Chemistry Library

Library of Congress Cataloging-in-Publication Data

Anisotropic organic materials : approaches to polar order / Rainer Glaser, editor, Piotr Kaszynski, editor.

p. cm.—(ACS symposium series ; 798)

Developed from a symposium sponsored by the Division of Organic Chemistry at the 218th National Meeting of the American Chemical Society, New Orleans, Louisiana, August 22-26, 1999.

Includes bibliographical references and index.

ISBN 0–8412–3689–5 (alk. paper)

1. Optical material—Congresses. 2. Organic compounds—Congressess. 3. Anisotropy—Congresses.

I. Glaser, Rainer 1957-. II. Kaszynski, Piotr, 1960-. III. American Chemical Society. Division of Organic Chemistry. IV. American Chemical Society. Meeting (218th: 1999 : New Orleans, La.). V. Series.

QC374.A55   2001
620.1'1795—dc21   2001033290

ACS SYMPOSIUM SERIES **798**

Anisotropic Organic Materials

Approaches to Polar Order

Rainer Glaser, Editor
University of Missouri–Columbia

Piotr Kaszynski, Editor
Vanderbilt University

American Chemical Society, Washington, DC

Foreword

The ACS Symposium Series was first published in 1974 to provide a mechanism for publishing symposia quickly in book form. The purpose of the series is to publish timely, comprehensive books developed from ACS sponsored symposia based on current scientific research. Occasionally, books are developed from symposia sponsored by other organizations when the topic is of keen interest to the chemistry audience.

Before agreeing to publish a book, the proposed table of contents is reviewed for appropriate and comprehensive coverage and for interest to the audience. Some papers may be excluded to better focus the book; others may be added to provide comprehensiveness. When appropriate, overview or introductory chapters are added. Drafts of chapters are peer-reviewed prior to final acceptance or rejection, and manuscripts are prepared in camera-ready format.

As a rule, only original research papers and original review papers are included in the volumes. Verbatim reproductions of previously published papers are not accepted.

ACS Books Department

Contents

Liquid and Crystals

Liquid Crystals

Preface

In 1998, the National Research Council published its report "Harnessing Light—Optical Science and Engineering for the 21st Century" (National Academy Press, Washington, D.C., 1998). One of the report's main thesis states that "although optics is pervasive in modern life, its role is that of a technological enabler: It is essential, but typically it plays a supporting role in a larger system." This insight leads to the issue as to how to ensure the future vitality of a field that lacks a recognized academic or disciplinary home? The committee addressed this issue as follows: "Underpinning the explosive growth of optics are investments in education and research. Although the field is growing rapidly and its impact is both pervasive and far-reaching, it remains a 'multidiscipline' with components in many university departments and government programs. The presence of optics in these diverse programs reflects its pervasiveness but also reveals its Achilles' heel. Trends and developments in optics can easily be missed in such a disaggregated enterprise. Educational and research organizations will need to pay close attention to ensure that the field develops in a healthy way that ensures benefits to society."

It is in this spirit, that we organized the symposium on "Anisotropic Organic Materials," which was held as part of the program of the Division of Organic Chemistry at the 218th National Meeting of the American Chemical Society (ACS), August 22–26, 1999, in New Orleans, Louisiana. The symposium brought together a group of ten internationally recognized experts —representing a broad range of disciplines—to speak on the topic of "Anisotropic Organic Materials." The two half-day sessions focused on "Crystals and Polymers" and "Liquid Crystals," and contributed posters were presented in the "State-of-the-Art" and again in the "Sci-Mix" poster sessions. The present ACS Symposium Series volume captures the spirit of this symposium and provides a more complete overview. Both, the symposium activities and content of this book should reveal current trends and developments in optics and, hopefully, attract organic chemists to enter and to strengthen this exciting multidisciplinary field.

The development of organic materials for optics clearly is one of the frontiers of modern organic materials chemistry. One of the key features that renders materials particularly useful concerns the issue of orientational order. Highly anisotropic materials are especially difficult to realize when the building blocks are polar. Most nonlinear optical materials are based on dipolar chromophores and the "dipole alignment problem" is a central issue in

ix

the development of advanced organic materials for optics applications. Only very recently strategies have emerged that allow for systematic approaches to polar order. With these strategies becoming more fully defined, the tuning of prototypes as well as the de novo design of new materials pose challenging problems to synthetic organic chemists. At the same time, the insights of physical and theoretical organic chemists are needed to more fully understand the structure–property relationships so that promising target molecules can be identified.

While the interest in polar organic materials has been tightly linked to applications in optics, organized polar materials are of interest in and of themselves. Aside from optical effects, many other phenomena depend on molecular anisotropy and parallel dipole alignment, and these effects include, for example, electrical conductivity, thermal conductivity, and ferroelectric and ferroelastic responses. From a fundamental point of view, polar materials allow for systematic studies of the effects of a polar environment on molecular properties. Polar alignment creates significant internal electric fields that can in turn affect the molecular properties. Because of the polarization, the properties of the molecule in such a polar solid can be quite different from the properties of the free molecule itself. Hence, the quest for "approaches to polar order" defines a fundamental issue that goes beyond the quest for anisotropy and that also goes beyond the realm of optics. Although many of the contributions in this book deal with issues in optics, the underlying science most frequently presents progress toward the higher goal: Polar alignment. We selected the title "Anisotropic Organic Materials: Approaches to Polar Order" to reflect this idea.

The book contains 21 chapters and they are grouped into five sections: Characterization of Polar Materials, Organic Thin Films, Molecular Materials and Crystals, Liquid and Crystals, and Liquid Crystals.

Characterization of polar materials contains two chapters. In Chapter 1, Grossmann, Weyrauch, and Haase discuss electroabsorption spectroscopy and the laser intensity modulation method for the experimental characterization of polar order in polymers. In Chapter 2, Geissinger describes the state-of-the-art of measuring electric fields in crystals with molecular and indeed atomic resolution.

Organic Thin Films contains two chapters. In Chapter 3, Cai et al. report the results of a collaboration of researchers from Zürich and Lausanne, which is aimed at the development of self-assembled organic thin films. In their exciting account, the authors report on the fabrication of highly anisotropic and polar films by molecular beam deposition. In Chapter 4, Metzger, Chen, and Baldwin describe the electrical conductivity of Langmuir–Blodgett thin films of hexadecylquinolinium tricyanoquinodimethanide. This chapter exemplifies well the importance of polar order as a requirement for the observation of electrical rectification.

Molecular Materials and Crystals contains six chapters. The first four chapters in this section address design issues of molecular organic materials for solid-state applications. Crystal engineering begins with the design of molecules with a view to the creation of an interesting electronic structure. Next, the design must consider the relation between the molecular structure and the consequential intermolecular interactions. And finally, crystal engineering is concerned with the mutual effects between the crystal packing and the electronic structure of the building blocks. Chapters 5–8 exemplify these three issues. In Chapter 5, Kaszynski presents computational and experimental studies of a novel kind of building block with interesting electronic properties which is based on *closo* boranes. Chapters 6 and 7 exemplify approaches to rational crystal engineering based on the synthon approach with focus on novel longitudinal and lateral synthons, respectively. In Chapter 6, Hanks, Pennington, and Bailey describe properties of the N⋯I charge transfer interaction and how this new longitudinal synthon can be employed in supramolecular synthesis. In Chapter 7, Lewis, Wu, and Glaser review arene–arene interactions and demonstrate the utility of so-called arene–arene double T-contacts in the crystal engineering of highly anisotropic materials including two near-perfectly dipole parallel aligned crystals. In Chapter 8, Staley, Peterson, and Wingert present a study of polarization effects in crystals of 6-arylfulvenes, which exemplifies the use of electrostatic potential maps for the characterization of intermolecular interactions. The last two chapters of this section address two kinds of dynamic aspects of crystals. In Chapter 9, Guha, Graupner, Yang, and the Chandrasekhars explore the response of single crystals of conjugated molecules to high pressure thereby revealing information about intra- and intermolecular bonding in the crystals. Finally, in Chapter 10, Gurney, Kurimoto, Subramony, Bastin, and Kahr apply optical probes to studies of crystal growth and the authors stress the specificity of intermolecular interactions between molecules and already formed anisotropic crystal regions.

Liquid and Crystals contains two chapters by the same authors. Oikawa, Kasai, and Nakanishi describe research in an area that they have pioneered and that they have named appropriately "Liquid and Crystals." Organic microcrystals occupy a mesoscopic phase between single molecules and bulk crystals and they are therefore likely to exhibit special optical and electrical properties. In Chapters 11 and 12, the authors describe the preparation, the properties, and some applications of organic microcrystals.

Liquid Crystals contains nine contributions covering recent developments in the chemistry and properties of nematic, smectic, and discotic materials. In the first chapter of this section, Chapter 13, Guittard and Geribaldi review some highlights of the fluorophobic effect on phase behavior of liquid crystals in the context of formation of tilted smectic phases. In Chapter 14, Reiffenrath and Bremer describe the first synthesis and characterization of a nematic with negative birefringence, a long sought after material for display applications. Vill, Bertini, and Sinou report supramolecular properties of a new class of chiral mesogens cleverly derived from carbohydrates in Chapter 15. The next four chapters are dedicated to chiral smectic phases with an emphasis on ferroelectric behavior of Sc mesogens and their practical applications. Thus in Chapter 16, Guillon, Cherkaoui, Sebastiao, Mery, Nicoud, and Galerne discuss their findings about the dependence of spontaneous polarization on molecular structure. In the next chapter, Lemieux, Dinescu, and Maly report the use of chiral and photoactive dopants for liquid crystalline optical switches and spatial light modulators. In Chapter 18, Anthamatten and Hammond describe phase transition phenomena in smectic C copolymers, whereas Shashidhar and Naciri describe ferroelectric properties of such high molecular weight smectogens in Chapter 19. In Chapter 20, Walba, Körblova, Shao, Maclennan, Link, Glaser, and Clark report results on the structure and electrooptical switching of chiral smectic phases formed from achiral bent-shaped molecules. Finally, in the last chapter, Janitz, Goldmann, Schmidt, and Wendorff describe exciting new results in the supramolecular assembly of discotic materials using donor-acceptor and hydrogen-bonding interactions.

Acknowledgments

The organization of this symposium on "Anisotopic Organic Materials" was stipulated by a conversation between Rainer Glaser and Joseph Gajewski at the Dallas National Meeting of the ACS in the Spring of 1998. In recognition of the growing importance of the materials chemistry and of its overlap with organic chemistry, Dr. Gajewski indicated that the ACS might view proposals for a symposium on organic materials favorably. We thank the ACS for their sponsorship and financial support of the New Orleans Symposium and we gratefully acknowledge the Petroleum Research Fund for a PRF–SE grant, which made possible a truly international symposium with speakers from the United States, Canada, the United Kingdom, France, Germany, and Japan. The symposium found a corporate sponsor in Rolic Research, Ltd. of Switzerland and we thank Dr. Martin Schadt for his generous contribution. We thank Dr. Stan Hall, the National Program Chair, for his guidance with the organization of the symposium and his superb organizational effort which ensured that the event went smoothly. The editors also want to thank their deans for travel fellowships and for their support. Jan Wagner, Jason Jett, and Chris Sanders of the Academic Support Center of the University of Missouri at Columbia created the symposium web site (http://web.missouri.edu/~chemrg/aomsymposium. html), posters and postcards. Finally and most importantly, we thank Anne Wilson and Kelly Dennis of the

ACS Books Department for their guidance in the process of getting the book together; it was a pleasure working with both of them.

Dr. Rainer Glaser
Department of Chemistry
University of Missouri at Columbia
Columbia, MO 65211

Dr. Piotr Kaszynski
Department of Chemistry
Vanderbilt University
Nashville, TN 37235

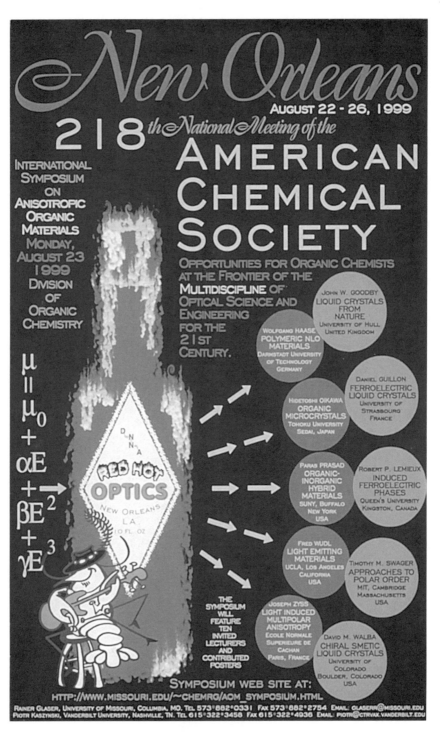

Courtesy of Jason Jett

Characterization of Polar Materials

Chapter 1

Characterization of NLO-Materials for Photonic Application

S. Grossmann, T. Weyrauch, S. Saal, and W. Haase

Institute of Physical Chemistry, Darmstadt University of Technology,
64287 Darmstadt, Germany

The use of electroabsorption (EA) spectroscopy and pyroelectric investigations by the laser intensity modulation method (LIMM) for the characterization of nonlinear optical (NLO) polymer systems is demonstrated. EA spectroscopy may be used to evaluate basic parameters of NLO chromophores and their polar order. The influence of the rotational mobility of the chromophores on the EA spectra and the application of this effect to characterize the relaxation of the chromophores is shown. EA spectroscopy may be applied to determine in situ the electric field in the electrooptic active layer of multi layer polymer stacks. Pyroelectric LIMM measurements are applied to characterize the polarization profile in double layer polymer systems.

Introduction

Second order nonlinear optical (NLO) polymers are promising materials for a large variety of integrated optical devices such as frequency converters, switches and electro-optic (EO) modulators (*1,2*). A lot of physical properties are responsible for the working efficiency of photonic elements. The important properties may be assigned to three major tasks: (*i*) Optimization of molecular properties, (*ii*) improvement of the polar orientation, which must be induced to the material, and its thermal and temporal stability, (*iii*) integration of materials into device structures. In order to accomplish tasks (*i*) and (*ii*) several experimental methods have been used and developed during the last 15 years (*2*). E. g. electric field induced second harmonic generation (EFISH) and Hyper-Rayleigh scattering (HRS) are suitable for the characterization of molecular nonlinear optical susceptibilities. Besides NLO methods such as second harmonic generation (SHG) and Pockels effect, a series of methods sensitive to polar properties have been applied for the study of the poling process and the poling stability. Here piezoelectricity and pyroelectricity as well as thermally stimulated depolarization techniques should be mentioned. The experimental methods for the work on task *(iii)* depend sensitively on the specific device and its requirement on the materials, e. g. the possibility to form 3-dimensional waveguide structures. Another point is the behavior of polymeric sandwich structures in electric field, a problem which is relevant for the poling procedure and during operation of different devices such as modulators based on Mach-Zehnder interferometers or polymeric light emitting diodes.

A method not mentioned in this paper until now is the electroabsorption (EA) or Stark spectroscopy, i. e. the measurement of absorbance changes of NLO chromophores in an electric field. One aim of this paper is to show the various possibilities of this multifunctional technique in order to get information with regard to all three tasks *(i-iii)* with comparable low experimental expenditure. It will be shown, that molecular parameters, polar order parameters and the study of voltage distributions in multilayer polymer films all are possible with one experimental setup only. The study of multilayer films is one aspect of the more general task of characterizing polarization profiles in poled polymers. Therefore the laser induced pressure pulse (LIPP) method or the laser intensity modulation method (LIMM) may be used. We used LIMM because the method is definitely non-destructive and the experimental setup (using a lock-in technique similar to EA spectroscopy with high signal-to-noise ratio) is much simpler in comparison to the LIPP method. However, the evaluation of polarization profiles has some peculiarities, which will be discussed within this paper.

Electroabsorption spectroscopy

Background

Orientationally Fixed Chromophores

Under the assumption of an orientationally fixed chromophore molecule characterized by two energy levels (the energy of the electronic ground state and the lowest excited state) the electroabsorption (i. e. the change of the absorbance band under the action of an external electric field) is due to the Stark effect. The absorption band $A(\tilde{\nu})$ is shifted according to the differences of dipole moments $\Delta\mu$ and polarizabilities $\Delta\alpha$ in the ground and the excited state. Experimentally one usually measures the effective value of the change in absorbance $\Delta A_{eff}(\omega)$ at fixed wave number $\tilde{\nu}$ under application of an ac electric field $E_{eff}(\omega)$ with frequency ω by lock-in technique. The quadratic effect (effective value of change in absorbance measured at the frequency 2ω) is given by

$$\Delta A_{eff}^{q}(2\omega) = \frac{E_{eff}^{2}(\omega)}{\sqrt{2}}\left[\frac{\Delta\alpha}{hc}\tilde{\nu}\frac{\partial(A/\tilde{\nu})}{\partial\tilde{\nu}} + \frac{(\Delta\mu)^{2}}{10h^{2}c^{2}}\tilde{\nu}\frac{\partial^{2}(A/\tilde{\nu})}{\partial\tilde{\nu}^{2}}\right] \qquad (1)$$

for the case of an isotropic or weak polar sample (3). The quadratic effect may also be measured at the frequency ω if a dc field E_0 is applied simultaneously. Thus

$$\Delta A_{eff}^{ql}(\omega) = 2E_0 E_{eff}(\omega)\left[\frac{\Delta\alpha}{hc}\tilde{v}\frac{\partial(A/\tilde{v})}{\partial\tilde{v}} + \frac{(\Delta\mu)^2}{10h^2c^2}\tilde{v}\frac{\partial^2(A/\tilde{v})}{\partial\tilde{v}^2}\right], \tag{2}$$

this may be referred to as the quasilinear effect. If the sample has a polar average orientation of the relevant chromophore the linear Stark effect

$$\Delta A_{eff}^{lin}(\omega) = \frac{<\cos\theta\sin^2\theta>}{<\sin^2\theta>}\frac{\Delta\mu E_{eff}(\omega)}{hc}\left[\tilde{v}\frac{\partial(A/\tilde{v})}{\partial\tilde{v}}\right] \tag{3}$$

will also be measured at the modulation frequency ω. Thus the measurable signal is in general a superposition of the linear and the quasilinear effect. However they are distinguishable due to the different spectral behavior of the linear effect in comparison to the quasilinear effect.

Orientationally Mobile Chromophores

If the NLO chromophore in the sample has an electric dipole moment and is free to reorient, the application of the modulation voltage in the electroabsorption experiment will vary the orientation of this molecule and thus the absorption of the sample. The most important effect is the change of the average angle between the transition dipole moment and the electric field vector of light. For the rod like molecules under investigation the direction of the transition dipole moment and the electric dipole moment can to be considered to be parallel to the molecular long axis. Because the molecules tend to be parallel to the modulation field the absorption will be lowest in case of highest field (independend of the direction of the field). Thus an additional orientational contribution to the electroabsorption signal at 2ω with

$$\Delta A_{eff}^{orient}(2\omega) = -\eta(T,\omega)\frac{1}{\sqrt{2}}\frac{\left(\mu_g E/kT\right)^2}{15}A(\tilde{v}) \tag{4}$$

will be expected for mobile chromophores in the case of the quadratic effect (eq 1). The term $\eta(T,\omega)$ describes the magnitude of the orientational mobility: In case of maximum reorientation of the chromophores, i. e. $\eta(T,\omega) = 1$, the contribution (eq 4) is limited by the Boltzmann distribution. If the chromophores are fixed in space the term $\eta(T,\omega)$ becomes zero.

Experiment

The electroabsorption was measured using a spectrometer (Figure 1) based on a monochromator (TOPAG LM-01) and a 100 W halogen lamp as light source. The ac and dc part of the intensity transmitted through the sample (ΔI and I, respectively) were measured by a photomultiplier, a lock-in amplifier (PAR 5210) and a digital multimeter (HP-34401A). The absorbance change was calculated from these measurements according to

$$\Delta A = -\Delta I / (I \ln 10).\tag{5}$$

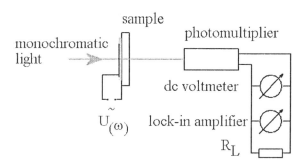

Figure 1. Scheme of the setup for the measurement of electroabsorption spectra.

Applications of Electroabsorption Spectroscopy

Electroabsorption spectroscopy offers the possibility to investigate different properties, which are important for the optimization of NLO materials.

Basic Physical Properties

The first important step in optimization of NLO properties is to select the used chromophores. The quadratic or quasilinear Stark effect allow to determine fundamental molecular properties of the chromophores according to the following procedure: From measurements of the absorption spectra $A(\tilde{v})$ the derivatives as given in eqs. 1 and 2 are calculated. These derivative spectra are used to fit eqs 1 and 2 to the experimental data. Here $\Delta\mu$ and $\Delta\alpha$ are independent fit parameters because of the different wavenumber dependencies of the related derivative spectra. After fitting, with the knowledge of those parameters, the hyperpolarizability of the chromophore can be estimated according to the 2-level model (*4*).

In order to use $\chi^{(2)}$ effects like SHG a polar orientation of the chromophores must be present. Therefore, the poling process tends to be the most important step during the preparation of polymeric systems for photonic applications. In those polar media the linear in field Stark effect is a useful method to investigate the poling efficiency. After determination of $\Delta\mu$ (using the quadratic effect) it is possible to calculate the first polar order parameter $<\cos\Theta \sin^2\Theta>/<\sin^2\Theta>$ (where Θ is the angle between the polar axis and the molecular dipole) by fitting eq. 3 to the experimental electrobsoprtion spectra of the linear effect. One advantage of this method is that uncertain parameters such as local field correction factors are cancelled out, because both, $\Delta\mu$ and $<\cos\Theta\ \sin^2\Theta>/<\sin^2\Theta>$, are determined by the same experimental method and for the same sample (5,6).

Multilayer Polymer Stacks

Aside from the characterization of basic physical properties the EA spectroscopy is also a tool for the investigation of internal field distributions, especially in case of multilayer polymer stacks. Due to the different electrical properties of the different layers, application of a voltage to a multilayer system does not produce a uniform electric field in the system, moreover the distribution has been shown to be time dependent (dc drift phenomenon) (7,8). No direct method exists for the selective measurement of the field (9), but it is possible to determine the electric field in the EO active layer via the NLO properties of this layer. However, during application of a dc electric field the apparent NLO properties are a sum of second order ($\chi^{(2)}$) and third order ($\chi^{(3)}$) effects. We have shown recently that electroabsorption spectroscopy offers the possibility to overcome this problem (10): Both, $\chi^{(2)}$ and $\chi^{(3)}$ effects, contribute to the EA spectra with different wavelength dependencies, both vanish at different wavelengths within the absorption band. Choosing one of these wavelengths in the electroabsorption experiment one of these contributions can be measured selectively. Applying this technique to a double layer system with one inactive layer, we could determine the transient behavior of the electric field in the EO active layer during switching on or off a dc electric field. The behavior could be understood in terms of a model considering the resistances and capacities of both layers. However, also single layer samples show a dc drift phenomenon, which gives evidence for formation of screening charges at the interfaces between electrodes and polymer (10,11,12).

Orientational Relaxation of Chromophores

As described above, in the case of orientationally mobile chromophores the EA spectra of the quadratic effect are different from that case where the chromophores don't reorient in the electric field. The most important change is described by the orientational contribution $\Delta A_{eff}^{orient}(2\omega)$ to the electroabsorption signal, which has a spectral dependency proportional to the absorbance (in contrast to the linear or the quadratic effect) and is thus well distinguishable from other contributions. Analysing the experimental EA spectra by using a fitting procedure gives the value of $\eta(T,\omega)$ and thus information about the reorientation mobility of the chromophores. Doing electroabsorption experiments at various modulation frequencies and at different temperatures the study of the relaxation process of the chromophores is possible (13).

The investigated samples possess a capacitor like structure. Thin films of dye doped polymer solution were spin coated onto ITO covered glass substrates. A polymeric matrix poly(methyl-methacrylate) (PMMA) was used. The azo dye 4 – [ethyl(hydroxy-ethyl)-amino]4-nitroazobenzene (DR1) was used as electrooptic active dopant. The dye concentration was 5 wt-%. The top electrode was a transparent gold layer sputtered onto the polymer film. Typical thickness of those samples was 1.7 μm.

To demonstrate the difference of EA spectra of mobile and immobile systems the quadratic effect was measured with two different modulation frequencies (2 Hz and 2 kHz) at the same temperature (Figure 2). At 2 kHz, the fitting procedures gives a value of $\eta(T,\omega) \approx 0$, i. e. the chromophores cannot follow the electric field and are immobile at this time scale. At 2 Hz however, the shape of the experimental spectrum has changed. The fitting procedure gives a significantly higher value of $\eta(T,\omega)$, indicating a certain mobility of the chromophores at the frequency of 2 Hz, but $\eta(T,\omega)$ doesn't reach the theoretical maximum.

The $\eta(T,\omega)$ values determined from EA measurements at various modulation frequencies at a fixed temperature allow to calculate the relaxation frequency of the chromophores, e. g. by fitting a Cole-Cole function to the experimental data. Figure 3 shows the result for two temperatures. The increase of the relaxation frequency with increasing temperature is observable. A more detailed study on various systems is presented in (13,14,15). It should be noted, that in comparison to other methods (e. g. dielectric relaxation spectroscopy), the application of EA spectroscopy is sensitive to the chromophores only and not to the matrix material, which allows for a comparison of matrix and chromophore behavior. Also measurements are possible even at very low dye concentrations.

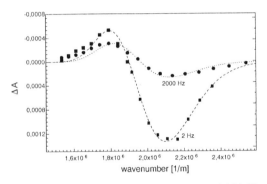

Figure 2. Electroabsorption spectra of DR1 in PMMA at 2000 Hz (rigid system) and 2 Hz (mobile system) modulation frequency.

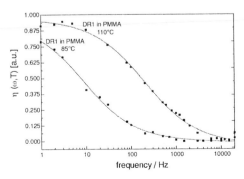

Figure 3. Frequency dependenc of $\eta(T,\omega)$ as function of temperature.

Pyroelectric effect – LIMM Technique

Theory

The LIMM technique and its application to NLO polymer films has been intensively reviewed by Bauer (16). The sample is embedded between two metal layers. The top electrode is irradiated by an intensity modulated laser beam. At the top electrode, which is the only light absorbing layer, a heat wave is generated. The penetration characteristic depends on the modulation frequency. The current $I(\omega)$ between the electrodes in dependence of the modulation frequency ω measurable in a closed loop is given by the convolution integral

$$I(\omega) = \frac{i\omega A}{d} \int_0^d p(z)\Delta T_\omega(z)dz \ , \tag{5}$$

where A is the electrode area, d is the film thickness, z is the coordinate along the film normal and $\Delta T_\omega(z)$ is the temperature oscillation amplitude. $p(z)$, the polarization function searched for, is given by

$$p(z) = \alpha_p P(z) + (\alpha_z - \alpha_\varepsilon)\varepsilon_0 \varepsilon E(z) \ . \tag{6}$$

α_p, α_z and α_ε are the temperature coefficients of the polarization, thermal expansion and the dielectric constant, respectively. The internal electric field $E(z)$ contains contributions from the polarization as well as from space charges (17).
The temperature oscillation amplitude $\Delta T_\omega(z)$ can be calculated for a polymer film on a glass substrate. Because the substrate is much thicker than the polymer film it may be considered as semiinfinite in z direction. Then

$$\Delta T_\omega(z) = \frac{\Delta T_\omega(0)\sinh k_p(d-z) + \Delta T_\omega(d)\sinh k_p z}{\sinh k_p d} \ , \tag{7}$$

$\Delta T_\omega(0)$ and $\Delta T_\omega(d)$ are the values at the interface between the absorbing electrode and the polymer ($z = 0$) and the interface between the polymer and the glass substrate ($z = d$) and are given by

$$\Delta T_\omega(0) = \frac{\eta J_\omega}{\kappa_s k_s} \frac{1 + \kappa_s k_s \tanh k_p d /(\kappa_p k_p)}{1 + \kappa_p k_p \tanh k_p d /(\kappa_s k_s)} \exp(i\omega t) \tag{8}$$

and

$$\Delta T_\omega(d) = \frac{\eta J_\omega}{\kappa_s k_s} \frac{1}{1 + \kappa_p k_p \tanh k_p d /(\kappa_s k_s)} \frac{1}{\cosh k_p d} \exp(i\omega t) \ . \tag{9}$$

κ_p and κ_s are the thermal conductivities, k_p and k_s are the thermal wave vectors of polymer and substrate, respectively. The latter are given by

$$k_p = (1+i)\sqrt{\omega/2D_p} \text{ and } k_s = (1+i)\sqrt{\omega/2D_s} \tag{10}$$

with the thermal diffusivities D_p and D_s of polymer and substrate, respectively ($D = \kappa/\rho c$, ρ = density, c = heat capacity).

Experiment

The experimental setup used for the pyroelectric LIMM measurements is shown in Figure 4. A 30 mW laser diode (λ = 690 nm) was modulated by a TTL signal, which is also connected to the reference input of a PAR 5210 lock-in amplifier. The electrodes of the sample are connected with the voltage input of the lock-in amplifier. The sample is placed in a metallic cover in order to achieve an electrical screening. Data recording is performed automatically by a personal computer, which also controls the output frequency of the signal generator and the preamplifier settings of the lock-in amplifier.

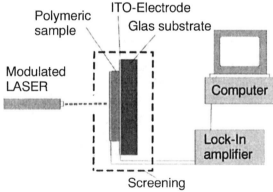

Figure 4. Experimental setup for LIMM measurements.

At low frequencies the resulting pyroelectric current signal is proportional to $\sqrt{\omega}$, i. e. the signal increases with increasing modulation frequency. Thus in the most interesting high frequency range signals are high. Therefore the measurement of the pyroelectric current is favorable. However, the current must be converted by a current-to-voltage-converter to detect it by a look-in-amplifier. The conversion characteristic (phase and amplitude) depends on the impedance of the sample especially in the high frequency range using samples with high capacity as in our case. To avoid any problems and misinterpretations we therefore used the voltage input of the lock-in-amplifier. In this case the input acts as an additional low frequency pass with a characteristic frequency $\omega_{LP} = R_{lock-in}C_{sample}$, given by the capacity of the sample and the input resistance of the lock-in amplifier. Thus we expect the following behavior (cf. Figure 5): In the low frequency range the signal should be proportional to $\sqrt{\omega}$ as in the case of current measurements. At ω_{LP}, the imaginary part of the signal changes its sign. In Figure 5 (and in subsequent plots) a double logarithmic plot of the absolute values of the signal is given, where the change of sign is visible as a drastic drop of the imaginary part to low values (in Figure 5 at around 20 Hz). Above the characteristic frequency of the low pass the voltage signal (real part) tends to be proportional to $1/\sqrt{\omega}$, thus the signal decreases with increasing frequency. The high frequency range is determined by the sample geometry and polarization profile. This will be discussed in more detail below. It should be noted, that this experimental arrangement requires a good electrical screening to reduce noise and disturbing signals in the most interesting high frequency range.

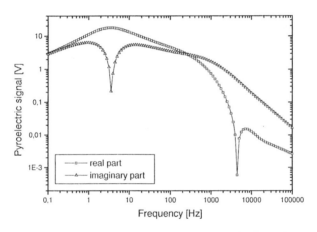

Figure 5. Pyroelectric voltage as function of frequency, calculated for a uniformly poled sample.

Signal Interpretation

As described in eq. 5, the measured signal is a convolution integral of the heat wave and the polarization function. One of the most important problems analyzing LIMM measurements is to find an appropriate deconvolution procedure in order to calculate the polarization function from the experimental data. An analytic solution of this problem is not possible, however several approximation methods have been developed and are discussed in literature (e.g. *18*). All these methods are very sophisticated and require a detailed knowledge of the procedure in order to avoid misinterpretations because of choosing inappropriate conditions for the deconvolution formalism. Another possibility to interpret the experimental data may be to suggest a first approximation of the polarization function $p(z)$. Then a least square fit may be used to obtain undetermined parameters.

In the case of double layer polymer stacks one may expect in a first approximation a constant polarization function within one layer, whereas the value changes from one layer to another. In the extreme case one layer is poled and the polarization is nonzero whereas the other layer is not polar and the polarization is zero. One additional problem of signal interpretation is that such a case is usually present also in single layer samples, because of the finite thickness of the absorbing electrode. To include the influence of the electrode it may be considered as a nonpolar layer (16), because the thermal waves don't start at the polymer/electrode interface but within the electrode or at the electrode/air surface (depending on the transmittivity of the electrode). An analysis of the influence of the electrode thickness on the pyroelectric signal demonstrates the general behavior of the double layer systems.

As one can see from Figure 5, the real part Re{U} of the pyroelectric signal of a uniformly poled sample shows a sign inversion in the high frequency range. The frequency ν_0 where Re{U} becomes zero (visible as a peak downwards if the absolute values are shown in logarithmic scale) depends on the thickness of the sample. As a general rule ν_0 is the lower the thicker the sample is. Additionally ν_0 depends on the thickness of the absorbing electrode, which has to be considered as an additional nonpolar layer. As shown in Figure 6 ν_0 is shifted to smaller frequencies with increasing electrode thickness.

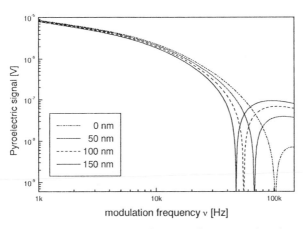

Figure 6. Frequency dependence of pyroelectric voltage signal as function of the electrode thickness.

The sensitivity of ν_0 to thickness changes of the unpolar layer is demonstrated in Figure 7. A layer thickness of 0.15 μm shifts ν_0 from more than 100 kHz (without unpolar layer) to less than 50 kHz.

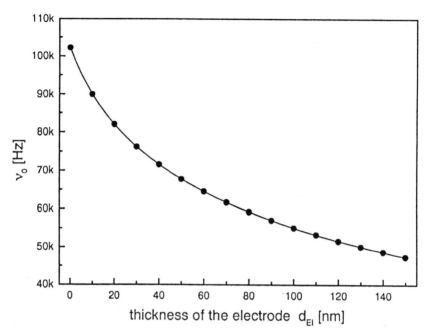

Figure 7. Frequency of ν_0 in dependence of the electrode thickness.

Sample Preparation

Double layer polymer stacks were prepared by spin coating the polymer on ITO coated glass substrates. Top electrodes were provided by sputtering a semitransparent gold electrode. The electrooptic active layer consisted of the dye p-N,N-dimethylamino-p'-nitroazobenzene (DMANA, Figure 8) diluted in a polycarbonate bisphenol-A (PC) matrix (2% of weight).

Figure 8. Chemical structure of p-N,N-dimethylamino-p'-nitroazobenzene (DMANA).

We used the polyimide PI 2566 (DuPont) as inactive layers. The top gold electrode was prepared by sputtering. The thickness of the samples was determined using an interference microscope. Absorbance spectra were taken using a Cary 17 spectrometer.

Experimental Results and Discussion

Influence of the Electrode Thickness on the Pyroelectric Signal

We compared measurements on samples with different effective electrode thicknesses to demonstrate the influence of the thickness of the absorbing top electrode on the pyroelectric signal. Figure 9a shows the measurement on a sample with the gold top electrode as the only absorbing layer. The value of the characteristic frequency v_0 of sign inversion is about 65 kHz, which is in accordance with the thickness of the electrode. For the measurements shown in Figure 9b the top gold electrode was covered with a 250 nm thick highly absorbing layer. A pronounced shift of v_0 to about 20 kHz was observed, in accordance with the calculations presented above (cf. Figure 7).

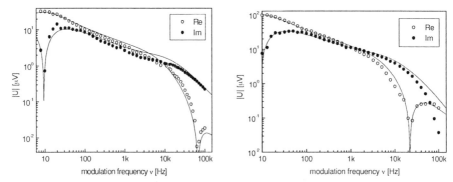

Figure 9. a) Pyroelectric voltage signal measured on a sample with a thin top electrode. b) Pyroelectric signal with a 250 nm thick absorbing layer.

Pyroelectric Measurements on a Double Layer System

For pyroelectric measurements on a double layer system a sample was prepared as follows: On the glass substrate with the ITO electrode a 5.8 μm thick layer of polyimide was spin coated and cured at 250 °C. The second layer was prepared from DMANA doped polycarbonate with a thickness of 2.7 μm. The top gold electrode's thickness was 30 nm. The sample was poled at 110 °C and subsequently cooled down to room temperature under application of 100 V to the electrodes. After the poling procedure the electrodes are short-circuited for 48 hours. Because the glass transition of the polyimide is much higher than 110 °C, one should expect, that using this procedure only the polycarbonate layer should receive a persistent polar order, which can be observed after poling. This was verified by our measurements (cf. Figure 10): Using a polarization function taking into account a 5.8 μm thick inactive layer and a 2.7 μm thick polar layer we can fit the experimental data. For comparison the curves calculated under the assumption of a uniform polarization are also shown in Figure 10. Here a much smaller value of the characteristic frequency v_0 in the real part of the

14

signal than observed experimentally is expected. Thus the measurements clearly indicate the inhomogeneously polar structure of the double layer system.

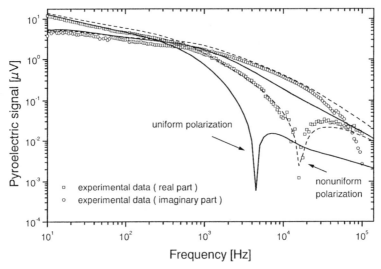

Figure 10. Pyroelectric signal on a double layer sample together with calculated curves for uniform and nonuniform polarization profile of the sample.

Summary and Conclusion

In this paper we presented a brief overview of the possibilities of electroabsorption spectroscopy and pyroelectric measurements for the characterization of NLO polymers for photonic application. E. g. the measurement of the rotational mobility and relaxation of NLO chromophores by EA spectroscopy with variable modulation frequency was demonstrated. It should be pointed out that this technique may be considered as a very suitable tool for the investigation of the orientational enhancement of the photorefractive effect in polymers (*19*). EA spectroscopy as well as LIMM pyroelectric measurements were used to characterize multilayer polymer stacks during and after poling. These methods give information about the voltage and polarization distribution across the different layers and may also be also used to characterize and optimize polymeric electroluminescence devices.

Acknowledgment

The authors gratefully acknowledge financial support by the Volkswagen-Stiftung.

References

1. Hornak, L. A.; Ed.; Polymers for Lightwave and Integrated Optics; Dekker, New York, 1992, p 287.

2 Bauer, S.; *J. Appl. Phys.* **1996**, *80*, 5531.

3 Barnik, M. I.; Blinov, L. M.; Haase, W.; Palto, S. P.; Tevosov, A. A.; Weyrauch, T. In *Polymers for Second-Order Nonlinear Optics*; Lindsay, G.A.; Singer, K.D., Eds.; ACS Symposium Series Vol. 601; American Chemical Society, Washington, D.C., 1995; p 288.

4 Paley, M. S.; Harris, J. M.; Looser, H.; Baumert, J. C.; Bjorklund, G. C.; Jundt, D.; Twieg, R. J.; *J. Org. Chem.* **1989**, *54*, 3774.

5. Blinov, L. M.; Palto, S. P.; Tevosov, A. A.; Barnik, M. I.; Weyrauch, T.; Haase, W.; *Mol. Materials.* **1995**, *5*, 311.

6 Grossmann, S.; Weyrauch, T.; Haase, W.; *J. Opt. Soc. Am. B.* **1998**, *15*, 414.

7. Shi, Y.; Olsen, D. J.; Bechtel, J. H.; Kalluri, S.; Steier, W. H.; Wang, W.; Chen, D.; Fetterman, H. R.; In *Organic Thin Films for Photonics Applications*, OSA Technical Digest Series Vol. 21, Optical Society of America, Washington, D.C., 1995; p 325.

8. Park, H.; Hwang, W.-Y.; Kim, J.-J.; *Appl. Phys. Lett.* **1997**, *70*, 2796.

9. Girton, D. G.; Anderson, W. W.; Marley, J. A.; Van Eck, T. E.; Ermer, S. In *Organic Thin Films for Photonics Applications*; OSA Technical Digest Series Vol. 21, Optical Society of America, Washington, D.C., 1995; p 470.

10 Grossmann, S.; Weyrauch, T.; Saal, S.; Haase, W.; *Opt. Mat.* **1998**, *9*, 236.

11 Grossmann, S.; Weyrauch, T.; Haase, W.; In *Electrical, Optical and Magnetic Properties of Organic Solid-State Materials* Reynolds, J. R.; Jen, A. K.-Y.; Rubner, M. F.; Chiang, L. Y.; Dalton, L. R., Eds.; Mat. Res. Soc. Symp. Proc. 488; Materials Research Society: Pittsburgh, PA, 1998, p 413.

12 Grossmann, S.; Weyrauch, T.; Haase, W. *Nonlinear Optics*, **1999**, *20*, 223.

13 Saal, S.; Haase, W.; *Chem. Phys. Letters* **1997**, *278*, 127.

14 Haase, W.; Saal, S.; Grossmann, S.; Weyrauch, T. In *Organic Photonic Materials and Devices,* Kippelen, B., Ed.; Proc. SPIE Vol. 3623; The Society of Photo-Optical Instrumentation Engineers: Bellingham, WA, 1999, p 159.

15 Saal, S.; Ph.D. thesis, Darmstadt University of Technology, Darmstadt, Germany, 1999.

16. Bauer, S.; *J. Appl. Phys.* **1994**, *75*, 5306.

17. Ploss, B., Emmerich, R.; Bauer, S.; *J. Appl. Phys.* **1992**, *72*, 5363.

18 Das-Gupta, D. K.; Hornsby, J. S.; *IEEE Transactions on Electrical Insulation* **1991**, *26*, 63.

19 Burland, D. M.; Devoe, R. G.; Geletneky, C.; Jia, Y.; Lee, V. Y.; Lundquist, P. M.; Moylan, C. R.; Poga, C.; Twieg, R. J.; Wortmann, R.; *Pure and Applied Optics* **1996**, *5*, 513.

Chapter 2

Quantitative Measurement of Internal Molecular Electric Fields in Solids

Peter Geissinger

Department of Chemistry, University of Wisconsin at Milwaukee, Milwaukee, WI 53201-0413

This chapter describes recent developments for the quantitative measurement of internal electric fields in solids at molecular and atomic resolution. These methods represent a significant advance, since previous models provided only indirect information on internal fields. Internal electric fields originate from the charge distributions of atoms and molecules. Therefore, they characterize the electrostatic properties of a material, e.g. the degree of polar order. Moreover, internal fields are involved in processes involving charge separation and transport in a variety of systems, particularly biological systems like proteins and enzymes. Results are presented for internal fields in molecular crystals and proteins.

Introduction

The purpose of this chapter is to highlight recent advances for the direct, quantitative measurement of internal electric fields in solids. Over the years, the term "internal electric field" has been used inconsistently, describing different contributions to the total electric field that atoms and molecules experience in a solid. Following refs. [1-4], we define the internal electric field as the electric field that is present in solids (and liquids) in the absence of an externally applied electric field. To illustrate the concept of internal electric fields we consider a molecular crystal, for example, an *n*-alkane crystal. The individual *n*-alkane molecules are charge neutral and carry no dipole moment. However, the electric fields generated by their charge distributions combine to form the internal electric field according to Coulomb's Law

$$\vec{E}_{int}(\vec{x}) = \sum_k \int \rho_k(\vec{x}_k') \frac{\vec{x} - \vec{x}_k'}{\left|\vec{x} - \vec{x}_k'\right|^3} d\vec{x}_k', \tag{1}$$

where $\rho_k(\vec{x}_k)$ is the charge distribution of molecule k in its crystal environment. Of course, in a high symmetry environment such as a crystal the electric fields from many molecules might cancel at certain sites. However, at typical condensed matter intermolecular distances these internal electric fields **can** be substantial! The summation in Eqn. (1) extends over all molecules that constitute the sample, excluding only the molecule whose charge distribution overlaps with the location $\vec{x}$ at which the internal field is to be evaluated [1]. In other words, the internal electric field at a molecule does not include a contribution from the molecule itself. It has to be emphasized that the internal fields considered here are intermolecular fields. These have to be distinguished from the intramolecular electric fields that determine the charge distribution on the individual molecules according to the Schrödinger equation!

It is convenient to model the charge distributions ρ_k with an appropriately chosen set of point charges. In this case the internal field can be written as

$$\vec{E}_{int}(\vec{x}) = \sum_k \sum_{a(k)} q_{a(k),k} \frac{\vec{x} - \vec{x}_{a(k),k}}{\left|\vec{x} - \vec{x}_{a(k),k}\right|^3}, \tag{2}$$

where $q_{a(k),k}$ are the point charges (located at $\vec{x}_{a(k),k}$) needed to model the charge distribution of molecule k.

Eqn. (1) is not limited to molecular crystals, but can also be applied to large molecular systems like proteins and enzymes. For example, in the protein hemoglobin ρ_k can be taken as the charge distribution of an entire subunit, the summation accounting for the four subunits. Alternatively, the individual amino acids could be regarded – in a coarse approximation – as independent. In this case, ρ_k are the charge distributions of the individual amino acids and the summation extends over all amino acids in the system (as well as possible cofactors). The latter approach is of interest if the amino acid charge distributions are represented by point charges (see refs. [5-7]), in which case Eqn. (2) applies.

Anisotropic organic crystals, polar surfaces, and a large number of biopolymers are examples for ordered systems that sustain strong internal and surface electric fields. Biological systems offer a lesson on how to engineer molecular structures with the goal of optimizing internal and/or surface electric fields at certain sites. Obviously, the route that nature takes is through very complex structures that ensure that crucial groups are positioned in a way to create the desired field. Not only from the synthetic point of view are internal and surface electric fields of interest. In biological systems electric fields and function are intricately related. In general, electric fields are likely to be of importance in processes involving charge separation and transport. Electric fields are thought of being responsible for the pathway selection for electron transfer in bacterial photosynthetic reaction centers [8-12]. Electric fields emanating from the surfaces of biological systems into the surrounding solution facilitate docking (e.g. cytochrome c_1 to cytochrome c oxidase [13]) and electrostatic steering and guidance of charge substrates to enzyme (e.g. acetylcholine to acetylcholinesterase [14], superoxide ions to superoxide dismutase [15]). Furthermore, electric fields regulate binding and reaction properties, for example the binding affinities of oxygen and carbon monoxide to the active site in myoglobin and hemoglobin [16,17].

In general, measuring internal electric fields allows for the electrostatic characterization of certain sites in a sample. Of obvious interest for the scope of this book is that internal electric fields can provide a means of measuring the degree of polar order that has been achieved in an anisotropic organic crystal.

Clearly, it is important to have experimental access to both magnitude and orientation of these fields. Moreover, since continuum dielectric approaches are unable to establish the link between electrostatics and function, internal fields need to be known at a molecular or even at an atomic level. So far only limited, often quite indirect information on internal fields has been extracted from experimental data. In this chapter two novel approaches for the quantitative measurement at molecular and atomic resolution are presented. Both methods rely on the effect of an externally applied electric field on the electronic transitions of molecules located at the sites where the internal fields are to be measured. The two approaches differ in the resolution that they offer for the internal electric field: the *Molecular Resolution Method* provides one value of the internal electric field at the site of the investigated molecule, while the *Atomic Resolution Method* yields one value of the internal field for each atom of the molecule studied. In order to better illustrate the impact of recent developments, previous strategies for obtaining information on internal fields are also described. In the following we will briefly introduce the general principles that allow for the measurement of internal electric fields from the Stark effect on electronic transitions. A brief summary of optical line narrowing techniques that are key to achieving high resolution for internal fields is followed by the sample analysis of Stark data in molecular crystals and proteins.

Measurement Principle

As described in the introduction, the goal is to measure internal electric fields quantitatively at molecular or even atomic resolution. In order to obtain truly local information, a specific site in the system under study has to be selectively addressed and probed for the internal field it experiences. Therefore, a suitable molecular or atomic detector is required. Or more precisely, the sample needs to be dissected into a region where the internal field is generated and a region where the internal field is detected. In the ideal case, the detection region contains only one molecule (or atom). For molecular resolution, we need a molecule in the detection region. For atomic resolution we either need an atom in the detection region or we need to employ individual atoms of a molecule as detectors. To obtain truly local information, two conditions have to be satisfied:

- First, the concentration of the detector molecules has to be sufficiently small such that the probability of two detector molecules coming with interaction distance is practically zero. Therefore, the sample consists of many detection regions, each containing only one detector molecule.

- Second, we have to be able to address experimentally only the detector molecules, that is, we have to employ a molecular characteristic that allows distinguishing detector and host molecules. This molecular characteristic has to be sensitive to the experimental parameter to be used, which is an externally applied electric field.

In our studies, we employ the electronic states of atoms and molecules, as they offer the highest sensitivity to electric field perturbations. High-resolution techniques for the detection of transition energy shifts have been developed in the past two decades that allow for the measurement of line shifts as small as 0.0005 cm^{-1} (see e.g [18]). Therefore, we select detector molecules whose optical absorption does not overlap with the absorption of the host molecules in the spectral region of interest. By tuning the light source to the absorption wavelength of the detector molecule, we can selectively address only the detector molecules.

Of course, a rigorous division of the sample in a detector and source region for internal fields is not possible, as detector and host molecules mutually polarize each other and thereby modify the internal electric fields. From an electrostatic point of view, the substitution of one or more host units proceeds in two steps:

(1) *Removal of the host units to create a cavity in the system.*
 The polarizing contribution of removed molecules is now missing, leading to modified charge distributions $\tilde{\rho}_k(\vec{x}_k)$ on the remaining molecules, particularly in those molecules surrounding the cavity. The internal field is now given by

$$\vec{E}'_{int}(\vec{x}) = \sum_k \int \tilde{\rho}_k(\vec{x}_k') \frac{\vec{x} - \vec{x}_k'}{|\vec{x} - \vec{x}_k'|^3} d\vec{x}_k' , \qquad (3)$$

 The summation extends over all remaining host molecules. Evaluating Eqn. (3) at the site of the removed molecule(s) yields the field in the empty cavity.
(2) *Insertion of the detector molecule.*
 The detector molecule will polarize its environment, modifying the charge distributions of the remaining host molecules to produce a different internal field

$$\vec{E}''_{int}(\vec{x}) = \sum_k \int \sigma_k(\vec{x}_k') \frac{\vec{x} - \vec{x}_k'}{|\vec{x} - \vec{x}_k'|^3} d\vec{x}_k' , \qquad (4)$$

 where σ_k is the charge distribution of molecule k. If the internal field at the detector molecule is to be calculated, the summation in Eqn. (4) extends over all remaining host molecules.

Since the measurement principle employed here relies on the presence of a detector molecule that is distinct from the host molecule, it is the field $\vec{E}''_{int}$ described by Eqn. (4) that is accessible experimentally. Since the goal is to characterize the electrostatic properties of the host material only, (that is, to measure either $\vec{E}_{int}$ or $\vec{E}'_{int}$) the detector molecule has to be chosen such that its effect on the surrounding host molecules is minimized. Moreover, since the polarizing field created by the detector molecule depends on its electronic state, so does the internal field at the site of the detector molecule! In the following we will assume that the internal field is independent of the electronic state of the detector molecule.

For the measurement of internal fields in polar crystals, a detector molecule has to be chosen that (1) will co-crystallize with the host molecules, (2) does not destroy the polar order locally, and (3) does not significantly modify the charge distributions of the host molecules. These conditions are likely to be met if the dipole moments and polarizabilities of detector and host molecules are similar, although the analysis of the experimental data is simplified when non-polar detector molecules are used.

The above equations describe the microscopic situation correctly, but are unwieldy. Over the past century, many models to account for cavity fields and reaction fields were developed based on a continuum dielectric approach. A point dipole is placed into a spherical or elliptical cavity, which in turn is surrounded by the homogeneous dielectric [1,19-22]. These models are strictly valid for cavity dimensions that are large compared to a molecular volume. However, since at the typical condensed matter separations the detector molecule "sees" anisotropic charge distributions (except for some atomic hosts), the continuum approach does not allow to obtain a truly local picture of internal and reaction fields.

Stark-Effect on Electronic Transitions

In the following we will discuss three basic methods that have been developed to describe the effect of electric fields on the electronic states of a molecule embedded in a host material – the classical analysis as well as the atomic and molecular resolution methods. The latter two, which were developed in the group of B.E. Kohler [9,23-26], allow for direct experimental access to internal electric fields.

In a Stark experiment an external electric field $\vec{E}_{ext}$ is applied to the sample. Due to shielding effects and reaction fields, the detector molecule is not exposed to the externally applied field $\vec{E}_{ext}$, but a local field $\vec{E}_{loc} = \bar{\bar{f}}\,\vec{E}_{ext}$ [27]. The tensor $\bar{\bar{f}}$ maps the macroscopic field $\vec{E}_{ext}$ onto the microscopic field $\vec{E}_{loc}$. The total electric field at the site of the detector molecule is the vector sum of the internal electric field and the local field:

$$\vec{E}_{tot} = \vec{E}_{int} + \vec{E}_{loc} \ . \tag{5}$$

To calculate the energy levels of a molecule under the influence of an electric field, we add the standard electric field perturbation term to the molecular Hamiltonian H_0:

$$H = H_0 - \vec{E} \cdot \hat{\vec{\mu}}, \tag{6}$$

where $\hat{\vec{\mu}}$ is the dipole moment operator. By writing the electric field perturbation term as $\vec{E} \cdot \hat{\vec{\mu}}$, we treat the detector molecule in a point dipole approximation, which is permissible if the electric field is homogeneous over the volume of a detector molecule.

Up to this point, we have not specified the meaning of the Hamiltonian H_0 and the electric field $\vec{E}$ in Eqn. (6). The standard interpretation has been to interpret $\vec{E}$ as the local electric field at the detector molecule site: $\vec{E} \equiv \vec{E}_{loc}$. In this case the Hamiltonian H_0 describes the molecular electronic states including all interactions between detector and host molecules, that is, diagonalizing H_0 yields the energies of the electronic states of the detector molecule in its host environment as observed in the spectrum. Eqn. (6) can be analyzed using perturbation theory to yield the external field dependent energies ε_i for the molecular electronic states i

$$\varepsilon_i = \varepsilon_i^0 - \vec{\mu}_i \cdot \vec{E}_{loc} - \frac{1}{2} \vec{E}_{loc} \vec{\alpha}_i \vec{E}_{loc} - \dots \tag{7}$$

where $\vec{\mu}_i$ and $\vec{\alpha}_i$ are dipole moment vector and polarizability tensor for the molecular state i, respectively. Higher order terms in Eqn. (7) containing hyperpolarizabilities can usually safely be neglected for typical external field strengths used in Stark experiments [28]. From Eqn. (7) we readily obtain the electric field induced energy shift for an electronic transition between molecular energy levels i and j

$$\Delta \varepsilon_{ij} = -\Delta \vec{\mu}_{ij} \cdot \vec{E}_{loc} - \frac{1}{2} \vec{E}_{loc} \Delta \vec{\alpha}_{ij} \vec{E}_{loc} \dots \, , \tag{9}$$

where $\Delta \vec{\mu}_{ij}$ and $\Delta \vec{\alpha}_{ij}$ now are the differences of dipole moment vector and polarizability tensor between the molecular states j and i, respectively. For a detector molecule with zero dipole moment in both states i and j, Eqn. (9) predicts that the Stark shift of the electronic transition between states i and j should depend quadratically on the external field strength. While this behavior is indeed found for some detector molecules in crystalline hosts, in the majority of cases the experimentally observed Stark shift is linear in spite of $\Delta \vec{\mu}_{ij} = 0$! This observation presents the opportunity – at least in principle – to measure the internal electric field at the site of the detector molecule: the linear Stark effect is caused by the internal electric field, which induces a dipole moment difference $\Delta \vec{\mu}_{ij,ind}$ in the detector molecule according to

$$\Delta \vec{\mu}_{ij,ind} = \Delta \vec{\alpha}_{ij} \vec{E}_{int} \, , \tag{10}$$

assuming that the internal electric field does not depend on the electronic state of the detector molecule. The effective dipole moment difference in Eqn. (9) is now given as the sum of the permanent and induced dipole moment differences, leading to the following expression for the Stark shift

$$\Delta \varepsilon_{ij} = -\left(\Delta \vec{\mu}_{ij,per} + \Delta \vec{\mu}_{ij,ind}\right) \cdot \vec{E}_{loc} - \frac{1}{2} \vec{E}_{loc} \Delta \vec{\alpha}_{ij} \vec{E}_{loc} \dots \. \tag{11}$$

Eqn. (11) shows that even in the absence of a permanent dipole moment difference ($\Delta \vec{\mu}_{ij,per} = 0$) the Stark effect has a linear contribution, which is consistent with the experimental observations. For this reason, only the linear term has been used for the analysis of Stark data in samples containing nonpolar detector molecules [29-31]. Models based on the linear term extract from the field dependent shift of the transition energies the value of the induced dipole moment as well as a parameter characterizing the width of its distribution. Eqn. (10) allows for the calculation of the internal electric field from the measured dipole moment difference, provided that the polarizability tensor difference $\Delta\vec{\alpha}_{ij}$ is known.

Three critical shortcomings of this approach are the following:
- The polarizability tensor difference $\Delta\vec{\alpha}_{ij}$ in Eqn. (10) characterizes the electronic states of the detector molecule in its host environment. This polarizability tensor is usually not known. In principle, $\Delta\vec{\alpha}_{ij}$ can be calculated from the second order contribution to the field induced energy shift, provided that sufficiently large external field strengths can be applied to measure the quadratic contribution. This has been done for some molecules [32] within the restrictions of the point dipole approximation for cases with a dominant quadratic Stark effect.
- The restriction to the term linear in the external field in Eqn. (11) can lead to erroneous results for samples in which the detector molecules are randomly oriented with respect to the external field. In this case, a group of detector molecules exists for which $\Delta\vec{\mu}_{ij,ind} \perp \vec{E}_{ext}$. The absorption line shift of this group of molecules is governed by the second and higher order terms in Eqn. (11), but is not zero as implied by the simple approach. The consequence of restricting the analysis to the first order term is that the width of the distribution of the values for $\Delta\vec{\mu}_{ind}$ will be massively overestimated. This leads to wrong conclusions about the structure of the local environment of the detector molecule.
- The induced dipole moment difference $\Delta\vec{\mu}_{ind}$ is a quantity that characterizes both the detector molecule and the host environment (see Eqn. (10)). The quantity that we are seeking to determine – the internal electric field – characterizes the host environment (subject to the limitations described by Eqns. (3) and (4)).

In spite of these limitations, this procedure can provide useful information. Particularly for nonpolar, centrosymmetric detector molecules in centrosymmetric host materials, the absence of a linear effect shows that the centrosymmetry of the host is maintained in the presence of the detector molecule. According to Eqn. (10), $\Delta\vec{\mu}_{ind} = 0$ implies that the internal electric field at the detector molecule site vanishes. This is the case, for example, for octatetraene in n-octane, where an upper limit of less than 1 kV/cm has been estimated for the internal field [33]. The molecular resolution method, which will be described in the following paragraphs, determines an internal field of 1.6 MV/cm at the octatetraene site in n-hexane [9]! The central feature of the molecular resolution method [9,23,26] is that it allows for the direct determination of internal fields without knowledge of values for the polarizability tensor difference $\Delta\vec{\alpha}_{ij}$. The starting point is the Hamiltonian (6), which implies that the detector molecule is again treated in a point dipole approximation. Now, however, we interpret the electric field in Eqn. (6) to be the total electric field at the detector molecule site. With $\vec{E} \equiv \vec{E}_{tot}$ we have

$$H = H_0' - \vec{E}_{tot} \cdot \hat{\vec{\mu}}, \qquad (12)$$

In this case the Hamiltonian H_0' describes the molecular electronic states including all interactions between detector and host molecules *except* for the electrostatic interaction. Therefore, H_0' produces the electronic state energies of the detector molecule in its host environment with the external and the internal electric field "turned off." These energies are not experimentally observable and therefore have to be treated as parameters in the analysis.

If the internal electric fields are smaller than the intramolecular fields, a perturbation treatment of H is still allowed. It is advantageous, however, to numerically diagonalize H for each value of the externally applied field in order to obtain the new external field dependent energy levels. Fitting the calculated energies to the measured line shifts allows extrapolating to the internal electric field. While the diagonalization and perturbation analyses both suffer from the limited number of states that can be used in the calculation (basis set problem), the diagonalization of the Hamiltonian is equivalent to infinite order perturbation theory and therefore provides a more reliable result.

The electric field in the Hamiltonian (12) always appears as an inner product with the transition dipole operator. This means that experimentally the projections of the internal electric field onto the relevant transition dipole moments are determined. Moreover, the method yields one value for the internal field for the entire detector molecule. This method will be applied below to proteins and *n*-alkane crystals.

The models described so far are based on the point dipole approximation, which implies a homogeneous electric field over the volume of the detector molecule. While this condition might be satisfied reasonably well for the local fields [25], internal fields are likely to be quite inhomogeneous. In order to account for the inhomogeneity of the internal field, Kohler et al. [25] developed a procedure that incorporates the effect of internal and external electric field on a detector molecule at an atomic level. The total electric field is calculated at the sites of the atoms of the detector molecule and the perturbation of the atomic orbitals is calculated. In the next step, the atomic orbitals are combined to form molecular orbitals. Their energies, which depend on the external and internal field strengths, are then compared to experimental data. The atomic resolution method requires knowledge of the following: (1) the structure of the local environment, either from molecular mechanics calculations or from X-ray or NMR-data (for biological systems, particularly); (2) a model for the charge distribution of the host molecules (preferably a point charge model); (3) a model for electronic structure of the detector molecule that is amenable to the calculation of atomic levels perturbations. Results are presented below.

Optical Absorption Spectra of Molecules in a Host Matrix

The width of the electronic absorption line is determined by the energy and phase relaxation processes. This width is known as the homogeneous line width. When a molecule is incorporated into a host matrix, the interaction of this molecule with the host matrix leads to a shift of the transition energy (and also to a broadening since the solid host matrix offers additional relaxation channels). The strength of this interaction and therefore the magnitude of the line shift depend on the relative separation and relative orientation of guest and host molecules. If the local structure is unique, that is, if each of the guest molecules has exactly the same microscopic local environment, all guest molecules will experience exactly the same shift of their electronic state energies. In reality, however, a distribution of local environments exists, which, in turn, leads to a distribution of transition energies, provided that the electronic states involved in the optical transition under study respond differently to the slight variations in the environment (which for transitions involving π-electron states is almost always the case). This results in an inhomogeneously broadened absorption band (from here on also called the inhomogeneous band), which is the envelope of a large number of homogeneous absorption lines that are shifted with respect to each other (see Figure 1a). The ratio Γ of inhomogeneous width to homogeneous width – sometimes referred to as the multiplexing factor – can exceed 10^5 [34]! For example, for one the systems that will be discussed in more detail below – octatetraene in n-hexane – this ratio is approximately 3×10^4.

Since the homogeneous line widths are now masked by the inhomogeneous distribution, we are faced with a severe loss of spectral resolution. All methods for the determination of internal fields are based on the application of an external electric field to the sample. In order to generate measurable changes to the entire inhomogeneous band, the external fields (just like other external perturbations like magnetic fields, pressure, or temperature) have to be substantial, of the order or even exceeding the magnitude of the internal fields. As a matter of fact, the external electric fields necessary to produce detectable shift of the entire inhomogeneous band are roughly a factor of Γ larger than those needed for measurable homogeneous absorption line shifts! Under the experimental conditions required for these experiments – they are carried out at liquid helium temperatures – fields of this magnitude are hard to sustain, as gas discharge effects lead to a breakdown of the fields. Moreover, high external fields modify the charge distributions in the host molecules substantially - we can no longer talk about performing the measurements in near equilibrium (meaning external field free) conditions.

In spite of these restrictions a number of Stark experiments on inhomogeneous absorption bands were carried out (see e.g. [35]). Electroabsorption spectra, which employ an electric field modulation technique to allow for higher peak external field strengths, were also recorded [10,11]. However, since the entire band consists of a multitude of local environments, the resulting induced dipole moment difference represents the average over a wide distribution of $\Delta \vec{\mu}_{ind}$-values. Stark experiments using the line-narrowing techniques described in the following section demonstrated a systematic variation of induced dipole moment difference (see e.g. refs. [36-39]) and internal electric field across the inhomogeneous band [23,40,41].

Line Narrowing Techniques

In order to overcome the limitations imposed by inhomogeneous line broadening, a number of line narrowing techniques have been developed. Collectively, these techniques have been referred to as site selection spectroscopy [42], since only certain molecular sites (i.e. a detector molecule and its local environment) in the sample are addressed in the experiment. However, the selection is made through the selective excitation of a subset of molecules within the inhomogeneous absorption band with monochromatic laser light. If site structure and detector molecule absorption energy are correlated, only a specific site is selected. This is not necessarily the case, since different sites can accidentally lead to identical transition energies.

The first such technique was fluorescence line narrowing spectroscopy (FLN) [43,44]. The emission of the subset of detector molecules that is selectively excited with a narrow band laser consists of vibronic fluorescence lines that are considerably narrower than the inhomogeneous bands. To achieve the maximum gain in resolution (that is, the largest narrowing effect), the laser line width has to be less than the homogeneous line width of a molecular transition. In this case, the gain in resolution is given roughly by the multiplexing factor Γ. This technique is reviewed in Ref. [42].

Shortly after the first demonstration of FLN, the line narrowing properties of persistent spectral hole burning were demonstrated [45,46]. The hole burning technique relies on the use of detector molecules that can undergo a phototransformation from the electronically excited state. Again, the initial step is the selective excitation of a subset of detector molecules at the laser frequency. Only the excited subset now undergoes a phototransformation (see Figure 1b). If the photoproducts absorb at different frequencies, the absorption at the laser frequency (the so-called "burning frequency") is depleted, leading to the formation of an indentation in the spectrum – the spectral hole. In the ideal case, the width of the spectral hole is twice the homogeneous width [47]. The lifetime of the spectral hole depends on the nature of the phototransformation mechanism and on the temperature. At liquid helium temperatures, a large number of detector/host systems exist that allow for a hole lifetime that greatly exceeds the experimental time scale. For reviews see refs. [34,47].

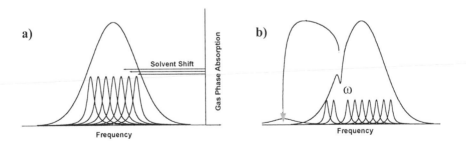

Figure 1: a) Inhomogeneous absorption band as superposition of shifted homogeneous lines. b) Spectral hole burnt at frequency ω; the photoproduct absorption is removed (arrow) from the original absorption band.

In our studies, the spectral hole serves as a narrow frequency marker, which responds with high sensitivity to external perturbations such as electric fields. Performing the Stark experiments on spectral holes rather than the entire inhomogeneous absorption band is advantageous as the external fields required to induce measurable changes of the spectral holes are smaller by typically several orders of magnitude smaller. The gain in resolution is again related to the multiplexing factor Γ.

Following the first observation of the Stark effect on spectral holes [48], theoretical models for the analysis of hole-shifts and -splittings were developed (see e.g. Ref. [31]). These models are usually based on an interaction term linear in the external field and are therefore subject to the restrictions described earlier. When analyzed with the advanced models [9,23-26] that are the focus of this chapter, the external electric field induced changes of spectral holes allow for a direct, quantitative determination of internal electric fields.

As mentioned above, a spectral hole usually cannot be associated with a particular molecular environment of the detector molecule, as different molecular environments can lead to identical detector molecule absorption energies. However, the reduction of sites compared to the number of different sites in an inhomogeneous band is substantial. Advances in molecular spectroscopy allowed for the ''ultimate'' increase in spectral resolution: the detection of single molecules embedded in solid host materials [49-51]. Single molecule spectroscopy is true site selection spectroscopy since now individual environments can be probed! Single molecules can be detected using their absorption using modulation techniques [49,50,52] or their fluorescence [51]. So far about 20 detector/host combinations are known that are suitable for single molecule spectroscopy [53]. Reviews can be found in refs. [53-56].

The first Stark experiments on single molecules were performed using centrosymmetric detector molecules in crystalline and amorphous hosts: pentacene in para-terphenyl [32] and terrylene in polyethylene [57]. The results confirmed earlier hole-burning results: in polyethylene, the amorphous nature of the local environments of the detector molecules manifested itself in strongly varying linear Stark effect for different molecules. In a crystalline para-terphenyl host the Stark effect was largely quadratic. The small linear contribution, which was estimated to correspond to an internal field of less than 10 kV/cm, verified that the centrosymmetry of the host crystal is maintained to a large degree. These results further elucidate the link between internal electric field and site geometry.

Clearly, the possibility of measuring internal fields for individual detector molecules presents exciting prospects. As mentioned above, however, the number of systems suitable for single molecule spectroscopy is rather limited [53]. This is likely to change in the future, particularly if absorption techniques for the measurement of single molecules are developed further [52]. At present, however, we believe that spectral hole burning still holds an advantage with respect to the requirements that the investigated systems have to meet and also with respect to the experimental requirements. Therefore, in our studies – some of which will be described below – we employ the hole-burning technique for the measurement of internal electric fields.

Applications: Internal Electric Fields in *n*-Alkanes and Proteins

Before presenting the first results obtained with the molecular and atomic resolution methods, it is necessary to comment briefly on the local electric field. Classical electrodynamics accounts for the screening effects and reaction fields in terms of a tensor $\bar{f}$ that links external and local fields. This tensor can be factored in a product of two tensors $\bar{D}$ and $\bar{Q}$, which reflect the polarization of the host material by the external field and the polarizability difference of guest and host molecules, respectively [27]. Due to the polarizability difference $\Delta \bar{\alpha}_{ij}$, the local field will also depend on the electronic state of the detector molecule. Usually the latter two contributions are neglected. In this case only the Lorentz correction $\bar{f} = (\varepsilon + 2)/3$ is used, which holds true for a virtual spherical cavity of dimensions large compared to the molecular volume surrounded by an isotropic continuum with dielectric constant ε. Also, this expression is only valid within the point dipole model. An improved treatment has to consider the host molecules close to the detector molecule as anisotropic polarizable charge distributions, while molecules far from the detector molecule can still be treated with a macroscopic polarization field [58]. This approach was employed in the first application of the atomic resolution method [25] (see below).

Molecular Resolution Analysis for Crystals and Proteins

So far, two different types of systems were analyzed using the molecular resolution method: octatetraene (OT) in *n*-hexane and *n*-heptane and the proteins myoglobin and cytochrome c. Both proteins already contain a "built-in" detector molecule, namely heme, an iron porphyrin. Its optical absorption is far removed from the protein backbone absorption. Also, heme, being the active site, is ideally positioned to investigate the electrostatics-function link. The hole-burning properties of heme being rather poor and to achieve maximum resolution gain through the hole-burning method, heme was replaced with its free-base equivalent protoporphyrin IX. This replacement will modify the internal field to some extent through the modified polarization and reaction field. Future experiments using metal porphyrins will shed light on the importance of this modification.

   Figure 2a shows a spectral hole, which was burnt into a $\pi\pi^*$-transition of myoglobin, at zero external field (bottom) and with an external field of 11 *kV/cm*. The splitting of the spectral hole shows that there is a well-defined internal electric field at the heme site – a field distribution would conceal the splitting. We performed the diagonalization of the Hamiltonian (12) for a basis set of 5 molecular states, the ground state and two pairs of excited states that are orthogonally polarized. Therefore, we obtain two orthogonal components of the internal field in the heme plane (see Table I). In octatetraene, only one component along its long axis can be obtained. The transition dipole moments are taken from Refs [59,60]. For myoglobin the components are 29.9 *MV/cm* and 38.6 *MV/cm*, respectively. These fields are substantially larger than the molecular resolution field found in nonpolar n-alkanes [9,25,26] (see Table I)! Therefore, charged groups in close proximity to the π-electron systems must be the primary sources of this field: it turns out the deprotonated propionic acid side chains of protoporphyrin are essentially the exclusive sources of the *in-plane* fields [23,24,61]. In cytochrome c the fields are somewhat smaller. The protonation state of the propionates in not clear, since heme is buried in cytochrome c. Also, heme in cytochrome c is surrounded by charged lysine groups, which will affect the field. An interesting fact is that the fields in *n*-heptane are significantly smaller than those in *n*-hexane, in spite of their almost identical (macroscopic) dielectric properties! This demonstrates again that continuum dielectric models cannot reflect the electrostatic properties at a local level.

Table I: Measured Internal Electric Fields in Proteins and *n*-Alkanes

	$E_{int,x}$ [MV/cm]	$E_{int,y}$ [MV/cm]	$E_{int,axis}$ [MV/cm]
Myoglobin	29.9	38.6	–
Cytochrome C	11	24	–
OT/*n*-hexane	–	–	2
OT/*n*-heptane	–	–	0.5

Atomic Resolution Analysis for Crystals

The atomic resolution method was first applied to the analysis of holes burnt into ππ*-transitions of octatetraene [18,26,62] in *n*-hexane and *n*-heptane. The required input data (see above) is obtained as follows: (1) the microscopic structure of the local environment is calculated using molecular mechanics simulations; (2) the *n*-alkane charge distributions – consisting of atom and bond centered charges – are taken from refs. [25,63,64], and (3) the electronic structure of octatetraene is modeled with a modified Hückel model [65].

Given the structure of the local environment and the location and magnitude of the point charges, the internal fields at the carbon atom sites of OT can be calculated with Eqn. (2). In order to calculate the external field induced hole shifts, the local field has to be added to the internal fields, also atomically resolved. This can be accomplished by treating the polarizability of a host molecule as the sum over the polarizabilities of its bonds [66]. The effect of the external field is to displace the bond centered point charges according to the bond polarizabilities, thereby inducing a bond dipole moment. The difference of the fields generated by the displaced charges and the charges in the original positions is then added to the Lorentz corrected external field to yield the local field. After these corrections, the local field – in contrast to the external field – is no longer homogeneous. Also, for some atomic sites the local field is less than the external field, which according to the commonly used Lorentz correction is not possible!

After the calculation of the total field perturbation for each *atomic* orbital, molecular orbitals are formed and their energies fitted to the data. A large number of hole spectra are analyzed simultaneously, such that the resulting internal electric field map is uniquely defined. Figure 2b shows the case of OT in *n*-hexane. The arrows represent the measured internal field at the carbon atom sites of OT. The internal fields vary dramatically in magnitude and orientation over the length of the molecule. It has to be pointed out that for each carbon atom of OT, all three components of the internal electric field vector are available (for details see Ref. [25]).

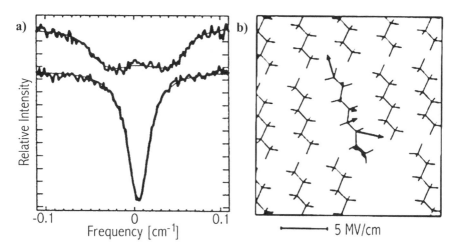

Figure 2: a) Bottom trace: spectral hole burnt in protoporphyrin IX substituted myoglobin; Top trace: the same hole in the presence of an external field of 11 kV/cm. b) Measured internal fields at the carbon atoms of octatetraene (in the center) – the internal electric field is strongly inhomogeneous in magnitude and orientation. (a) Adapted from Ref. [23]. Copyright 1995 American Chemical Society. b) Adapted from Ref. [25]. Copyright 1995 American Institute of Physics.)

Summary and Conclusions

Two novel methods were introduced that offer unprecedented access to the electrostatic properties of the environment of detector molecules in solids. As proof of principle, these methods were applied to two very different systems: molecular crystals and proteins. The molecular resolution method, despite its approximations and its simplicity, produced internal electric fields that are consistent with expectation. For example, in myoglobin, the internal field in the plane of the porphyrin detector molecule are created by the closest charged groups – the propionic acid side chains. In *n*-hexane and *n*-heptane the results were surprising: the measured internal fields at the octatetraene detector molecule differ by a factor of four, even though the macroscopic dielectric properties of both hosts are almost identical. This shows the importance of the local geometry for the internal fields. Another feature of this method is that the structure of the local environment of the detector molecule is not required.

More detailed information is provided by the atomic resolution method, which was only applied to octatetraene in *n*-hexane and *n*-heptane. In both cases an array of eight values for the internal field can be determined – one value for each of the carbons of octatetraene. Moreover, this method also provides a microscopic description of the shielding of the detector molecule from external fields using bond polarizabilities of the host molecules. One result was that the almost universally employed Lorentz correction for the local field is quantitatively and qualitatively incorrect in this case.

We feel that there is great potential for these methods. Once the basic ingredients are in place, many different types of systems are open for study. More precisely, once a suitable model for the electronic states of the detector molecule that allows for the incorporation of electric field effects at an atomic level is obtained, this detector molecule can – in principle – be inserted in many different systems. For biological systems the outlook is equally promising: since many biological systems contain porphyrins in their native forms, they are all open to investigation with these models.

References

1. Böttcher, C. J. F. *Theory of Electric Polarization*, 2nd ed.; Elsevier Scientific Publishing Company: Amsterdam, 1973; Vol. 1.
2. Bogner, U.; Schätz, P.; Seel, R.; Maier, M. *Chem. Phys. Lett.* **1983**, *102*, 267-271.
3. Greiner, W. *Klassische Elektrodynamik*; Harri Deutsch: Thun & Frankfurt am Main, 1986; Vol. 3.
4. Maier, M. *Appl. Phys. B* **1986**, *41*, 73-90.
5. Chipot, C.; Maigret, B.; Rivail, J. L.; Scheraga, H. A. *J. Phys. Chem.* **1992**, *96*, 10276-10284.
6. Chipot, C.; Angyan, J.; Maigret, B.; Scheraga, H. A. *J. Phys. Chem.* **1993**, *97*, 9788-9796.
7. Chipot, C.; Maigret, B.; Rivail, J. L.; Scheraga, H. A. *J. Phys. Chem.* **1994**, *98*, 1518-1518.
8. Boxer, S. G.; Franzen, S.; Lao, K. Q.; Lockhart, D. J.; Stanley, R.; Steffen, M.; Stocker, J. W. Electric Field Effects on the Quantum Yields and Kinetics of Fluorescence and Transient Intermediates in Bacterial Reaction Centers. In *The Photosynthetic Bacterial Reaction Center II*; Breton, J., Verméglio, A., Eds.; Plenum Press: New York, 1992.
9. Gradl, G.; Kohler, B. E.; Westerfield, C. J. *J. Chem. Phys.* **1992**, *97*, 6064-6071.
10. Middendorf, T. R.; Mazzola, L. T.; Lao, K. Q.; Steffen, M. A.; Boxer, S. G. *Biochim. Biophys. Acta* **1993**, *1143*, 223-234.
11. Steffen, M. A.; Lao, K. Q.; Boxer, S. G. *Science* **1994**, *264*, 810-816.
12. de Silva, A. P.; Rice, T. E. *Chem. Commun.* **1999**, *Jan 21*, 163-164.

13. Voet, D.; Voet, J. G. *Biochemistry*; John Wiley & Sons: New York, 1995.
14. Gilson, M. K.; Straatsma, T. P.; McCammon, J. A.; Ripoll, D. R.; Faerman, C. H.; Axelsen, P. ; Silman, I.; Sussman, J. L. *Science* **1994**, *263*, 1276-1278.
15. Getzoff, E. D.; Cabelli, D. E.; Fisher, C. L.; Parge, H. E.; Viezzoli, M. S.; Banci, L.; Hallewell, R. A. *Nature* **1992**, *358*, 347-351.
16. Springer, B. A.; Sligar, S. G.; Olson, J. S.; G.N. Phillips, J. *Chem. Rev.* **1994**, *94*, 699-714.
17. Borman, S. A Mechanism Essential to Life. In *Chem. Eng. News*, 1999; pp 31-36.
18. Kohler, B. E.; Personov, R. I.; Woehl, J. C. Electric Field Effects in Molecular Systems Studied Via Persistent Hole Burning. In *Laser Techniques in Chemistry*; Myers, A. B., Rizzo, T. R., Eds.; John Wiley & Sons: New York, 1995; Vol. XXIII; pp 283-323.
19. Onsager, L. *J. Am. Chem. Soc.* **1936**, *58*, 1486.
20. Bordewijk, P. *Physica* **1970**, *47*, 596-600.
21. Chen, F. P.; Hanson, D. M.; Fox, D. *J. Chem. Phys.* **1975**, *63*, 3878-85.
22. Chen, F. P.; Hanson, D. M.; Fox, D. *J. Chem. Phys.* **1977**, *66*, 4954-60.
23. Geissinger, P.; Kohler, B. E.; Woehl, J. C. *J. Phys. Chem.* **1995**, *99*, 16527-16529.
24. Geissinger, P.; Kohler, B. E.; Woehl, J. C. *Synthetic Metals* **1997**, *84*, 937-938.
25. Kohler, B. E.; Woehl, J. C. *J. Chem. Phys.* **1995**, *102*, 7773-7781.
26. Kohler, B. E.; Woehl, J. C. *Mol. Cryst. Liq. Cryst.* **1996**, *291*, 119-134.
27. Hanson, D. M.; Patel, J. S.; Winkler, I. C.; Morrobel-Sosa, A. Effects of Electric Fields on the Spectroscopic Properties of Molecular Solids. In *Spectroscopy and Excitation Dynamics of Condensed Molecular Systems*; Agranovich, V. M., Hochstrasser, R. M., Eds.; North-Holland Publishing Company: Amsterdam, 1983; Vol. 4; pp 621-679.
28. Brunel, C.; Tamarat, P.; Lounis, B.; Woehl, J. C.; Orrit, M. *J. Phys. Chem. A* **1999**, *103*, 2429-2434.
29. Kador, L.; Haarer, D.; Personov, R. I. *J. Chem. Phys.* **1987**, *86*, 53000-5307.
30. Meixner, A.; Renn, A.; Wild, U. P. *Chem. Phys. Lett.* **1992**, *190*, 75-82.
31. Schätz, P.; Maier, M. *J. Chem. Phys.* **1987**, *87*, 809-820.
32. Wild, U. P.; Güttler, F.; Pirotta, M.; Renn, A. *Chem. Phys. Lett.* **1992**, *193*, 451-455.
33. Kohler, B. E., personal communication.
34. *Persistent Spectral Hole-Burning: Science and Applications*; Moerner, W. E., Ed.; Springer-Verlag: Berlin, 1988; Vol. 44, pp 315.
35. Barker, J. W.; Noe, L. J.; Marchetti, A. P. *J. Chem. Phys.* **1973**, *59*, 1304-13.
36. Altmann, R. B.; Renge, I.; Kador, L.; Haarer, D. *J. Chem. Phys.* **1992**, *97*, 5316-5322.
37. Altmann, R. B.; Haarer, D.; Renge, I. *Chem. Phys. Lett.* **1993**, *216*, 281-285.
38. Altmann, R. B.; Haarer, D.; Ulitsky, N. I.; Personov, R. I. *J. Luminesc.* **1993**, *56*, 135-141.
39. Altmann, R. B.; Kador, L.; Haarer, D. *Chem. Phys.* **1996**, *202*, 167-174.

40. Geissinger, P.; Kohler, B. E.; Woehl, J. C. *Mol. Cryst. Liq. Cryst* **1996**, *283*, 249-254.

41. Geissinger, P.; Kohler, B. E.; Woehl, J. C. *Mol. Cryst. Liq. Cryst.* **1996**, *283*, 69-74.

42. Personov, R. I. Site Selection Spectroscopy of Complex Molecules in Solutions and Its Applications. In *Spectroscopy and Excitation Dynamics of Condensed Molecular Systems*; Agranovich, V. M., Hochstrasser, R. M., Eds.; North-Holland Publishing Co.: Amsterdam, 1983; Vol. 4; pp 555-619.

43. Szabo, A. *Phys. Rev. Lett.* **1970**, *25*, 924-6.

44. Personov, R. I.; Al'shits, E. I.; Bykovskaya, L. A. *Opt. Commun.* **1972**, *6*, 169-173.

45. Kharlamov, B.; Personov, R. I.; Bykovskaja, L. A *Opt. Commun.* **1974**, *12*, 191-193.

46. Gorokhovskiî, A.; Kaarli, R.; Rebane, L. *JETP Lett.* **1974**, *20*, 216-218.

47. Friedrich, J.; Haarer, D. *Angew. Chem., Int. Ed. Engl.* **1984**, *23*, 113-140.

48. Marchetti, A. P.; Scozzafava, M.; Young, R. H. *Chem. Phys. Lett.* **1977**, *51*, 424-426.

49. Moerner, W. E.; Kador, L. *Anal. Chem.* **1989**, *61*, A1217-A1223.

50. Moerner, W. E.; Kador, L.; Ambrose, W. P. *Mol. Cryst. Liq. Cryst.* **1990**, *183*, 47-57.

51. Orrit, M.; Bernard, J. *Phys. Rev. Lett* **1990**, *65*, 2716-2719.

52. Kador, L.; Latychevskaia, T.; Renn, A.; Wild, U. P. *J. Chem. Phys.* **1999**, *111*, 8755-8758.

53. Moerner, W.E.; Orrit, M. *Science (Washington, D.C.)* **1999**, *283*, 1670-1676.

54. Moerner, W. E.; Basche, T. *Angew. Chem., Int. Ed. Engl.* **1993**, *32*, 457-476.

55. Plakhotnik, T.; Donley, E. A.; Wild, U. P. *Annu. Rev. Phys. Chem.* **1997**, *48*, 181-212.

56. Xie, X. S.; Trautman, J. K. *Annu. Rev. Phys. Chem.* **1998**, *49*, 441-480.

57. Orrit, M.; Bernard, J.; Zumbusch, A.; Personov, R. I. *Chem. Phys. Lett.* **1992**, *196*, 595-600.

58. Jackson, J. D. *Classical Electrodynamics*, 2nd ed.; John Wiley & Sons: New York, 1975.

59. Baker, J. D.; Zerner, M. C. *Chem. Phys. Lett.* **1990**, *175*, 192-196.

60. Soos, Z. G.; Ramasesha, S. *J. Chem. Phys.* **1989**, *90*, 1067-1076.

61. Geissinger, P.; Kohler, B. E.; Woehl, J. C. "Measuring Internal Electric Fields in Proteins Using Hole-Burning Spectroscopy"; International Conference on Lasers '96, 1996, Portland, Oregon, U.S.A.

62. Kohler, B. E.; Woehl, J. C. *J. Phys. Chem. A* **1999**, *103*, 2435-2445.

63. Jolly, W. L.; Bakke, A. A. The Study of Molecular Electron Distribution By X-Ray Photoelectron Spectroscopy. In *Electron Distributions and the Chemical Bond*; Coppens, P., Hall, M. B., Eds.; Plenum Press: New York, 1982; pp 431-445.

64. Chipot, C.; Ángyàn, J. G.; Ferenczy, G. G.; Scheraga, H. A. *J. Phys. Chem.* **1993**, *97*, 6628-6636.

65. Kohler, B. E. *J. Chem. Phys.* **1990**, *93*, 5838-5842.

66. Denbigh, K. G. *Trans. Faraday Soc.* **1940**, *36*, 936-947.

Organic Thin Films

Chapter 3

Self-Assembly Growth of Organic Thin Films and Nanostructures by Molecular Beam Deposition

Chengzhi Cai[1,3], Martin Bösch[1], Christian Bosshard[1], Bert Müller[1,4], Ye Tao[1,5], Armin Kündig[1], Jens Weckesser[2], Johannes V. Barth[2], Lukas Bürgi[2], Olivier Jeandupeux[2], Michael Kiy[1], Ivan Biaggio[1], Ilias Liakatas[1], Klaus Kern[2], and Peter Günter[1]

[1]Nonlinear Optics Laboratory, Institute of Quantum Electronics, Swiss Federal Institute of Technology (ETH), CH-8093 Zurich, Switzerland
[2]Institut de Physique Expérimentale, EPFL, PHB-Ecublens, CH-1015 Lausanne, Switzerland
[3]Current address: Department of Chemistry, University of Houston, Houston, TX 77204-5641
[4]Current address: Biocompatible Materials Science and Engineering, ETH Zürich, Switzerland
[5]Current address: Institute for Microstructural Sciences, National Research Council Canada, M-50 Montreal Road, Room 178, Ottawa, Ontario K1A 0R6, Canada

The concept of supramolecular assemblies based on strong and directional hydrogen bonding has been applied to organic molecular beam deposition (OMBD) for growth of anisotropic nanostructures and thin films. Aligned nanostructures were generated on Ag(111) surfaces. Thin films with a thickness of 100–400 nm were grown on glass substrates by oblique incidence OMBD, and studied by second harmonic generation experiments, indicating that the average direction of the dipolar molecules in the films was parallel to the projection of the molecular beam on the substrate surface. This intriguing result is rationalized by a proposed mechanism considering the fundamental processes of self-assembly on surfaces.

Introduction

Organic thin films and nanostructures have attracted increasing interest for their potential applications in a variety of advanced technologies, including nonlinear optics (NLO), microelectronics, nanotechnology, light emitting devices, field-effect transistors, liquid crystals, sensors, and solar cells (1-7). For many of these applications, the device performance is crucially dependent on the orientation of the functional molecules in the film or the nanostructure (1-3). Therefore, methods for alignment of molecules in the structures are of great technological interest. The alignment can be centrosymmetric or non-centrosymmetric. Non-centrosymmetric alignment is more challenging, because the acentric molecules need to be aligned with a directional preference. It is, however, the basic requirement for organic second order nonlinear optics (1,2) which is our primary interest.

Several techniques have been developed for the growth of anisotropic organic thin films (1-7). For the alignment of molecules perpendicular to the substrate surface, most common techniques include Langmuir-Blodgett (LB) film formation, high electric field poling, and self-assembly using the layer-by-layer methodology (4). For aligning molecules parallel to the substrate surface (in-plane), the anisotropy can be induced by an anisotropic substrate surface (7). However, if an isotropic substrate surface is required, the surface alone cannot induce an in-plane preferential alignment over a large area. To circumvent this problem, external physical means may be applied to impose a preference to the alignment. For example, we have demonstrated that, during LB film deposition, the dipping direction can be used to define the alignment direction of some second order NLO chromophores (8).

Organic Molecular Beam Deposition (OMBD)

In general, organic thin films can be deposited in solution or gas phase. Solution-based deposition is relatively easy to set up, and has been intensively studied. When we became involved in this field, we were, however, interested in an instrumentally sophisticated technique based on ultrahigh vacuum, referred to as organic molecular beam deposition (OMBD) (7). It is an offshoot of the physical vapor deposition (PVD) technique that has been widely used in microelectronic and optical industries for the deposition of inorganic thin films (9).

The OMBD process is illustrated in Figure 1. The materials in the effusion cells (d, Figure 1) are evaporated, and the gas molecules rush out of the small hole of the cell into the ultrahigh vacuum chamber, forming a molecular beam (f). Some of them stick on a relatively cold substrate (a) while the others are absorbed by the liquid nitrogen shroud (g). The ultrahigh vacuum ($< 10^{-8}$ mbar) is generated by the turbo molecular pump (h) and the liquid nitrogen shroud (g). The film thickness can be monitored with monolayer sensitivity by the quartz thickness monitor (b) or the ellipsometer (c). The deposition (on/off) can be controlled by the shutters (e).

As compared to the solution-based growth techniques, OMBD has several practical advantages. It is carried out in ultrahigh vacuum that is an ultraclean environment. It enables precise and in-situ control of substrate temperature, growth rate, and film thickness. Because the molecules in the beam under ultrahigh vacuum are too far away to interact with each other, they fly straight. If the dimensions of the substrate are much shorter than the distance between the effusion cell and the substrate, the approaching direction of the gas molecules should be approximately the same over the whole film area. As discussed later, this "beam-like" feature can be used to align molecules in the film. From a practical point of view, it also allows easy fabrication of well defined and miniature patterns using masks in front of the substrate. In addition, OMBD can combine the deposition of metals, organics, and semiconductors in the same chamber with a predetermined sequence, composition, and layer thickness, for preparation of well defined hetero-layer structures. Moreover, high growth rates in the order of micrometers per hour are possible. Combining all these advantages, OMBD appeared to be an ideal tool for the fabrication of *e.g.* integrated optical and electrical circuits based on conjugated organic molecules.

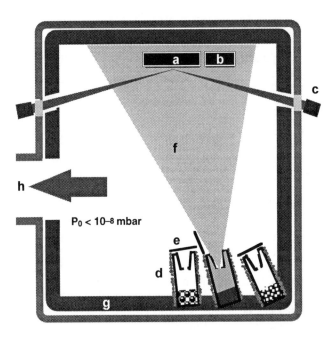

Figure 1. Illustration of Organic Molecular Beam Deposition (OMBD). *a: substrate; b: quartz crystal thickness monitor; c: ellipsometer; d: effusion cells; e: shutters; f: molecular beam; g: liquid nitrogen shroud; h: turbo molecular pump.*

Supramolecular Assemblies as Anisotropic Materials for OMBD

Despite the above advantages of OMBD, the technique has been far less developed for growth of anisotropic organic thin films as compared to the solution-based techniques. The main reason is probably that the OMBD technique is neither familiar nor accessible to many organic chemists. The shortage of input from organic chemistry hampers the development of materials for this technique.

Our main interest was to grow NLO thin films where the dipolar molecules are aligned in the same direction. In general, NLO materials can be grouped into two types: the low molecular weight crystalline materials and the polymers (1,2). It is extremely difficult to grow a large single crystalline film by OMBD. It has been demonstrated that an anisotropic polycrystalline film of 4'-nitrobenzylidene-3-acetamino-4-methoxy-aniline (MNBA), an efficient NLO material, can be grown on a centrosymmetric organic single crystal of ethylenediammonium terephthalate (10). Here, the anisotropy of the films is induced by the lattice-matched substrate surface (heteroepitaxy). However, it is difficult to apply this method to many practical inorganic substrates that are either amorphous or have a distinct lacttice constant from that of organic crystals. In fact, many well known low molecular weight NLO materials have been tried to grow on silicon and glass substrates, but at best resulting in films consisting of randomly oriented and μm-sized microcrystals that cause high scattering losses (11). On the other hand, polymers containing NLO chromophores can be easily processed to thin films by spin coating, and the dipolar chromophores can then be aligned by high electric field poling at the glass transition temperature of the polymer (1,2). Such films have a much better optical quality than the polycrystalline films of conventional low molecular weight materials. However, the non-volatile polymers are not suitable for OMBD, and the above mentioned advantages of OMBD are difficult to be realized with the polymer materials.

We reasoned that a compromise might be found by developing a new type of materials for OMBD based on supramolecular assemblies (12) where the molecules are linked to each other via strong and directional interactions such as hydrogen bonds (H-bonds). If the strong intermolecular bonds can be broken at elevated temperatures while the molecules are intact, then the materials might be suitable for OMBD. In addition, the grain boundary in such materials can be reduced (13) and the molecular alignment stabilized. Based on this idea, we designed a series of molecules having a pyridyl group at one end, and a carboxy group at the other, such as 4-[(pyridin-4-yl)vinyl]benzoic acid (**1**) and 4-[(pyridin-4-yl)ethynyl]benzoic acid (**2**) (Figure 2). They are expected to form strong and linear head-to-tail intermolecular H-bonds in the solid states (14-17).

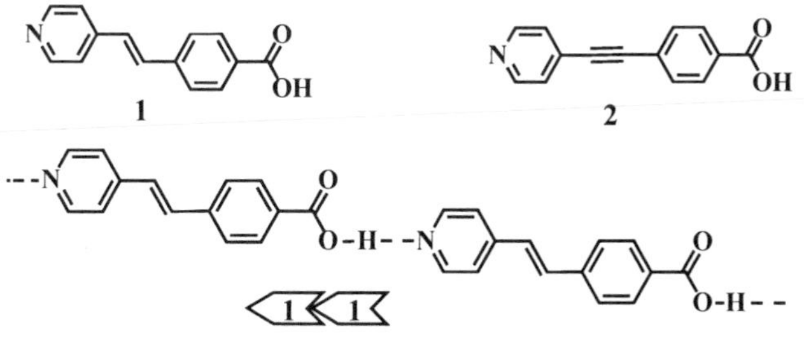

Figure 2. Molecules that can form supramolecular assemblies in the solid states.

Indeed, in the solid state ^{15}N-NMR spectra, the pyridyl ^{15}N signals of **1** and **2** appear at –105 and –106 ppm relative to that of $CH_3^{15}NO_2$ (0 ppm), while that of pyridine at –63 ppm, and that of the methyl ester of **2** at –67 ppm (15,16). The large upfield shift can only be attributed to the strong H-bonding to the pyridyl N-atom. Since there is only one peak found in the region of 0–200 ppm, it is unlikely that the solid state materials contain significant amount of short oligomers or dimers of the carboxylic acids. Otherwise, additional ^{15}N signals of the free pyridyl groups at the end of the oligomers and dimers should be present. In accord with the strong head-to-tail H-bonding, these low molecular weight materials have a high melting point (350°C for **1**, and 300°C for **2**), while the methyl ester of **1**, lacking intermolecular H-bonding, melts at 105–107°C. Compounds **1** and **2** can be sublimated at 250–220°C/0.01 mbar without decomposition, and thus are suitable for OMBD.

As revealed by the high resolution STM image (Figure 3B), where individual molecules of **1** were clearly resolved, each line actually consisted of polymer twin chains of **1** hydrogen-bonded in a linear and head-to-tail fashion along the chain direction (18). An analysis of STM contours lines demonstrates that the molecules within the twin chains were oriented anti-parallel to each other, as illustrated in the corresponding model in Figure 3C (18). When the films were grown into monolayer coverage, the perfect one-dimensional ordering was retained as shown in Figure 3D.

Since the twin chains were anti-parallel to each other and the films consisted of three rotational (120°) domains due to the threefold symmetry of Ag(111), such films are centrosymmetric in the bulk form. Nevertheless, the results are highly interesting in relation to nanotechnology and molecular electronics (5). They demonstrate that supramolecular assemblies might be processed into molecular wires by OMBD.

Growth of Multi-Layer Films with an In-Plane Directional Order

For nonlinear optic applications, it is necessary to grow macroscopically ordered films with a thickness larger than tens of nanometers. Here the key problem is how to maintain the same degree of order over a large area and through increasing thickness. We have shown that **1** can self-assemble into long and linear supramolecular polymers in a head-to-tail fashion. But we needed to find a way to orient each polymer chain in the same direction.

Initially, we reasoned that if a substrate surface is functionalized so that it bonds only one end of the molecules **1** and **2**, the head-to-tail H-bonding of the continuously arriving molecules should lead to a molecular alignment in the direction that is perpendicular to the substrate surface (19). Following this idea, we silylated glass substrates with 2-(4-chlorosulfonylphenyl)ethyltrichlorosilane to provide a surface of sulfonic acid groups that prefer to bond the pyridyl group of *e.g.* **1** (Figure 4A). Films of **1** and **2** were then grown on the silylated and bare glass or quartz substrates for a comparison.

In our home-built OMBD chamber (Figure 1), the distance between the beam source and the substrate ($2x2$ cm^2) was 26 cm, hence the molecular beam direction over the whole substrate surface is almost constant (Figure 4B). The deposition angle, defined as the angle between the molecular beam and substrate surface normal, was ~26°. Before OMBD, **1** and **2** were ground into fine powder, and degassed at

$100–120°C/10^{-9}$ mbar overnight. The substrates were washed with acetone in ultrasound for 5 min, and then dried at 120°C and 10^{-6} mbar for 0.5 h. During deposition, the base pressure was 5 x 10^{-9} mbar, the evaporation temperature was 230°C, and the substrate temperature was varied between 30 and 100°C. The deposition rate was about 5 nm/min. SHG experiments were performed using a BMI Nd:YAG laser at 1064 nm (7 ns pulses, 10 Hz repetition rate).

To our surprise, all experimental results (15-21) indicated that *1 and 2 in the multilayer films were preferentially lying flat on the substrate surface.* In addition, *they had a preferential in-plane direction that was parallel (or antiparallel) to the X_3 axis, defined as the projection of molecular beam direction on the substrate surface* (Figure 4B). Moreover, the preferential molecular direction was the same over the *whole* large film area (2x2 cm^2), and was **not** dependent on the different types of glass substrates no matter whether they were functionalized or not (19). This means that the in-plane alignment direction can be chosen simply by rotating the substrate around the X_2 axis, because the alignment direction is only defined by the molecular beam direction. This conclusion was drawn from the following SHG experiments.

In the first experiment, the polarizations of the input and output laser beam were set parallel to the X_3 axis. We rotated the substrate around the X_1 axis and recorded the output second harmonic signal as a function of the laser incident angle, that is, the angle between the X_3 and the polarization axes. The result reproduced in Figure 4C (15) shows that the SHG signal reaches the maximum when the X_3 axis is parallel to the polarization. This indicates that the dominant second harmonic susceptibility tensor of the film is parallel to the substrate surface. According to our semiempirical calculations (AM1), the second-order polarizability of the linear assemblies of **1** and **2** is dominated by its tensor component along the long molecular axis. Therefore the results indicates that the molecules are lying flat on the surface.

To find out the in-plane alignment direction, we rotated the sample and recorded the second harmonic signal as a function of the angle between the X_3 and the polarization axes. As shown in Figure 4D (15), when the X_3 axis is parallel to the polarization axis, the SHG intensity reaches its maximum, and when X_3 is perpendicular to the polarization, it drops to zero. This indicates that the molecules are preferentially aligned along the X_3 axis.

The above results hold for films of **1** and **2** grown on silylated and bare glass, quartz, and indium tin oxide (ITO) substrates at substrate temperatures ranging from 30° to 100°C. In addition, the SHG intensities at different places of the films varied within 10% which was the experimental error, indicating the same degree of order over the whole large film area (2x2 cm^2). The most noteworthy result is that *the same degree of order was obtained with different thickness from 100 to at least 400 nm,* as shown by the quadratic relationship (22) of the SHG intensities with the film thickness (Figure 4D) (15). In contrast, we are not aware of any other self-assembled films that could maintain their initial order beyond a thickness of 100 nm. In general, the disorders tend to accumulate during the growth, although this problem might be overcome by the laborious and time consuming layer-by-layer methodology (4).

The films grown on glass below 100°C were transparent and homogeneous. As revealed by scanning electron microscope (Figure 4E), films of **1** grown even at 100°C still had a featureless surface. The surface roughness (~ 5 nm) measured by atomic force microscopy was smaller than the roughness of the substrate surface. In contrast, the other low molecular weight NLO materials we examined readily formed μm-sized crystallites as observed by light microscope. The low tendency for **1** to form

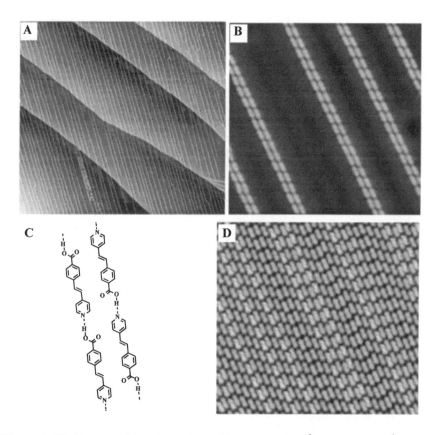

Figure 3. STM images of a submonolayer (A: 400 x 400 nm^2, B: 18 x 18 nm^2, C: model) and a monolayer (D, 20 x 20 nm^2) film of **1** grown on Ag(111) at 300K. Individual molecules of **1** are clearly resolved in the high resolution images B and D. The data were obtained at 77K. B is reproduced with permission from ref. 18.

large crystallites can be attributed to the directional head-to-tail H-bonding which dominates the other intermolecular interactions. The strong H-bonding is also expected to stabilize the polar order. Indeed, the SHG intensity of the films decreased only slightly before reaching 190°C for **1** and 180°C for **2**.

Among the above results, the most intrigue one is that the molecular alignment direction is defined by the projection of the obliquely incident molecular beam on the substrate surface (X_3 axis). In fact, it took us a long time to come to this conclusion and to be convinced that the alignment was not due to the possible substrate anisotropy. But then how can we explain this? The molecules in the molecular beam can rotate freely although they fly in the same direction, then why should they preferentially align along the X_3 axis? In addition, why the same degree of order can be kept for hundreds of layers without accumulation of errors?

Self-Assembly of Nanostructures by OMBD

We reasoned that the superstructures of ultra-thin films of **1** and **2** on a flat metal surface might be observed directly by scanning tunneling microscopy (STM). A series of ultra-thin films of **1** were then grown on the Ag(111) surface by OMBD, and characterized by in situ STM (18). Since Ag is a noble metal, the absorbate/substrate interactions are weak, and in view of the smoothness of the close-packed (111) geometry, it was expected that the intermolecular interaction could be reflected by the molecular arrangement at the surface. Indeed, we found that upon deposition at 300 K **1** self-assembled into linear lines (the bright lines in Figure 3A) with a length of up to several μm and a width of only about 1 nm. The lines oriented along <112> directions of the Ag-lattice with mesoscopic ordering at the μm scale, only weakly affected by the atomic steps of the substrate surface. This corresponds to a one-dimensional nanograting, noting that the distance between the parallel lines was about 10 nm.

Proposed Mechanism of Self-Assembly During OMBD

To answer the above questions, we need to consider details of the thin film growth process (17). First of all: why can the molecules in ultrahigh vacuum be deposited on a surface? Obviously, that is because the surface molecules bond the incoming molecules. This process can occur when the free energy of bonding is negative ($\Delta G = \Delta H - T\Delta S < 0$), that is, the enthalpy of bonding ($-\Delta H$) is larger than the entropy term ($-T\Delta S$) which favors dissociation particularly in ultrahigh vacuum. For **1** and **2**, the intermolecular interactions include H-bonding, van der Waas forces, and π-π stacking interaction. Without H-bonding, the other two bonding interactions appear too weak to keep the molecules from dissociating at room temperature. In fact, without a H-bond, the methyl ester of **1** has a dramatically lower melting point than **1** (106 vs 350°C). Materials with a melting point lower than 130°C usually cannot be

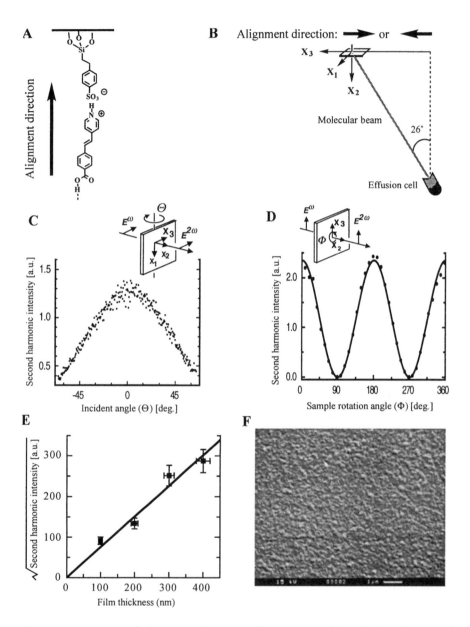

Figure 4. A: proposed alignment direction of **1** grown on silylated glass. B: setup for Oblique Incidence OMBD and the observed alignment direction of **1** and **2** on bare and silylated glass substrates. C: sample rotation angle dependent SHG. D: Incident angle dependent SHG. E. Thickness dependent SHG. F: SEM image of a film of **1** grown on glass at 100°C, reproduced with persission from reference 15.

deposited at 27°C by OMBD (11). The fact that films of **1** and **2** can be easily deposited even at 100°C should be due to the H-bonding.

Thermodynamic Aspects of the Hydrogen Bonding

For **1** and **2**, the strongest H-bonds are the tail-to-tail bonding and the head-to-tail bonding (Figure 5), and the former having two OH···O bonds is stronger than the latter having only one H···N bond. However, if many molecules are involved, the head-to-tail bonding that leads to supramolecular assemblies can be thermodynamically more favored, considering that each molecule in the chain has also two H···N bonds (Figure 5) that should be stronger than the H···O bonds. This is in accord with the above solid state ^{15}N-NMR studies that indicated the dominance of head-to-tail H-bonds. The enthalpy (ΔH) for dimerization of carboxylic acid derivatives is typically -15 kcal/mol, and the entropy (ΔS) is -36 cal/mol (23). Hence the dissociation temperature (when $\Delta G = 0$) for a tail-to-tail bond is 144°C. This is considerably higher than the desorption temperature of the film (128°C) measured by in situ ellipsometry (17). It suggests that the desorption may involves sequentially breaking the head-to-tail H-bonds of the surface molecules.

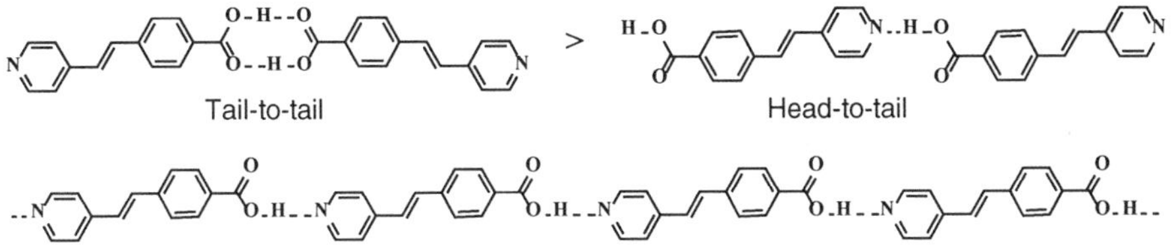

Tail-to-tail > Head-to-tail

Figure 5. The strongest H-bonds for 1.

Directional Requirement for Hydrogen Bonding

We have assumed that thin film growth of **1** and **2** is mainly due to the head-to-tail and tail-to-tail H-bonding. Now we need to consider the kinetics of the bonding. These bonds are most likely to form when two molecules collide in the ways shown by the large arrows in Figure 6, that is, the carboxy H-atom approaches another molecule along the axes of the non-bonding electron pairs at the O– or the N–atoms (24). If the collision happens in the other directions (the small arrow in Figure 6), the chance for the bonding is lower.

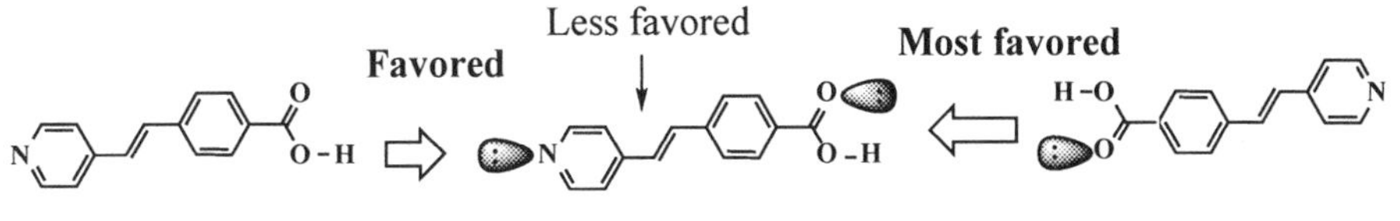

Favored   Less favored   Most favored

Figure 6. Illustration of the directional preference of H-bonding of 1.

44

As mentioned before, during OMBD, the molecules in the beam translate to the film surface in the same direction. *In this case, the H-bonding probability should be influenced by the orientation of the surface molecules.* This assumption was supported by the growth of **1** on Pd(110) surface at 27°C (25). Unlike Ag(111) surface, the Pd(110) surface bonds **1** strongly with an estimated bonding energy of ~65 kcal/mol (25). It forces **1** to lay flat on the Pd surface as shown by the in situ STM measurement (Figure 7). Interestingly, when the substrate surface was covered by a monolayer of **1** (Figure 7B) the growth stopped, that is, the incoming molecules could no longer bond to the already adsorbed molecules. This is probably due to the unfavorable orientations of the surface molecules that are determined by the substrate surface. This orientation provides the lowest chance for the surface molecules to hydrogen bond the incoming molecules (Figure 6), while the other intermolecular forces alone are too weak to keep additional molecules on the film surface. At low temperatures, those forces contribute more to the bonding of the arriving molecules, but they are far less directional than H-bonding, and the randomness is expected to increase. Indeed, the second harmonic intensity of the films of **1** grown at −190°C was only about 10% of those grown at 30°C (17).

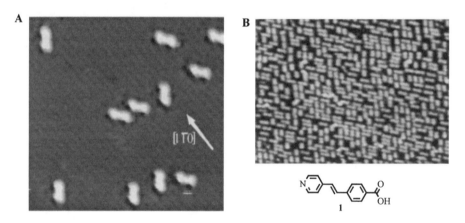

Figure 7. STM images of sub-monolayer(A) and monolayer(B) films of 1 on Pd(110). Reproduced with permission from reference 17.

Self-Correcting Effect

So far, we arrive to the conclusion that the highest growth rate is obtained when the end group (the pyridyl or carboxy) of the surface molecules tilts towards the molecular beam direction. The following discussion gives a plausible answer to how it is possible to achieve a directional alignment that remains over hundreds of layers (17). When the first layer of the molecules are deposited on a glass substrate, both the pyridyl and carboxy groups can tilt towards the molecular beam direction to capture the arriving molecules (Figure 8). Because the pyridyl groups prefer to bond to the carboxy groups, after the bonding, the molecular direction will be preserved for the surface molecules with their pyridyl group facing to the incoming molecules (Figure 8, left). For those orienting their carboxy groups towards the incoming molecules, although head-to-tail bonding to the pyridyl groups of the arriving molecules is also possible, for both thermodynamic and kinetic reasons (Figure 6), they are more likely to capture the arriving molecules through the tail-to-tail bonding. Hence, after the bonding, more than half of the carboxy groups on the surface will be changed to the pyridyl groups (Figure 8, right). Accordingly, after growth of n layers, the ratio of pyridyl vs carboxy groups on the surface will be larger than $2^n/1$ (Figure 8). This mechanism allows **1** and **2** to "self-correct" errors occurring during the growth, and hence keep the same degree of order over hundreds of layers. It also explains why the same results were obtained for multi-layer films of **1** and **2** grown on different glass substrate surfaces. A similar example of this effect on polar inclusion compounds of channel-type hydrocarbon crystals and NLO guests was provided by Hulliger et al (26). However, why is the alignment direction in our case not parallel to the molecular beam direction but to its projection on the substrate surface? This can be rationalized by the self-shadowing effect (17).

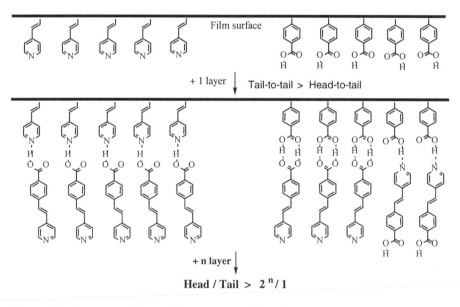

Figure 8. *Illustration of the "self-correcting effect" during OMBD with 1.*

Self-Shadowing Effect

The self-shadowing effect is a kinetic phenomenum happening at non-equilibrium conditions (low substrate temperatures). When a molecule arrives at the film surface, it should choose the nearest and less crowded bonding site on the film surface. After bonding to the surface, it then blocks the subsequent molecules from reaching its shadowing area. This leads to void formation in the film. The effect is well known for oblique incidence molecular beam deposition of inorganic thin films. However, to our knowledge, it has not yet been reported for OMBD. Considering the large size of organic molecules, the shadowing effect should be more significant for OMBD. Formation of voids in the outmost layers during the growth of **1** is illustrated in Figure 9. The smallest void diameter should be larger than the molecular length. The highest surface density of the pyridyl groups and the optimal H-bonding geometry are provided to the arriving molecules only when the surface molecules and hence the voids are tilted towards the molecular beam direction. The sum of the weak van der Waals and π–π stacking interactions between the tilted chains increases with the chain length. These interactions attract the molecules to fill the voids for a close packing. The most probable way to fill the voids, while keeping the chain parallel to the molecular beam direction, is that the molecules in the inner layers lie in the X_3 direction on the substrate surface through a two-center hydrogen-bonded intermediate illustrated in the gray box in Figure 9. For voids along the X_1 axis, they can be filled through the migration of the tilted molecular chains along the X_1 axis (17).

Is the Orientation of the Arriving Molecules Random?

When the incoming molecules are far from the film surface, their orientation should be random. However, when they get close to the film, they start to feel the long range interactions from the film, especially the electro-static interaction. If the film is anisotropic, such interactions are expected to impose an orientation preference to the arriving molecules. This is more likely to happen in our system, where the films are not only anisotropic but also dipolar. The distance-dependent electro-static field as a sum over the dipolar molecules in the film appears to be quite large, although calculations and quantitative measurements remain to be done to evaluate the field strength as compared to the rotational and translational kinetic energy of the arriving molecules which favors randomness. At first sight, the arriving molecules seemly prefer to orient anti-parallel to the surface molecules. However, it should be noted that **1** and **2** have a very different dipole moment when they are single molecules in the gas phase and when they are part of the supramolecular assemblies in the films. According to our AM1 calculations, the dipole moment of the arriving molecules is about 1 Debye. It mainly arises from the carboxy group and hence is perpendicular to the molecular axis as shown in Figure 9. For those molecules in the film, their dipole moment was calculated to be around 3 Debye and oriented along the molecular axis as illustrated by the upper large arrows in Figure 9. Therefore, the preferred orientation of the arriving molecules should be the one with the carboxy group facing to the film surface (Figure 9). That is the perfect orientation for H-bonding (Figure 6).

Summary

OMBD has many technological advantages over solution-based methods for deposition of organic thin films and nanostructures. It calls for materials designed to take these advantages and to meet the requirements of specific applications. In this work, we present a new concept to grow anisotropic nanostructures and dipolar multi-

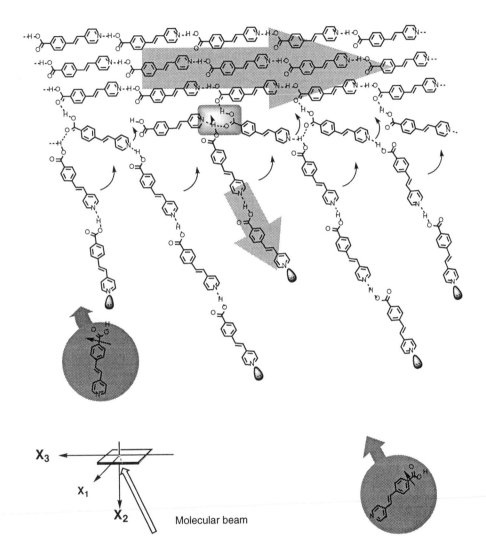

Figure 9. *Proposed mechanism for alignment of 1 along the X₃ axis, the projection of the molecular beam direction on the film surface.*

layer thin films by OMBD. It began with a simple assumption that the use of rigid rod molecules such as **1** and **2** that can form strong and linear head-to-tail H-bonding may reduce randomness. The directional interaction dominates the other non-directional interactions, and thus an anisotropic order could be generated. This idea was supported by the formation of centrosymmetric anisotropic nanostructures consisting of supramolecular assemblies of **1**. Surprisingly, a macroscopic polar order can be generated in multi-layer films of **1** and **2** on amorphous glass substrates by oblique incidence OMBD. The polar order direction is parallel to the projection of the molecular beam direction on the substrate surface. This is remarkable for a molecular beam incident angle of only 30°. In addition, the films were grown out of one source rapidly and continuously. Moreover, the polar order was independent of the film thickness in the range of 100–400 nm. To account for these intriguing results, we proposed that the growth of **1** and **2** is mainly due to the H-bonding of the arriving molecules to the surface molecules. The bonding probability is determined by the orientation of the surface molecules relative to the well-defined approaching direction of the arriving molecules. Besides this effect, the self-correcting and self-shadowing effects that are associated with the shape and bonding features of **1** and **2**, as well as the long range directional interactions of the film with the arriving molecules, all are believed to contribute to the in-plane directional alignment of the molecules.

There are still several open questions. The nonlinear optical coefficients of the films are rather low, around 0.5–1 pm/V. This is expected for **1** and **2** which have a weak donor and acceptor. The degree of the ordering is still not clear, since the molecular nonlinearity cannot be measured. We are trying to address this question by synchrotron X-ray diffraction and scattering studies. The proposed mechanism also needs more quantitative theoretical support and further experimental verification.

Although the nonlinearities obtained with the prototype molecules are still too low for practical applications, we have demonstrated that anisotropic films with a polar order can be obtained by oblique incidence OMBD with supramolecular assemblies based on strong and directional intermolecular interactions. Our results and proposed mechanisms would lead to a better understanding of the assembly of organic molecules on substrates. In the future better nonlinear optical thin films should be prepared by improving both the material design and the processing conditions. The work presented here also demonstrates the need for a multidisciplinary interaction between organic chemists, materials scientists, and phycisists.

References

1. Bosshard, C.; Sutter, K.; Prêtre, P.; Hulliger, J.; Flörsheimer, M.; Kaatz, P.; Günter, P. *Organic Nonlinear Optical Materials*, Gordon & Breach. Amsterdam, 1995.
2. *Organic Thin Films for Waveguiding Nonlinear Optics*; Kajzar, F.; Swalen J. D.; Eds.; Gordon & Breach: Amsterdam 1996.
3. Tredgold, R. H. *Order in Thin Organic Films*, Cambridge Univ. Press: Cambridge, 1994.
4. Ulman. A. *Chem. Rev.* **1996**, *96*, 1533–1554.
5. Reed, M. A. *Proceedings of the IEEE*, **1999**, *87*, 652.

6. Miyata, S.; Nalwa, H. S. Eds.; *Organic Electroluminescent Materials and Devices*, Gordon & Breach: Amsterdam, 1997.
7. Forrest, S. R. *Chem. Rev.* **1997**, *97*, 1793.
8. Decher, G.; Tieke, B.; Bosshard, C.; Günter, P. *Ferroelectrics* **1989**, *91*, 193.
9. Ohring, M. *The Materials Science of Thin Films*; Academic Press: Boston, 1992.
10. Dietrich, T.; Schlesser, R.; Erler, B.; Kündig, A.; Sitar, Z.; Günter, P. *J. Crystal Growth*, **1997**, *172*, 473.
11. Schlesser, R. Ph.D. Dissertation, ETH-Zürich, Switzerland, 1996.
12. Lehn, J.-M. *Supramolecular Chemistry*, VCH: Weinheim, 1995.
13. Sijbesma, R. P.; Beijer, F. H.; Brunsveld, L.; Folmer, B. J. B.; Hirschberg, J.; Lange, R. F. M.; Lowe, J. K. L.; Meijer, E. W. *Science*, **1997**, *278*, 1601.
14. The one-dimensional H-bond networks are shown in the crystal structure of isonicotinic acid, a lower analogue of **1** and **2**. Takusagawa, F.; Shimada, A. *Acta Crystallogr.* **1976**, *B32*, 1925.
15. Cai, C.; Bösch, M.; Tao, Y.; Müller, B.; Gan, Z.; Kündig, A.; Bosshard, C.; Liakatas, I.; Jäger, M.; Günter, P. *J. Am. Chem. Soc.* **1998**, *120*, 8563.
16. Cai, C.; Bösch, M.; Müller, B.; Tao, Y.; Kündig, A.; Bosshard, C.; Gan, Z.; Biaggio, I.; Liakatas, I.; Jäger, M.; Schwer, H.; Günter, P. *Adv. Mater.* **1999**, *11*, 745.
17. Cai, C.; Müller, B.; Weckesser, J.; Barth, J. V.; Tao, Y.; Bösch, M.; Kündig, A.; Bosshard, C.; Biaggio, I.; Günter, P. *Adv. Mater.* **1999**, *11*, 750.
18. Barth, J. V.; Weckesser, J.; Cai, C.; Günter, P.; Bürgi, L.; Jeandupeux, O.; Kern, K. *Angew. Chem.* Submitted.
19. Cai, C.; Bösch, M.; Tao, Y.; Müller, B.; Kündig, A.; Bosshard, C.; Günter, P. *Polymer Preprint*, **1998**, *39*(2), 1069.
20. Müller, B.; Cai, C.; Bösch, M.; Jäger, M.; Bosshard, C.; Günter, P.; Barth, J. V.; Weckesser, J.; Kern, K. *Thin Solid Films* **1999**, *343-344*, 171.
21. Müller, B.; Cai, C.; Kündig, A.; Tao, Y.; Bösch, M.; Jäger, M.; Bosshard, C.; Günter, P. *Appl. Phys. Lett.*, **1999**, *74*, 3110.
22. Lin, W.; Lin, W.; Wong, G. K.; Marks, T. J. *J. Am. Chem. Soc.* **1996**, *118*, 8034.
23. Curtiss, L. A.; Blander, M. *Chem. Rev.* **1988**, *88*, 827.
24. Legon, A. C.; Millen, D. J. *Acc. Chem. Res.* **1987**, *20*, 39.
25. Weckesser, J.; Barth, J. V.; Cai, C.; Müller, B.; Kern, K. *Surf. Sci.*, **1999**, *431*, 168.
26. Roth, S. W.; Langley, P. J.; Quintel, A.; Wubbenhorst, M.; Rechsteiner, P.; Rogin, P.; Konig, O.; Hulliger, J. *Adv. Mater.* **1998**, *10*, 1543.

Chapter 4

Unimolecular Rectification between 370 K and 105 K and Spectroscopic Properties of Hexadecylquinolinium Tricyanoquinodimethanide

Robert M. Metzger, Bo Chen, and Jeffrey W. Baldwin

Laboratory for Molecular Electronics, Department of Chemistry, University of Alabama, Tuscaloosa, AL 35487-0336

Hexadecylquinolinium tricyanoquinodimethanide (1) is a unimolecular rectifier of electrical current with a large hypsochromic shift in the absorption spectrum, due to an electronically allowed transition between a large ground-state dipole moment (43 ± 8 D) and a small excited-state dipole moment (3 to 8 D). Two fluorescent emissions were observed: one in the visible region (quantum yield $\phi \approx 0.01$, not solvatochromic) and one in the near infrared spectrum (weakly solvatochromic). The core-level N1s XPS spectrum shows the three expected N valence states, while the valence-level XPS spectrum can be correlated with theory. Simultaneous cyclic voltammetry and electron spin resonance of the radical anion of 1 shows that the spin density in the LUMO of 1 is concentrated on the tricyanoquinodimethanide portion of the anion. The molecule is clearly zwitterionic in the ground state, both in LB films and in solution, and is undissociated in its first excited state. The through-film electrical conductivity of Langmuir-Blodgett monolayer films of hexadecylquinolinium tricyanoquinodimethanide was studied as a function of temperature. Electrical rectification was observed between 370 K and 105 K. The rectification ratios at different temperatures do not change significantly, although the current is usually smaller at low temperature. The rectification can very probably be ascribed to an intramolecular electron transfer.

Introduction

The unimolecular rectifier was first proposed by Aviram and Ratner in 1974, who suggested that a molecule could be a rectifier of electrical current, meaning that current may flow through the molecule preferentially in one direction, if this molecule has a D-σ-A structure, where D is a strong electron donor, A is a strong electron acceptor, and σ is covalent (saturated) linker [1]. In the past two decades, several D-σ-A molecules were synthesized, but none had been unequivocally proven to be rectifiers of electrical current [2-5]. Ashwell and co-workers discovered a new class of zwitterions based on TCNQ [6], one of which is hexadecylquinolinium tricyanoquinodimethanide, $C_{16}H_{33}Q$-3CNQ (1) [7].

1, $C_{16}H_{33}Q$-3CNQ
high-moment ground state

1', $C_{16}H_{33}Q$-3CNQ
low-moment excited state

This molecule forms Z-type multilayer Langmuir-Blodgett (LB) films with an extremely high, resonance-enhanced second-order non-linear optical susceptibility ($\chi^{(2)}$ = 180 pm V^{-1} at 532 nm) [8]. As explained below, this molecule has a zwitterionic D^+-π-A^- (1) structure in its ground state, and is neutral (undissociated) in the first excited state with a D^0-π-A^0 (1') structure [9]. Molecule 1 was shown to be a multilayer LB rectifier [7], and was definitely proven to be a unimolecular rectifier [9]. Here we review the estimates of the excited-state dipole moment, using the solvatochromism of the absorption spectrum of 1 [10]. Fluorescence emission spectra for 1 have been reported [10], and the spin distribution of the radical anion of 1 was measured by electron spin resonance [10]. The X-ray core-level and valence-band photoelectron spectra of 1 gave information about the nature of the ground state, and about the valence-level molecular orbitals [10]. The measurements of current as a function of voltage at room temperature for all Al | LB monolayer of 1 | Al "sandwiches" (some of which do not rectify) were also reported [11]. Finally, the rectification of the LB films of 1 at temperatures ranging from 105 K to 370 K is reviewed [12]. The primary results have been reviewed elsewhere [5], [13], [14], but receive renewed attention below.

Ground and Excited States

It was suggested that 1 may be a twisted internal charge transfer (TICT) molecule [15]; there was also some concern that rectification may be caused by some thermally activated mechanism. These questions had to be investigated in detail [10, 12] to buttress the claim for molecular rectification.

Since no crystal structure could be obtained for 1, the twist angle θ between the quinolinium ring and the ethylene bridge is not known experimentally. If $\theta = 0°$, then 1 could be either a zwitterion (D^+-π-A^-), i.e. 1, or a neutral, undissociated, state (D^0-π-A^0), i.e. 1'. If $\theta = 90°$, then 1 would be a zwitterion with a TICT state [15] with no possibility of an intervalence transfer band (IVT): this would leave the molecule colorless. Since the molecule has a blue color in solution, this is not likely. Steric hindrance between the quinolinium ring and the cyano group most likely hinders the molecule from having θ = 0° and probably also hinders free rotation. The crystal structure of a very similar molecule had θ = 30° [6]. AM1 and PM3 molecular orbital calculations suggest that the "gas phase" ground state should have θ = 9° -11° [15], with a dipole moment about 10 D (Debye). The excited state was calculated to have θ ≈ 90°, with a dipole moment near 45 D [15]. The solution dipole moment of 1 was measured to be 43 D in CH_2Cl_2 at infinite dilution [9]. For such a high dipole moment to be measured in solution, the molecule must be zwitterionic in the ground state. Due to the large hypsochromic shift described below, the molecule's excited state must be neutral, i.e. undissociated.

Instrumentation and experimental details

A Shimadzu UV-1600 spectrophotometer was used to measure visible-UV spectra. Fluorescence spectra in the visible region were determined using a SPEX Fluoromax-2 spectrometer. Near-infrared fluorescence spectra were obtained using a specially equipped Fluoromax-2 at Furman University, access to which was kindly provided by Prof. N. A. P. Kane-Maguire. X-ray photoelectron spectroscopy (XPS) of LB monolayers and multilayers were obtained using a Kratos Analytical Axis 165 Scanning Auger / X-Ray Photoelectron Spectrometer. Monochromatized Al K_α photons (E = 1486.6 eV) were used as the exciting radiation, giving an intrinsic spectrometer resolution of < 0.2 eV. The spectra were fit using a proprietary program from Kratos Analytical, which applied Lorentz-Gaussian peak shapes and instrument-specific fitting parameters.

Silicon substrates were treated in "piranha" solution (a mixture of 30% H_2O_2 and concentrated H_2SO_4, 30:70 / v:v), and heated at 90°C for 30 minutes. The substrate thus acquires a hydrophilic surface. Gold was evaporated onto a silicon substrate by using an Edwards 306A evaporator with an oil diffusion pump (a few nanometers of chromium was coated on silicon surface before gold was evaporated without breaking the vacuum). Langmuir-Blodgett films were obtained using a microcomputer-controlled Nima Model 622D2 trough connected to a Lauda constant-temperature bath (5 - 30° C), in a room with HEPA-filtered air, and high resistivity water (Millipore Milli-Q, >14 MΩ cm).

A "sandwich structure" of Al (100 nm) | 1 LB monolayer of 1 (Z-type) | Al (100 nm) was fabricated in order to measure the rectification of the monolayer as a function of temperature. A microscope glass slide was cut into small pieces, treated with "piranha" solution, and heated at 90°C for 30 minutes. A layer of aluminum was evaporated onto the surface in a vacuum evaporator. The fresh substrate was then immersed in water subphase of 14°C in a LB trough, the solution of 1 in CH_2C_2 was spread onto the air/water interface, and the barriers were compressed immediately to a film pressure of 25 mN m^{-1}. One upstroke of the dipper gave a Z-type monolayer LB film of 1. The film was thereafter kept in a vacuum desiccator containing P_2O_5 for two days before it was taken out to apply the top electrode of aluminum (100 nm). The sample was then mounted in a Janis Varitemp 100R dewar, fitted with a Lakeshore 330 dual temperature controller, controlled by a Gateway 2000 P5/60 microcomputer. A eutectic (alloy of Gallium and Indium, liquid at room temperature) was used to connect Au wires (diameter of 0.0127 mm) to both electrodes. Two heaters were used to heat the samples up to 370 K.

Equations for solvatochromism were written using a Digital Visual FORTRAN 5.0 compiler on a IBM Thinkpad 365 XD PC microcomputer. Molecular orbitals were calculated using the CaCHE programming system on a Macintosh PowerPC 8100 microcomputer: a trial geometry was first optimized by molecular mechanics, then minimized again using the PM3 algorithm at the single-determinant restricted Hartree-Fock (RHF) level for the ground state, and the first singlet excited state.

Absorption and fluorescence spectra in solution

The absorbance and fluorescence spectra in Fig. 1 show the visible fluorescence from the excitation of the short-wavelength absorption maxima as well as the long-wavelength absorption band in acetonitrile, dichloromethane, and chloroform [10]. The spectra in solvents with lower dielectric constants have two peaks in the absorption band, while those in solvents with the higher dielectric constant only have one peak. The two closely spaced absorption peaks in the lower polarity solvents are probably due to vibronically resolved features.

It was previously determined that the long-wavelength visible absorption maximum, the intervalence transfer band (IVT), of 1 obeys Beer's law in CH_3CN and in CH_2Cl_2, and only partially deviates from it in $CHCl_3$ [9]. To check further for a possible monomer-oligomer equilibrium in solution, the UV-visible absorption spectrum was measured as a function of temperature and of concentration, but no isosbestic points were found in any of several solvents, e.g. in CH_2Cl_2 (-50°C to 30°C), or in CH_3CN (5°C to 50°C) [10]. Therefore the spectra in Figs. 1 and 2 were attributed to the monomer of 1 [10].

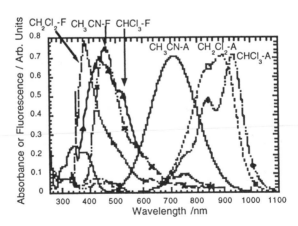

Figure 1. The absorption (-A) and fluorescence (-F) spectra (exciting wavelength = 300 nm) of solutions of 1 in chloroform ($CHCl_3$), dichloromethane (CH_2Cl_2), and acetonitrile (CH_3CN) at room temperature. A solvent Raman line is seen at 320 nm in CH_2Cl_2. (Reproduced from reference 10, Copyright 1999 American Chemical Society)

54

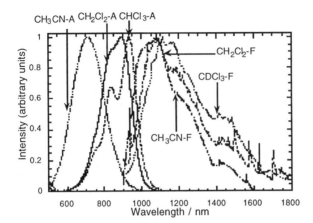

Figure. 2. Absorption (-A) and fluorescence (-F) spectra (exciting wavelength = 850, 850, and 720 nm) of solutions of 1 in chloroform (CHCl₃) or deuterochloroform (CDCl₃), dichloromethane (CH₂Cl₂), and acetonitrile (CH₃CN) at room temperature. Spikes due to muon absorption were eliminated from the fluorescence emission spectra. (Reproduced from reference 10, Copyright 1999 American Chemical Society)

The visible emission band in CH_2Cl_2 (excitation at 300 nm) includes a strong Raman emission at 320 nm [10]. The quantum efficiency ϕ of the visible emission was low, $\phi \approx 0.01 \pm 0.02$. The near-infrared emission spectra in Fig. 2 were noisy, and spikes caused by atmospheric muons were observed (but are not shown in Fig. 2) [10].

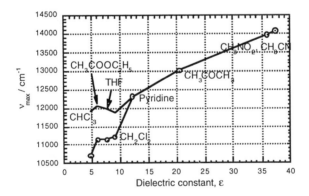

Figure 3. Hypsochromic (blue) shift of long-wavelength absorption maxima of 1 with increasing dielectric constant of the solvent. (Reproduced from reference 10, Copyright 1999 American Chemical Society)

The long-wavelength absorption bands are strongly solvatochromic: they are hypsochromic (blue shifted), and shift almost linearly with dielectric constant, as seen in Figure 3 [10]. This clearly indicates that the electronically excited state has a lower dipole moment than the ground state.

There are two sets of fluorescent emission lines: emissions below 600 nm, and emissions in the 1000 - 1400 nm region [10]. The emissions in the near IR region are very weakly hypsochromic. It is a violation of Kasha's rule that two fluorescent emissions exist for 1: normally, within a set of states of the same spin multiplicity, a fluorescent emission should occur only from the lowest excited state; other, higher excited states usually decay, internally and without radiative emission, into the lowest excited state, which then emits the photon. If, however the higher-energy emission is weak (here $\phi \approx 0.01$), and is not re-absorbed, then emission from a higher-energy state becomes possible [16]. Alternatively, molecule 1 may have different rotameric states: molecules of 1 with $\theta \approx 90°$, with no IVT, would emit in the visible, while molecules with a lower θ and a "blue" IVT will emit in the near infrared region [10]. The excited state dipole moment was estimated from Stokes shifts and from general theories of solvatochromism [10]: The estimated excited state dipole moment is a very reasonable 3 - 9 D, depending on the method used [10].

Absorbance and Emission of a LB monolayer

A single monolayer of 1 was transferred onto a quartz slide by the LB technique: weak absorption maxima were measured at 340 nm and at 565 nm (Figure 4) [10]. The same monolayer, excited at 350 nm, produces a clear emission peak at 492 nm,

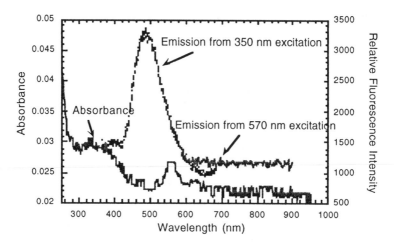

Figure 4. Visible absorption and emission spectrum of a 1-monolayer LB film of 1 on quartz. The lower curve is the absorbance (maxima at 340 nm and 565 nm); the upper curve is the fluorescence emission, which peaks at 492 nm (for excitation at 350 nm). The short curve on the right is the fluorescence emission (for excitation at 570 nm, i.e. at the maximum of the absorbance). (Reproduced from reference 10, Copyright 1999 American Chemical Society)

56

but no emission (from 570 nm to 900 nm) when excited at 570 nm [10]. An 11-monolayer film has an aborbance maximum at 565 nm [9]. An LB film of 1 on quartz was checked for changes in absorbance between room temperature and 77 K: the films stayed blue-violet at both temperatures, thus arguing for no large change in the twist angle θ between 300 K and 77 K.

Nitrogen 1s core-level XPS of Langmuir-Blodgett films

Figure 5 shows the XPS spectrum of an LB multilayer of 1 adsorbed on a Si substrate precleaned with "*piranha*" solution. Only C (1s, 286 eV), O 1s (533 eV, due to the substrate), and N (1s, ≈ 400 eV) peaks were obtained.

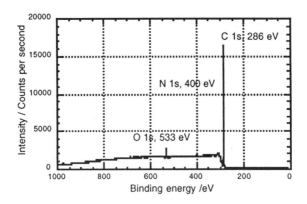

Figure 5. Wide-scan X-ray photoelectron spectrum of an LB multilayer on Si, corrected from reference 10.

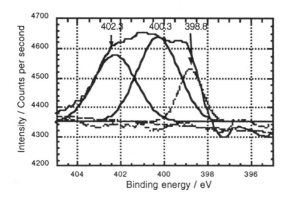

Figure 6. Nitrogen N1s X-ray photoelectron spectrum of an LB multilayer of 1 on Si, with Gaussian peak fits, and the error in the fit (small circles). (Reproduced from reference 10, Copyright 1999 American Chemical Society)

Figure 6 shows details of the N 1s region [10]. The feature associated with nitrogen is clearly composed of three peaks, which have been fit using standard techniques. The dominant feature at 400.5 eV is in the range for N in an organic matrix (398.9 - 401.0 eV [17]), exactly what one would expect from this molecule, given its chemical formula. However, there are features at higher (402.3 eV) and lower (398.8 eV) binding energies, which indicate the chemical shifts within the molecule. The lowest energy feature (398.8 eV) can be associated with nitrogen in the negatively charged C≡N moiety of ionic salts like KCN (N 1s 398.3-399.8 eV [18]). The highest binding energy feature (402.3 eV) is in the range associated with positively charged N in ammonium salts, e.g. $N^*H_4NO_3$ at 402.3 eV [19]. The shift to lower binding energy reflects reduction, or the increase in local electron density, and the shift to higher binding energy reflects oxidation, or a decrease in local electron density [10]. These results are totally consistent with the zwitterionic depiction of **1**. The N peak centered at 400.3 eV represents an undifferentiated N signal from the organic layer [10]. Progressive degradation of the films under X-irradiation may also explain why the peak-fit integrated areas do not scale with the ratio of different N atoms in **1**, which would be 1:1:2 for the features assigned to 402.3, 400.3, and 398.8 eV, respectively, in Figure 6.

Valence band XPS of Langmuir-Blodgett films

A valence band scan of a multilayer of **1** on Si (Figures 7, 9) and on Au (Figure 8) reveal several peaks at binding energies 3 eV to 25 eV [10]. Strong signals from the Au substrate in the 3 to 8 eV range (not shown) are much attenuated when the LB film covers it, but the signals from the LB film in this region may still be Au signals, so this region of the spectrum is not shown in Fig. 8. For Si substrates, there are features in the 2 - 6 eV range, which do not diminish after Ar^+ sputtering, but they are weaker, so the valence band spectrum for the LB film is shown in full in Figure 9.

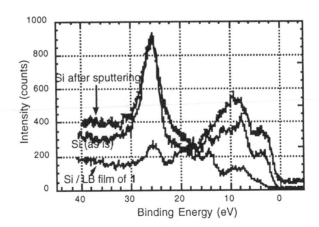

Figure 7. Valence-band XPS spectrum of an LB multilayer of 1 on Si (lowest scan), of the highly degenerate Si substrate, used as received, and of the Si substrate after Ar^+ sputtering (top scan). (Reproduced from reference 10, Copyright 1999 American Chemical Society)

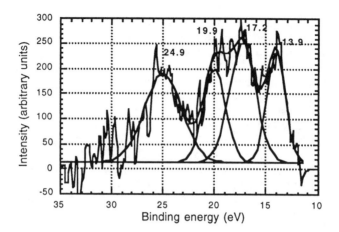

Figure 8. Valence-band XPS spectrum of an LB multilayer of 1 on Au (difference between LB film and Au substrate), with Gaussian line-fits. (Reproduced from reference 10, Copyright 1999 American Chemical Society)

A fit with Gaussian peaks was performed; the binding energies of the centers of these Gaussians are reported on Figures 8 and 9. There is strong agreement between the LB film peaks seen on Au and those seen on Si. To compare with theory the valence-band XPS spectral peaks, the values of the experimental peaks were adjusted to the vacuum

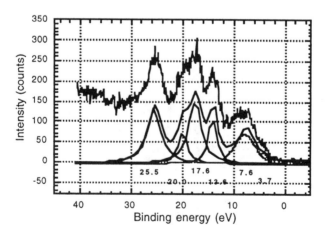

Figure 9. Valence-band XPS spectrum of an LB multilayer of 1 on Si, with Gaussian line-fits. (Reproduced from reference 10, Copyright 1999 American Chemical Society).

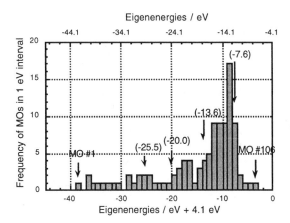

Figure 10. Histogram of occupied molecular orbitals, from PM3/RHF calculation of ground state of 1. The bottom abscissa axis gives the eigenvalues, shifted by 4.1 eV for better comparison with Figure 9; the numbers in parentheses show the XPS peaks from Figure 9. The upper abscissa axis gives the eigenvalues relative to the vacuum level. (Reproduced from reference 10, Copyright 1999 American Chemical Society)

level (E = 0) by adding to the binding energies the work functions ($\approx$ Fermi levels) of 4.1 eV for Si, and 4.58 eV for Au, and changing the sign [20]; they are then compared with a geometry-optimized PM3/RHF semiempirical molecular orbital calculation of the ground state of **1**. Where the data exist, there is strong agreement between theory and experiment. The fit between the onset of the valence-band XPS at 3.7 eV on Si (-7.8 eV versus E(vacuum)) and the calculated HOMO at -7.98 eV is remarkable. The experimental XPS spectra can be resolved into 4 to 6 broad peaks (Figs. 8, 9), while the PM3 calculation has 107 closely spaced occupied molecular orbitals (Fig. 10). The escape probabilities from each MO would have to be computed to improve the fit.

However, the same calculation for the ground state of **1** obtains a dipole moment of only 10.7 D. A PM3/RHF calculation for the excited singlet ($\theta \approx 90°$) does not agree well, although it yields a large moment of 45.4 D.

Temperature Dependence

It was shown that a single monolayer of **1**, sandwiched between Al electrodes (covered by a defect-rich native Al_2O_3) rectifies at room tmeperature, with a rectification ratio of up to 26 at 1.5 V [9]; this rectification ratio decreases upon cycling; the enhanced forward current is 0.33 electrons molecule^{-1} s^{-1} [9]. Under the same conditions, a centrosymmetric multilayer of arachidic acid does not rectify [9]. Rectification was also seen by STM (scanning tunneling spectroscopy) [9]. Not all junctions measured rectify, but there are plausible explanations for the IV characteristics of the "aberrant" Al | monolayer of **1** | Al sandwiches.

It was important to determine whether the rectification current was due to a thermally activated process [12]. Although the ultimate aim is to measure the current at 4.2 K, we present here the results obtained between 370 K and 105 K [12]. All the samples measured at low temperatures showed rectification at room temperature, before the temperature was decreased [12]. Results could easily be obtained when temperature was higher than 250 K, but below 250 K, either open circuits or permanent or intermittent short circuits happened very often: only a few reliable results were obtained at low temperatures [12]. We discuss below four samples that showed rectification (samples A, B, C, and D) [12].

Figures 11 and 12 show the I-V plots for sample A at 195 K and 105 K, respectively. The rectification ratio increased from 18 at 195 K to 48 at 105 K. There is a turning point at a positive bias of about 1 V, which means that the sample rectifies at the applied potential > 1 V (at 295 K, 195 K, or 105 K) [12]. On the other hand, some instability of the measured system can be seen from the hysteresis. Figures 13 and 14 are I-V plots of sample B at 290 K and 150 K [12]. At the lower temperature, the current at both positive and negative bias is much smaller than that at room temperature, and the rectification ratio drops from 14.9 to 2.7 [12]. Figures 15 and 16 show the I-V plots of sample C at 290 K and 370 K: the current at higher temperature increased as much as 3.2 times at 1.5 V and 2.8 times at -1.5 V, the rectification ratio didn't change much (from 10.7 to 12.5). Finally, Figure 17 gives the rectification of sample D at 120 K. The enhanced current increases with increasing temperature (for samples B, C, but not for sample A). There is a fair variation in this current from sample to sample [12].

The plots of the logarithms of the current versus the applied bias (Figure 18) are roughly linear with voltage far enough away from V = 0, but with steeper slopes at positive bias [12]. The current drops to a minimum I_{min} at $V_{min} = 0$ for sample D at 370 K, and for sample B at 290 K; I_{min} is displaced to $V_{min} = -0.1$ to -0.4 V for samples B at 150 K, D, and particularly C [12]. There is some hysteresis for samples B and C [12]. The data for sample A were too scattered to be included in Figure 18.

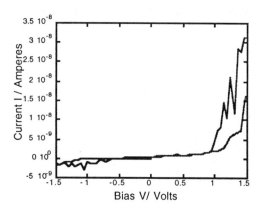

Figure 11. DC current I versus applied voltage V for sample A (pad area 2.0 mm²) at 195 K. (Reproduced from reference 12, Copyright 1999 American Chemical Society)

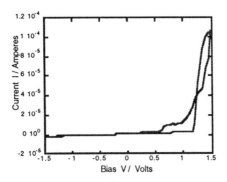

Figure 12. DC current I versus DC applied voltage V for sample A at 105 K. (Reproduced from reference 12, Copyright 1999 American Chemical Society)

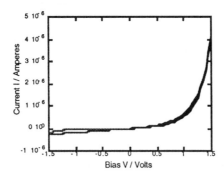

Figure 13. DC current I versus applied voltage V for sample B (pad area 2.0 mm² at 290 K. (Reproduced from reference 12, Copyright 1999 American Chemical Society)

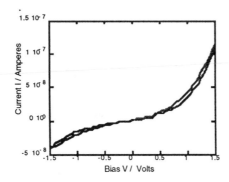

Figure 14. DC current I versus applied voltage V for sample B at 150 K. (Reproduced from reference 12, Copyright 1999 American Chemical Society)

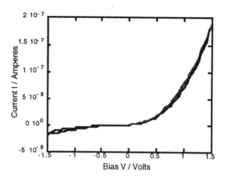

Figure 15. DC current I versus applied voltage V for sample C (pad area 2.5 mm²) at 290 K. (Reproduced from reference 12, Copyright 1999 American Chemical Society)

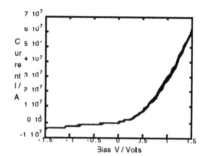

Figure 16. DC current I versus applied voltage V for sample C at 370 K. (Reproduced from reference 12, Copyright 1999 American Chemical Society)

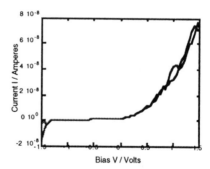

Figure. 17. DC current I versus applied voltage V for sample D (pad area 1.5 mm²) at 120 K. (Reproduced from reference 12, Copyright 1999 American Chemical Society)

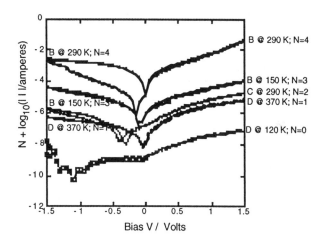

Figure 18.   Logarithm of the current I versus voltage V, for samples B, C, and D, displaced vertically by N = 4,3,2,1, or 0 for clarity. (Reproduced from reference 12, Copyright 1999 American Chemical Society)

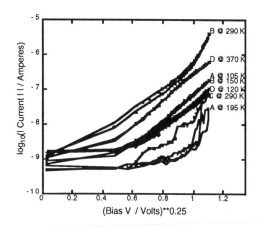

Figure 19.   Plot of log$_{10}$I versus V$^{0.25}$ in the forward regime (I > 0) for samples A, B, C, and D: there are several regimes, with different slopes. (Reproduced from reference 12, Copyright 1999 American Chemical Society)

Plots of log$_{10}$|I| versus both V$^{0.5}$ and versus V$^{0.25}$ show several regimes, a low-voltage regime, (0 < V < 0.5) with almost zero slope, and a high-voltage regime with several large and varying slopes [12]. The plot of $\log_{10}|I|$ versus $V^{0.25}$ (Figure 19) shows clearly that, above about 0.5 V bias, the current through a single monolayer of 1 is due neither to scattering by mobile defects ($\log j \propto V^{0.25}$), nor to simple tunneling through vacuum, nor to Poole-Frenkel nor to Schottky processes ($\log j \propto V^{0.5}$) [12]. The asymmetric I-V curves were obtained for monolayers of 1 between symmetrical electrodes between 1 V and 1.5 eV in the temperature range 370 K to 105 K. Since rectification by a single monolayer between symmetrical electrodes has now been observed as a function of temperature, therefore the possibility that rectification is caused by some thermally activated mechanism can be discounted. A molecular explanation, such as a modification [9] of the Aviram-Ratner intramolecular electron transfer mechanism [1], is reinforced [12]. Alternately, the large current under forward bias ("rectification") can be attributed to a large electronically allowed dipole moment change between two adjacent electronic energy levels of a molecule which can be ordered as a monolayer [5], [14]. This immediately suggests the design of further molecules which can exhibit such electronic behavior, and may usher in real progress in unimolecular electronics.

Conclusion

We have proved that the excited-state dipole moment of 1 is much smaller than its ground-state moment, and confirmed its zwitterionic ground state and undissociated excited state. The electrical conductivities across a monolayer of 1 at different temperatures (370 K to 105 K) confirm a molecular mechanism (modified Aviram-Ratner mechanism) of the rectification, or enhanced conductivity of 1 under positive bias.

References

1. Aviram, A.; Ratner, M. A. *Chem. Phys. Lett.* **1974**, *29*, 277-283.
2. Metzger, R. M.; Panetta, C. A. *New J. Chem.* **1991**, *15*, 209-221.
3. Metzger, R. M. In *Molecular and Biomolecular Electronics*, R. R. Birge, Ed. ACS Adv.in Chem. Ser. 240 (American Chemical Society, Washington, DC 1994) pp. 81-129.
4. Metzger, R. M. *Mater. Sci. & Engrg.* **1995**, *C3*, 277-285.
5. Metzger, R. M. *Acc. Chem. Res.* accepted and in press.
6. Metzger, R. .M.; Heimer, N. E.; Ashwell, G. J. *Mol. Cryst. Liq. Cryst.* **1984**, *107*, 133-149.
7. Ashwell, G. J.; Sambles, J. R.; Martin, A. S.; Parker, W. G.; Szablewski, M. *J. Chem. Soc. Chem. Commun.* **1990**, 1374-1376.
8. Ashwell, G. J. In Ashwell, G. J.; Bloor, D. Eds. *Organic Materials for Nonlinear Optics*, (Royal Soc. of Chem., Cambridge, 1993), pp. 31-39.
9. Metzger, R. M.; Chen, B.; Höpfner, U.; Lakshmikantham, M. V.; Vuillaume, D.; Kawai, T.; Wu, X.; Tachibana, H.; Hughes, T. V.; Sakurai, H.; Baldwin, J. W.; Hosch, C.; Cava, M. P.; Brehmer, L.; Ashwell, G. J. *J. Am. Chem. Soc.* **1997**, *119* : 10455-10466.

10. Baldwin, J. W.; Chen, B.; Street, S. C.; Konovalov, V. V.; Sakurai, H.; Hughes, T. V.; Simpson, C. S.; Lakshmikantham, M. V.; Cava, M. P.; Kispert, L. D.; Metzger, R. M. *J. Phys. Chem. B* **1999**, 103, 4269-4277.
11. Vuillaume, D.; Chen, B.; Metzger, R. M. *Langmuir* **1999**, *15,* 4011-4017.
12. Chen, B.; Metzger, R. M. *J. Phys. Chem. B* **1999**, *103,* 4447-4451.
13. Metzger, R. M., *Adv. Mater. Optics & Electronics* **1998**, *8,* 229-245.
14. Metzger, R. M. *J. Materials Chem.* **1999**, *9,* 2027-2036.
15. Broo, A.; Zerner, M. C. *Chem. Phys.* **1996**, *196,* 423-426.
16. Geldof, P. A.; Rettschnick, R. P. H.; Hoytink, G. J. *Chem. Phys. Lett.* **1969**, *4,* 59-61.
17. Chastain, J.; King, R. C., Eds., *Handbook of X-ray Photoelectron Spectroscopy* (Physical Electronics, Inc., Eden Prarie, MN, 1985).
18. Vannerberg, N. G.*Chem. Scripta* **9**: 122 (1976).
19. Burger, K.; Tschismarov, F.; Ebel, H. *J. Electron Spectroscopy and Relat. Phenom.* **1977**, *10,* 461.
20. Gray, D. E., Ed., *American Institute of Physics Handbook,* 2nd Edition (McGraw-Hill, New York, 1963) p. 5-123.

Molecular Materials
and Crystals

Chapter 5

closo Boranes as π Structural Elements for Advanced Anisotropic Materials

Piotr Kaszynski

Organic Materials Research Group, Department of Chemistry, Vanderbilt University, Nashville, TN 37235

Steric and electronic interactions between 6-, 10-, and 12-vertex *closo* boranes and their substituents are examined using experimental data and quantum-mechanical methods. The electronic spectra and structural studies supported by ZINDO and ab initio calculations demonstrate that the strength of electronic interactions between clusters and their π substituents decreases in the order: 6-vertex > 10-vertex > 12-vertex. The role of these interactions in the design of new electronic materials is studied computationally and supported with experimental results. Two such classes of materials containing *closo* boranes are examined in detail: conjugated carboraneacetylenes and NLO materials.

Introduction

A design of molecular and polymeric materials for photonic and electronic applications relies on intramolecular electronic communication and mobility of π electrons.[1-5] Typically, the π molecular system is composed of several molecular fragments the majority of which are unsaturated hydrocarbons (e.g. benzene, ethylene, acetylene) and their nitrogen analogs. Structural components based on elements other than C and N are rare and may not possess the molecular orbital symmetry and energy necessary for efficient electronic interactions.

Boron *closo* clusters are highly polarizable[6,7] and unusual inorganic ring systems regarded as 3-dimensional sigma-aromatic compounds.[8-11] Owing to their molecular geometry, the *closo* clusters are attractive building blocks for anisometric materials such as liquid crystals[12] and molecular scale construction sets.[13,14] The unique stereoelectronic structure of these compounds has attracted much attention and inspired numerous studies of cluster-substituent electronic interactions with the intent of developing novel materials. The vast majority of the work has concentrated on the most accessible 12-vertex carboranes, while other clusters have been largely neglected.

Besides the 12-vertex boranes **1** and **2**, are the 6- and 10-vertex *closo*-boranes represented by the borate dianions, **3** and **5**, and carboranes, **4** and **6** (Figure 1). Their O_h, D_{4d} and D_{4h} molecular symmetry provide molecular orbital manifolds particularly suitable for interactions with π substituents. Therefore we began systematic studies to understand the stereoelectronic interactions of these clusters with substituents and to utilize these electronic

68

structures in designing of new materials in which the clusters would play an active role and interact with π systems.

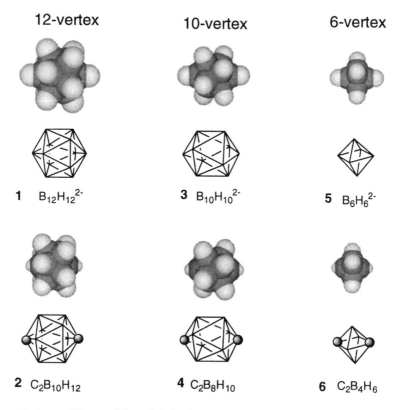

12-vertex **10-vertex** **6-vertex**

1 $B_{12}H_{12}^{2-}$ **3** $B_{10}H_{10}^{2-}$ **5** $B_6H_6^{2-}$

2 $C_2B_{10}H_{12}$ **4** $C_2B_8H_{10}$ **6** $C_2B_4H_6$

Figure 1. Space filling models and skeletal representations for closo boranes. Each vertex corresponds to a BH fragment and the circle represents a CH group.

Recently, we reviewed the chemistry of *closo* boranes[15] and properties of their liquid crystal derivatives.[12] Here we focus on electronic interactions between boron *closo* clusters **1-6** and π substituents in the context of anisotropic materials.

Molecular Properties of boron clusters and their interactions with substituents

Geometry

The molecular dimensions and symmetry of the clusters are significantly different than those for benzene. The 6-vertex *closo* p-carborane (**6**) is the smallest among the clusters with the calculated apex-to-apex separation of 2.16 Å. This compares to 3.04 Å for **2** and 3.68 Å for decaborate (**3**). The diameter of the cylinders of rotation defined by the hydrogen atoms in clusters **1-6** is in the range of 4.7-5.2 Å as shown in Table I. Analogous values for benzene are generally smaller and are 2.77 Å for the C(1)···C(4)

Table I. Molecular Dimensions, Frontier Orbital Energy Levels and Hybridization.[a]

compound	L (Å)	D (Å)	HOMO (eV)	LUMO (eV)	ΔE (eV)	sp^{x} [b]
1, $B_{12}H_{12}^{2-}$	3.415	5.205	-1.48 (g_u)	13.59 (g_g)	15.07	3.63
2, $C_2B_{10}H_{12}$	3.040	5.031	-10.97 (e_{2u})	3.68 (e_{1g})	14.65	1.91
3, $B_{10}H_{10}^{2-}$	3.725	4.895	0.16 (e_1)	13.12 (e_3)	12.96	1.46
4, $C_2B_8H_{10}$	3.331	4.773	-10.98 (e_2)	2.59 (e_3)	13.57	1.96
5, $B_6H_6^{2-}$	2.459	4.897	1.94 (t_{2g})	16.14 (t_{1u})	14.20	-
6, $C_2B_4H_6$	2.155	4.768	-9.60 (b_{2g})	4.80 (e_u)	14.40	1.71
C_6H_6	2.772	4.264	-9.00 (e_{1g})	4.07 (e_{2u})	13.07	2.37
C_5H_5N	2.773	4.279	-9.36 (a_2)	3.45 (b_1)	12.81	2.84
C_2H_2	1.185	na	-11.01 (π_u)	6.05 (π_g)	17.06	1.10

[a] L is apex-to-apex distance and D is a diameter of the cylinder of rotation carved out by the centers of the hydrogen atoms. MO energies are in eV. All values are derived from HF/6-31G* calculations. [b] x defines the apical atom hybrid used for the exocyclic bond.

separation and 4.26 Å for the diameter of the cylinder of rotation. The calculated values shown in Table I and molecular symmetries shown in Figure 1 correspond well to the experimental data.

Clusters **1-6** have high order rotational axes. The 6- and 10-vertex boranes exhibit the C_4 rotational symmetry with a 45° intra-cage twist in the latter, while the 12-vertex cluster has 5-fold rotational axes and 36° intra-cage twist (Figure 2). These symmetry properties and the resulting distribution of conformational minima are unique to the clusters and their derivatives and complement those available among the organic rings.

An effective demonstration of the differences between the symmetry of the clusters and organic rings is shown in Figure 2 for a series of homostructural oligomers substituted with alkyl chains at the terminal positions. The extended Newman projection along the long molecular axes shows the terminal positions of the oligomer and the relative orientation of the planes containing the alkyl substituents. The symmetry of the 12-vertex p-carborane (**2**) and cubane allows for the terminal alkyl chains to adopt antiperiplanar conformational minima in the entire series of oligomers regardless of the number of the rings in the core (Figure 2a and 2e). In contrast, the terminally disubstituted oligomers of 6-vertex p-carborane (**6**) and bicyclo[2.2.2]octane exhibit an "odd-even" effect where the antiperiplanar arrangement of the substituents is available only for the odd number of the boron cages or even number of bicyclo[2.2.2]octane rings (Figure 2c and 2d). The series of oligomeric 10-vertex p-carborane (**4**) derivatives is unique and the substituents are always offset from coplanarity by 45° forming chiral ground state conformers (Figure 2b). Similar relative orientation of substituents can be expected for other substituents.

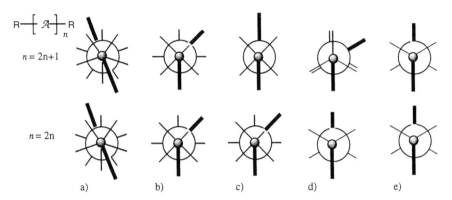

Figure 2. Extended Newman projection along the long molecular axes showing the terminal positions of odd- (upper row) and even-membered (lower row) oligomers of (a) 12-vertex p-carborane 2, (b) the 10-vertex p-carborane 4, (c) 6-vertex p-carborane 6, (d) bicyclo[2.2.2]octane, and (e) cubane. The bold lines represent the substituent plane.

The stereochemistry of the clusters have important implications for the properties of materials. The distribution and the depth of the conformational minima of cluster derivatives affect molecular packing in the solid state and also the dynamic molecular shape and hence mesophase stability.[16] Single crystal X-ray analyses show that 10-vertex derivatives exhibit frequent molecular disorders since the terminal substituents cannot adopt a coplanar orientation. For instance, while the heptyl and heptynyl substituents in the 12-vertex derivative 7 are antiperiplanar, the substituents in the crystal structure of the 10-vertex analogs 8 are not coplanar, and one is significantly disordered.[17] The difficulties with close packing are also evident from the intermolecular separation in the crystal of 8 (6.3 Å) which is larger than that found in 7 (6.1 Å) despite the smaller core width for the former. Liquid crystalline properties of 10-vertex derivatives are also generally inferior to the 12-vertex analogs.[12]

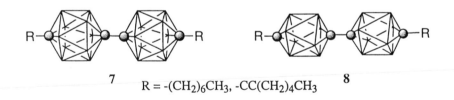

7     $R = -(CH_2)_6CH_3, -CC(CH_2)_4CH_3$ 8

Electronic Structures

According to Wade's rules,[18] an n-atom *closo* cluster contains n+1 electron pairs of delocalized electrons on 2n tangential p orbitals that account for bonding among the skeletal atoms.[19] Thus the 6-, 10- and 12-vertex *closo*-boranes contain 14, 22 and 26 delocalized electrons satisfying Hückel's 4n+2 rule for aromaticity. The resulting stabilization energy, estimated at 1.763, 1.145 and 0.844 β for 1, 3 and 5,[9] is revealed in chemical,[20,21] electrochemical,[22,23] and thermal stability of the clusters. In addition, their reactivity towards electrophiles is reminiscent of π-aromatic compounds.[15]

72

The largely unhybridized tangential p orbitals form molecular orbitals of π symmetry. The apical carbon atoms in carboranes contribute their AOs to the LUMO and HOMO-1 and HOMO-2 (Figure 3), among other MOs, but not to the HOMO. The calculated HOMO energy levels decrease in the order 6-, 10- and 12-vertex for both anions and the isoelectronic carboranes (Table 1). The HOMO of the latter series lies lower than that of benzene. The generally larger HOMO-LUMO gap for the boron clusters **1-6** than for benzene results in their marginal electronic absorption above 200 nm.[21,24,25]

The hybridization of the exocyclic bond to hydrogen formed by the apical atoms in the clusters is close to $sp^{2.0}$ (Table 1), with the noticeable exception of the B-H bond in anion **1**, which has a calculated hybridization of $sp^{3.63}$.

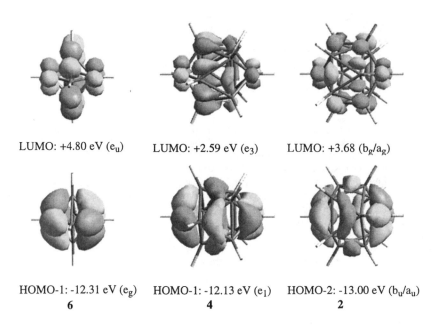

LUMO: +4.80 eV (e_u)   LUMO: +2.59 eV (e_3)   LUMO: +3.68 (b_g/a_g)

HOMO-1: -12.31 eV (e_g)   HOMO-1: -12.13 eV (e_1)   HOMO-2: -13.00 eV (b_u/a_u)

6   **4**   **2**

Figure 3. Graphical representations of the HF/6-31G* MO's of 6-, 10- and 12-vertex p-carboranes.

Electronic interactions with substituents

The symmetry and energy levels of the MOs in the *closo* clusters (Table 1) facilitate electronic interactions with π substituents. Spectroscopic and, more recently, computational studies suggest that the interactions are stronger for the 10-vertex than for 12-vertex analogs,[26] but there is little data for 6-vertex *closo* boranes.[27]

Generally, *p*-carboranes act as bathochromic and hyperchromic auxochromes, modifying principal aromatic absorption bands in their aryl derivatives.[28] For instance, replacing the bicyclo[2.2.2]octyl moiety in **9a** with 10-vertex *p*-carborane group in **9b** results in a 22 nm red shift and a 12% increase of intensity.[29] These results are consistent with previous findings for 12-vertex carboranes[30,31] and azaborane[32] in which the borane cages exert an inductive effect on the π substituents.

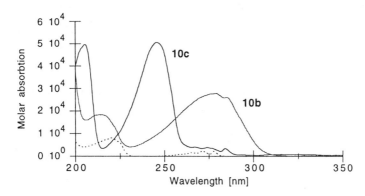

C_5H_{11}—A—⟨phenyl⟩—⟨phenyl⟩—CN   RO—⟨phenyl⟩—A—⟨phenyl⟩—OR

9   **10**

$C_7H_{15}O$—⟨pyridine⟩N—⟨cage⟩—N⟨pyridine⟩—OC_7H_{15}

11

a $A =$ 1,4-bicyclo[2.2.2]octadiyl, **b** $A =$ 1,10-$C_2B_8H_8$, **c** $A =$ 1,12-$C_2B_{10}H_{10}$

Occasionally, electronic interactions between *closo* boranes and their aryl substituents result in new and strong absorption bands such as those observed for carborane derivatives **10b** and **10c** (R=C_7H_{15}) at 279 and 246 nm, respectively (Figure 4).[29] INDO/2 calculations for **10** (R=H) reproduced the two bands remarkably well, showing that they originate primarily from the HOMO to LUMO+4 (**10b**) and HOMO to LUMO+3 (**10c**) transitions with longitudinal transition moments. The HOMO is largely localized on the phenyl rings while the unoccupied orbitals extend over the entire molecule. The calculations also account for the shapes of the absorption bands.

*Figure 4. Electronic absorption spectra for **10b**, **10c** (R=C_7H_{15}) and anisole (dashed line).*

Pyridine derivatives of 10-vertex *closo* borane such as **11**, an isoelectronic analog of **10b**, have a strong absorption band at about 320 nm.[33,34] The absorption is ascribed to the symmetry-allowed HOMO to LUMO or cluster-to-ring electronic transition with a longitudinal transition moment. Compound **11** exhibits strong fluorescence with a relatively large Stokes shift of 1.29 eV.[34] A similar charge-transfer phenomenon is observed for tropylium derivatives of **1** and **3**.[35]

A clear spectroscopic signature of cage-substituent electronic interactions is observed in a series of isoelectronic derivatives of 10-vertex *closo* clusters substituted with triple bonded functional groups **12**. Although the individual structural components show no or little absorption above 200 nm, **12a-12e** exhibit strong absorption bands.[26,33]

The spectroscopic results are well reproduced by INDO2//HF/6-31G* calculations, which ascribe the long wavelength absorptions to symmetry-allowed HOMO-LUMO excitations with a longitudinal transition moment. Qualitatively, they are identical to those in terephthalonitrile **13c**.[26] The transition in **12d** (R=H) is calculated at 236 nm and observed at 233 nm (R=TMS) with the oscillator strength f=0.19. The same transition in the analogous 6-vertex derivative **15d** is calculated to appear at 184 nm (f=0.69) and in the 12-vertex analog it is at 182 nm (f=0.11). Judging by the oscillator strengths, the order of electronic interactions with the acetylene group is 6-v >> 10-v > 12-v.

a: X = B, Y = N, Z = N
b: X = B, Y = N, Z = C-R
c: X = C, Y = C, Z = N
d: X = C, Y = C, Z = C-R
e: X = C, Y = N, Z = C

Further information about the electronic interactions within acetylene derivatives of *p*-carboranes is provided by ab initio calculations supported with X-ray analysis of the 10-vertex (**12d**, R=TMS)[26] and 12-vertex (**14d**, R=TMS) derivatives.[36] In diacetylene **14d**, the experimental C-C (1.452(2) Å) and C≡C (1.193(3) Å) bond lengths are typical for alkylacetylenes (1.466(10) Å and 1.181(14) Å, respectively),[37] while the analogous distances in the 10-vertex analog **12d** (1.436(2) Å and 1.199(2) Å, respectively) are typical for arylacetylenes (1.434(6) Å and 1.192(14) Å, respectively).[37] Although the difference in bond lengths between **12d** and **14d** is crystallographically insignificant, the trend and the values are well reproduced by ab initio calculations (Table II). Moreover, ab initio results

Table II. Comparison of π Bonds and 6-, 10- and 12-Vertex *p*-Carborane Derivatives.[a]

$$HC≡C-\mathcal{A}-C≡CH$$

	$\mathcal{A}$	Bond Length		π–π overlap[b]	
		C-C	C≡C	C-C	C≡C
14d	$1,12\text{-}C_2B_{10}H_{10}$	1.4493 $(1.452)^{c,d}$	1.1847 $(1.193)^{c,d}$	0.055	1.936
12d	$1,10\text{-}C_2B_8H_8$	1.4390 $(1.436)^{c,e}$	1.1853 $(1.199)^{c,e}$	0.058	1.935
15d	$1,6\text{-}C_2B_4H_4$	1.4325	1.1858	0.067	1.928
13d	$1,4\text{-}C_6H_4$	1.4414	1.1881	0.069	1.929
	$1,2\text{-}C_2$	1.348 $(1.378)^{e,f}$	1.1880 $(1.205)^{e,f}$	0.115	1.883

[a] HF/6-31G* calculations. [b] MNDO//HF/6-31G* calculations. [c] Experimental value for the TMS derivative. [d] Pakhomov S.; Kaszynski, P.; Young, V. G. Jr., unpublished results. [e] Pakhomov, S.; Kaszynski, P.; Young, V. G. Jr. *Inorg. Chem.* **2000**, *39*, 2243. [f] McNaughton, D.; Bruget, D. N. *J. Mol. Struct.* **1992**, *272*, 11.

suggest that an even larger distortion towards the allenic resonance form should be expected for the 6-vertex carborane derivative in which the bond lengths approach the values calculated for diacetylene.

The differences in the C-C bond lengths reflect differences in hybridizations of the cage carbon atom. For instance, the calculated C-C distance in **15d** is significantly shorter than that in benzene derivative **13d** (Table II) which coincides with the higher s character of the cage carbon atom in **6** than that in benzene ($sp^{1.71}$ vs. $sp^{2.37}$; Table I).

Despite the bond length differences, both **15d** and **13d** have almost equal calculated $\pi-\pi$ overlaps of about 0.07, which is the largest among the carborane derivatives and more than a half of that calculated for diacetylene (Table II). The lowest bond order for the C-C bond (1.055) is observed for the 12-vertex p-carborane derivative **14d** which again demonstrates the weakest electronic interactions between the cage and the substituent. With the increase of the bond order for the C-C bond in the series, the C≡C bond is proportionally depopulated.

A chemical consequence of the high degree of the $\pi-\pi$ overlap in **12-15** is the relative stability of the dinitrogen derivatives **12a**, **14a** and **15a** compared to that of benzenediazonium cation (**16**). Calculations (HF/6-31G*) show that the heterolytic cleavage of the B-N bond and loss of N_2 has a similar endotherm for the 10-vertex, **12a** (22.8 kcal/mol) and 6-vertex, **15a** (20.2 kcal/mol) derivatives, close to that calculated (22.1 kcal/mol) and observed (25.5 kcal/mol)[38] for **16**, while the 12-vertex derivative **14a** is predicted to be much less stable with an endothermicity of only 12.9 kcal/mol.[39] The decompositions proceed with negligible activation barriers.

$$N\equiv N-\mathcal{A}-N\equiv N \longrightarrow N\equiv N-\mathcal{A} + N_2$$

Like many benzenediazonium salts, 1,10-bis(dinitrogen)decaborane (**12a**) is a stable compound[33] that can be sublimed below 100 °C under vacuum.[40] Although the 6-vertex analog **15a** is still unknown, the peculiar dinitrogen derivative **17**[41] strongly suggests the relatively high stability of such a B-N≡N link in 6-vertex cluster derivatives.

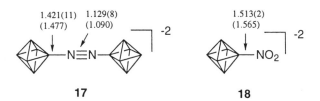

17   **18**

Figure 5. Experimental and calculated (in parentheses) selected bond lengths for 17 and 18.

Among a handful of other derivatives of 6-vertex *closo* boranes with π substituents is the nitrohexaborate **18**.[42] Both compounds **17** and **18** are reported to be absorbing in the visible (orange and yellow, respectively), indicating strong cage-substituent electronic interactions. The long wavelength absorption bands are calculated (INDO2//HF/6-31G*) for **17** and **18** at 356 nm ($f = 0.003$) and 351 nm ($f = 0.003$), respectively with transverse transition dipole moments. The charge transfer transitions with the longitudinal moment are predicted to be at 314 nm ($f = 2.42$) and 256 nm ($f = 0.06$) for **17** and **18**, respectively.

Single crystal X-ray analysis for **17** and **18** shows that the B-N bonds are calculated (HF/6-31G*) to be too long, while the N-N bond in the former compound is calculated to be too short. This suggests that in reality there is much greater cage-substituent π–π overlap than the calculations account for.[41] These and other compounds in the 6-vertex series are still awaiting complete theoretical analysis.

New Materials

The relatively high degree of electronic communication between boron clusters and π substituents makes them, especially the 6- and 10-vertex series, attractive structural elements for a variety of electronic materials typically based on benzene. Here we discuss the application of 6- and 10-vertex clusters in two classes of compounds: conjugated oligomers and non-linear optical materials.

Conjugated polymers

Polymers such as PPV in which aromatic or heteroaromatic rings alternate with ethylene or acetylene linkers[43] are an important class of materials used in a variety of electronic devices such as organic conductors, organic light emitting diodes,[4] sensors, or as linkers providing electronic communication between terminal substituents.[44,45] Boron clusters offer a rare opportunity to design inorganic-organic hybrids or purely inorganic conjugated polymers such as **19a[n]** and **19b[n]**. To assess the effect of carborane cage in the polymer, several oligomers were analyzed using ab initio and ZINDO methods and the results were compared to those for polycarbyne (**19c[n]**).

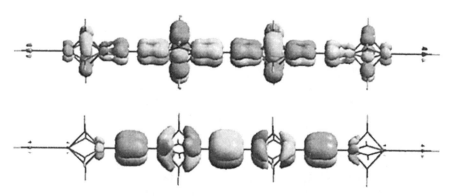

Analysis of electronic structures for both *p*-carboraneacetylene oligomers **19a[n]** and **19b[n]** shows extensive delocalization of the molecular orbitals, as demonstrated for **19b[4]** in Figure 6.

Figure 6. Calculated (INDO2 // HF/3-21G) HOMO and LUMO+2 (top) contours for 19b[4].*

ZINDO calculations for the series of oligomers **19[n]** (n=1-4) show a progressive bathochromic shift of the long wavelength absorption with the increasing length of conjugation. To establish the λ_{max} limit for each infinite polymer, the calculated transition energy for the L band was plotted as a function of reciprocal length (Figure 7). For comparison experimental values for α,ω-dimethyl-**19c[n]** (n = 2,[46] n = 3÷6[47]) and α,ω-di-*t*-butyl-**19c[7]**[48] are also plotted showing that the calculated transition energies were overestimated by about 1.3 eV. Extrapolation of the best fit quadratic functions (y = ax^2+bx+c) gives λ_{max} for an infinite polymer of 210 nm for **19b[n]** and 269 nm for **19a[n]**, which compares to 354 nm (exp.: 574 nm) for polycarbyne **19c[n]**. It is interesting to note that the parabolas for the carborane polymers lie above the vertex (a>0) while those found for polycarbyne and *p*-polyphenylacetylene lie below the vertex (a<0).[26] This suggest that the electronic interactions in the carborane derivatives are different than those in the hydrocarbon polymers. All curves shown in Figure 7 have high correlation value r > 0.999 except for **19a[n]** (r = 0.996) which appear to quickly level off at about 262 nm for n=3.[26]

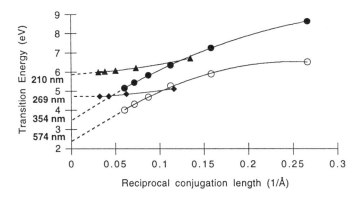

Figure 7. π–π* (L) transition as a function of the reciprocal total length calculated (INDO2 // HF/3-21G*, CI=40x40) for 19a[n] (diamonds), 19b[n] (triangles), and 19c[n] (full circles). Open circles represent experimental data for α,ω-dimethyl-19c[n] (see text). The correlation factor is r > 0.999 except for 19a[n] (r = 0.996).

The computational results indicate that the electronic interactions are stronger in 6-vertex poly-carboraneacetylene (**19b[n]**) than in the 10-vertex analog **19a[n]** but weaker than those in polycarbyne **19c[n]**. It can be expected that the carborane oligomers **19a[n]** and **19b[n]** will be less conductive of electronic effects than their hydrocarbon analogs.

Nonlinear Optical Materials
        The cage-to-ring CT transition observed in the highly quadrupolar dipyridine derivative **12**[34] and other bis-betaine compounds[35] offers interesting and unique possibilities for designing new materials with NLO properties.[49]   When the decaborate cage **3** is substituted with only one pyridine ring and the second onium group Q is an "electronically isolated" sigma substituent, a directional CT transition should significantly lower polarity of the pyridine-cage bond and result in a substantial increase of the net molecular dipole moment as shown in Figure 8.   In consequence the material should exhibit a significant first hyperpolarizability β.

Indeed, ZINDO calculations for decaborane derivative **I** substituted with pyridine and ammonium group as Q show a relatively large β value which is four (static) or seven (dynamic) times larger than that calculated for p-nitroaniline (PNA) (Table III). Similar

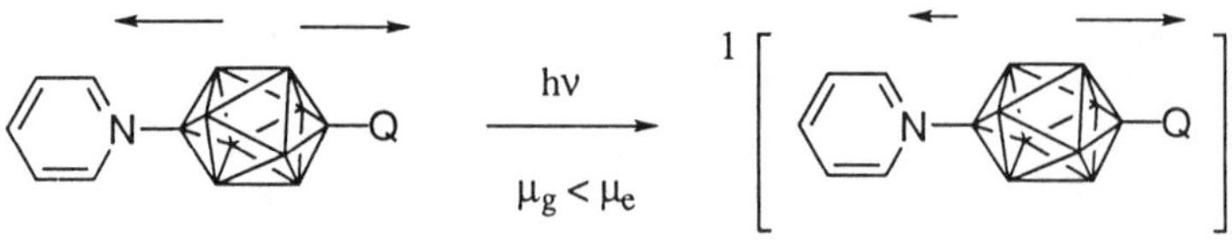

Figure 8. Directional photoexcitation in the 1,10 derivative of decaborate containing one π (pyridine) and one $\sigma(Q)$ substituent. The arrows represent local dipole moments.

calculations for analogous derivatives of the two other *closo* borates **1** and **5**, give almost identical high hyperpolarizability for the 6-vertex derivative and relatively small value for the 12-vertex compound comparable with that for PNA. The calculated λ_{max} for the decaborane **I** is overestimated by about 90 nm.

Table III. A Comparison of Calculated Dipole Moments and Molecular Hyperpolarizabilities for I and II.[a]

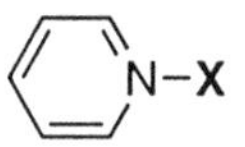

	X	μ_g	$\beta_{vec} \times 10^{-30}$		CT band
		D	0 μm	1.06 μm	λ_{max} (osc)
I	1,6-B$_6$H$_4$-NH$_3$	0.76	18.4	54.4	404 (f=0.28)
	1,10-B$_{10}$H$_8$-NH$_3$	1.15	18.7	53.6	395 (f=0.30)
	1,12-B$_{12}$H$_{10}$-NH$_3$	1.67	6.7	11.4	303 (f=0.29)
II	1,6-C$_2$B$_4$H$_4$	5.37	-0.5	-0.7	218 (f=0.16)
	1,10-C$_2$B$_8$H$_8$	5.23	-1.2	-1.9	278 (f=0.08)
	1,12-C$_2$B$_{10}$H$_{10}$	4.67	0.15	0.06	214 (f=0.02)
	1,4-C$_6$H$_4$ (PNA)[b]	7.31	-4.7	-7.8	293 (f=0.46)

[a] INDO2 // HF/3-21G* CI=30x30, TDA method used for β_{vec}. [b] Exp μ_g = 6.2 D, β_{vec} = 34.5x10^{-30} esu; Oudar, J. L.; Chemla D. S. *J. Chem. Phys.* **1977**, *66*, 2664.

Calculations for carborane analogs of PNA **II** uniformly show very low hyperpolarizabilities with the lowest values obtained for the 12-vertex carborane derivative (Table III). These results are consistent with previous findings on NLO properties of some 12-vertex carborane derivatives.[50,51] More recent calculations show that the magnitude of the NLO response in derivatives of **1** and **2** can be significantly increased by using strong donor-acceptor systems such as pentadienylide-tropylium.[52]

The structure of NLO chromophore **I** lends itself to further modification and incorporation into a liquid crystalline framework. Thus, using quinuclidine or thiacyclohexane ring as the "sigma" onium fragment Q and addition of alkyl substituents in the antipodal positions yields elongated molecules such as **20** and **21** with liquid crystalline properties.[53]

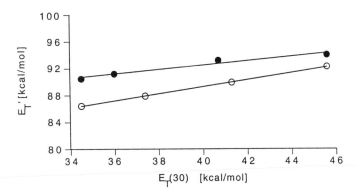

The quinuclidine derivative **20** exhibits a highly ordered smectic phase with the clearing point above 250 °C. The substitution of the thiacyclohexane ring for quinuclidine in **21** lowers the clearing point and changes the smectic into a nematic mesomorph. Both compounds absorb at about 320 nm and are transparent in the visible region.

The predicted increase of the dipole moment upon excitation in **20** and **21** should give rise to a positive solvatochromic effect whose magnitude is a reflection of the dipole moment difference.[54] Our studies found, however, moderate negative solvatochromic effects (Figure 9), indicating the dominant role of the local rather than overall molecular polarity in the effect observed in **20** and **21**.[53]

Figure 9. Correlation between solvent polarity ($E_T(30)$ scale) and the charge transfer transition energy E_T' for 1-(4-heptyloxy-1-pyridine)-10-(4-pentyl-1-quinuclidine)-closo-decaborane (20, O, slope=0.523) and 1-(4-heptyloxy-1-pyridine)-10-(4-pentyl-1-thiacyclohexane)-closo-decaborane (21, ●, slope=0.325).

The two pyridine derivatives represent materials that are unique from both liquid crystal and NLO points of view. To our knowledge, they are the first examples of liquid crystalline bis-betaines, which also makes them rare as NLO chromophores.[55] The compounds exhibit low dipole moment and are highly quadrupolar in the ground state but upon excitation they become highly dipolar.

The mesogenic properties of these bis-betaines combined with the low ground state dipole moment of about 1-2 D should permit parallel molecular alignment. The parallel orientation of the chromophores can be reinforced further by application of a weak electric field since the compounds are expected to exhibit a positive dielectric anisotropy. Finally, the significant $\Delta\mu$ combined with marginal absorption in the visible range is very desirable for NLO applications, and it is characteristic for π-betaines.[56]

The design for NLO materials can, in principle, be extended to the corresponding monocarbaborates, although in contrast to the bis-betaines, mono-betaines possess high ground state dipole moments.[57] This may limit the density of parallel packing of the chromophores.

Prospects

The unusual structural and electronic features of *closo* boranes make them attractive structural elements for anisotropic materials such as liquid crystals, conjugated polymers and NLO chromophores. The particular application depends chiefly on the degree of electronic interactions with the substituent which appears to be the strongest for the 6-vertex and weakest for the 12-vertex clusters. The highly symmetrical and weakly electronically interacting 12-vertex carborane are particularly attractive for photochemically and thermally stable liquid crystals. The strongly interacting 6- and 10-vertex borates appear well suitable for design of NLO chromophores while the 6-vertex carborane gives polymers with the highest degree of conjugation yet almost complete UV transparency. These considerations can be extended to monocarbaborates.

Besides applications mentioned above, the *closo* boranes might serve as conduits for electron transfer systems and facilitate spin-spin interactions. Further progress in design and experimental studies of such materials requires advances in synthetic and structural chemistry of these species especially the 6-vertex *closo* boranes. The best studied system to date is the 12-vertex, however it appears as the least promising for electronic applications.

Acknowledgment

This project was supported by the NSF CAREER (DMR-9703002) and ONR (331/99/0237) grants.

References

1. *Molecular Electronics*; Ashwell, G. J., Ed.; Wiley & Sons: New York, 1992.
2. *Conjugated Conducting Polymers*; Kiess, H. G., Ed.; Springer Verlag: New York, 1992.
3. *Handbook of Organic Conductive Molecules and Polymers*; Nalwa, H. S., Ed.; Wiley & Sons: New York, 1997.
4. Kraft, A.; Grimsdale, A. C.; Holmes, A. B. *Angew. Chem. Int. Ed.* **1998**, *37*, 402.
5. Shirota, Y. *J. Mater. Chem.* **2000**, *10*, 1.
6. Kaczmarczyk, A.; Kolski, G. B. *J. Phys. Chem.* **1964**, *68*, 1227.
7. Kaczmarczyk, A.; Kolski, G. B. *Inorg. Chem.* **1965**, *4*, 665.
8. King, R. B. *Russ. Chem. Bull.* **1993**, *42*, 1283.

9.   Aihara, J. *J. Am. Chem. Soc.* **1978**, *100*, 3339.
10. Gimarc, B. M.; Zhao, M. *Inorg. Chem.* **1996**, *35*, 825.
11. Schleyer, P. v. R.; Najafian, K. *Inorg. Chem.* **1998**, *37*, 3454.
12. Kaszynski, P.; Douglass, A. G. *J. Organomet. Chem.* **1999**, *581*, 28.
13. Muller, J.; Baše, K.; Magnera, T.; Michl, J. *J. Am. Chem. Soc.* **1992**, *114*, 9721.
14. Hawthorne, M. F.; Mortimer, M. D. *Chem. Brit.* **1996**, *April*, 33.
15. Kaszynski, P. *Collect. Czech. Chem. Commun.* **1999**, *64*, 895, and references cited therein.
16. Douglass, A. G.; Both, B.; Kaszynski, P. *J. Mater. Chem.* **1999**, *9*, 683.
17. Czuprynski, K.; Kaszynski, P. *Liq. Cryst.* **1999**, *26*, 775.
18. Wade, K. *Adv. Inorg. Chem. Radiochem.* **1976**, *18*, 1.
19. King, R. B.; Rouvray, D. H. *J. Am. Chem. Soc.* **1977**, *99*, 7834.
20. Todd, L. J. In *Progress in Boron Chemistry*; R. J. Brotherton and H. Steinberg, Eds.; Pergamon Press: New York, 1970; Vol. 2; pp 1-35.
21. Muetterties, E. L.; Balthis, J. H.; Chia, Y. T.; Knoth, W. H.; Miller, H. C. *Inorg. Chem.* **1964**, *3*, 444.
22. Schmitt, A. P.; Middaugh, R. L. *Inorg. Chem.* **1974**, *13*, 163.
23. Yarosh, M. V.; Baranova, T. V.; Shirokii, V. L.; Erdman, A. A.; Maier, N. A. *Russ. J. Electrochem.* **1994**, *30*, 366.
24. Thibault, R. M.; Hepburn, D. R. J.; Klingen, T. J. *J. Phys. Chem.* **1974**, *78*, 788.
25. Wright, J. R.; Klingen, T. J. *J. Inorg. Nucl. Chem.* **1970**, *32*, 2853.
26. Pakhomov, S.; Kaszynski, P.; Young, V. G. Jr. *Inorg. Chem.* **2000**, *39*, 2243.
27. Preetz, W.; Peters, G. *Eur. J. Inorg. Chem.* **1999**, 1831.
28. Fox, M. A.; MacBride, J. A. H.; Peace, R. J.; Wade, K. *J. Chem. Soc., Dalton Trans.* **1998**, 401.
29. Pakhomov, S.; Douglass, A. G.; Kaszynski, P., unpublished results.
30. Grimes, R. N. *Carboranes*; Academic Press: New York, 1970, and references cited therein.
31. Harmon, K. M.; Harmon, A. B.; Thompson, B. C.; Spix, C. L.; Coburn, T. T.; Ryan, D. P.; Susskind, T. Y. *Inorg. Chem.* **1974**, *13*, 862.
32. Fendrich, W.; Harvey, J. E.; Kaszynski, P. *Inorg. Chem.* **1999**, *38*, 408.
33. Knoth, W. H. *J. Am. Chem. Soc.* **1966**, *88*, 935.
34. Kaszynski, P.; Huang, J.; Jenkins, G. S.; Bairamov, K. A.; Lipiak, D. *Mol. Cryst. Liq. Cryst.* **1995**, *260*, 315.
35. Harmon, K. M.; Harmon, A. B.; MacDonald, A. A. *J. Am. Chem. Soc.* **1969**, *91*, 323.
36. Kaszynski, P.; Pakhomov, S.; Young, V. G. Jr., unpublished results.
37. Allen, F. H.; Kennard, O.; Watson, D. G.; Brammer, L.; Orpen, A. G.; Taylor, R. *J. Chem. Soc., Perkin Trans. II*, **1987**, S1.
38. Kuokkanen, T. *Acta Chim. Scand.* **1990**, *44*, 394. For recent discussions of the bonding in ion **16** see Glaser, R.; Horan, C. J.; Lewis, M.; Zollinger, H. *J. Org. Chem.* **1999**, *64*, 902.
39. Kaszynski, P., unpublished results.
40. Whelan, T.; Brint, P.; Spalding, T. R.; McDonald, W. S.; Lloyd, D. R. *J. Chem. Soc. Dalton Trans.* **1982**, 2469.
41. Franken, A.; Preetz, W. *Z. Naturforsch.* **1995**, *50B*, 781.
42. Franken, A.; Preetz, W.; Rath, M.; Hesse, K.-F. *Z. Naturforsch.* **1993**, *48B*, 1727.
43. Bunz, U. H. F. *Chem. Rev.* **2000**, *100*, 1605.
44. Mayr, A.; Yu, M. P. Y.; Yam, V. W.-W. *J. Am. Chem. Soc.* **1999**, *121*, 1760.

45. Creager, S.; Yu, C. J.; Bamdad, C.; O'Connor, S.; MacLean, T.; Lam, E.; Chong, Y.; Olsen, G. T.; Luo, J.; Gozin, M.; Kayyem, J. F. *J. Am. Chem. Soc.* **1999**, *121*, 1059.

46. Price, W. C.; Walsh, A. D. *Trans. Faraday Soc.* **1945**, *41*, 381.

47. DMS *UV Atlas of Organic Compounds*; Plenum Press: New York, 1966.

48. Bohlmann, F. *Chem. Ber.* **1953**, *86*, 657.

49. Kaszynski, P.; Lipiak, D. In *Materials for Optical Limiting*; R. Crane, K. Lewis, E. V. Stryland and M. Khoshnevisan, Eds.; MRS: Boston, 1995; Vol. 374; pp 341-347.

50. Murphy, D. M.; Mingos, D. M. P.; Forward, J. M. *J. Mater. Chem.* **1993**, *3*, 67.

51. Murphy, D. M.; Mingos, D. M. P.; Haggitt, J. L.; Powell, H. R.; Westcott, S. A.; Marder, T. B.; Taylor, N. J.; Kanis, D. R. *J. Mater. Chem.* **1993**, *3*, 139.

52. Newlon, A. E.; Rudd, G. P.; Taylor, J. W.; Caruso, J. D. III; Allis, D. G.; Spencer, J. T. Boron USA VII, Pittsburgh, June 2000.

53. Harvey, J. E.; Kaszynski, P., unpublished results.

54. Reichardt, C. *Solvents and Solvent Effects in Organic Chemistry*; VCH: New York, 1988.

55. The low-dipole quadrupolar NLO materials were first reported by Zyss (J. Zyss, Chemla, D.S.; Nicoud, J. F. *J. Chem. Phys.* **1981**, *74*, 4800), and more recently the concept has been explored extensively for azines (Glaser, R.; Chen, G. S. *J. Comp. Chem.* 1998, 19, 1130, and references cited therein).

56. Abe, J.; Shirai, Y. *J. Am. Chem. Soc.* **1996**, *118*, 4705.

57. Grüner, B.; Janoušek, Z.; King, B. T.; Woodford, J. N.; Wang, C. H.; Všetečka, V.; Michl, J. *J. Am. Chem. Soc.* **1999**, *121*, 3122.

Chapter 6

Iodine and Organoiodide Templates in Supramolecular Synthesis

T. W. Hanks[1], William T. Pennington[2], and Rosa D. Bailey[2]

[1]Department of Chemistry, Furman University, 3300 Poinsett Highway, Greenville, SC 29613
[2]Department of Chemistry, Clemson University, Clemson, SC 29634

The halogen bond between I_2 or an organoiodide and a nitrogen heterocycle can be used to control crystal packing. The interaction is relatively strong, highly directional and well-described by quantum mechanical calculations. This allows it to be reliably used for crystal engineering projects. Examples include the resolution of chiral perfluorohaloalkanes, the control of polymorph formation and the control of diacetylene solid state polymerization. An unique feature in these applications is the ability to remove one component of the co-crystal by thermolysis or solvolysis.

The rational construction of supramolecular architectures is greatly facilitated by strong, highly directional intermolecular interactions. Hydrogen bonding sites, for example, can be programmed into a molecule and used to build complex assemblies, both in solution and in the solid state. While as Perlstein points out, "the H-bond is only one vector in a multitude of other noncovalent vectors with which it competes", (1) chemists have cleverly built structures in which this interaction clearly dominates crystal packing. Likewise, metal ions coordinate organic ligands in well-defined patterns that have been used to great effect by both nature and bench chemists. (2)

Weaker interactions can also be used as the principal synthons for the engineering of solids. (3) While it is easier to overwhelm them, such synthons offer advantages, particularly in applications where a "molecular scaffolding" must be removed after serving a templating function. The key to this approach is to use highly directional interactions that are sufficiently strong for the purpose at hand, but are weak enough (or can be made weak enough) to be broken when required. Applications that could take advantage of such removable templates include: the control of polymorphism, the separations of enantiomers, the construction of "organic zeolites" and the fabrication of molecular sensors.

The N···I_2 Charge-Transfer Interaction

It has long been known that when iodine is dissolved in solvents such as ethanol or water a brown color results, while the violet color of the halogen is retained in other solvents. The possibility that this difference is due to the formation of a molecular complex between nucleophilic solvents and I_2 was recognized more than a century ago, (4) but decades of study and great advances in bonding theory were required before a satisfactory understanding of the phenomenon could be developed. (5) The language used to describe the interaction of an electron donor and a halide has evolved from terms such as "adhesion by stray feeler lines of force" through "electron clutching" to more modern terms such as "donor-acceptor" or "charge-transfer" interactions. (6) We are particularly partial to the recently advanced term "halogen bond", (7,8) because it emphasizes the strength (~10 Kcal/mol for the N···I interaction) and directionality observed in a variety of systems. (9)

A simple orbital description of the interaction between an aromatic nitrogen heterocycle such as pyridine and iodine is that the lone pair sp^2 orbital on nitrogen overlaps a σ^* anti-bonding orbital on I_2. Of course, this weakens the I-I bond and can eventually break it, leading to ion formation. (10) In many cases (though not with pyridine) the intermediate heterocycle-I_2 complex can be isolated from solution as a meta-stable solid. The N-I-I bond angle in these systems is very close to 180°. The N-I bond distances in 1:1 adducts range from about 2.3 to 2.6 Å, while the I-I distance increases from 2.67 Å to between 2.75 and 2.83 Å. (6)

Figure 1 shows a quantum mechanical depiction of the donor-acceptor complex between pyridine and iodine. (11) It can be seen that both the HOMO (a) and the LUMO (b) are located primarily on the I_2 moiety, with the LUMO lying along the N-I-I axis. Not surprisingly, electrostatic potential isosurfaces show that the iodine takes on a partial negative charge (Figure 1c, solid surface), while the pyridine becomes partially positive (Figure 1c, mesh surface). It is interesting to note the shape of the two potential isosurfaces near the terminal iodine. Despite the overall increase in electron density on this atom, there is a void in the negative isosurface directly opposite the pyridine. This suggests that under the right conditions, a second Lewis base might be capable of associating at this position, leaving a bridging I_2. For example, Figure 1d shows the results of the same calculation with pyrazine as the donor. Pyrazine is a weaker electron donor, leading to less electron transfer onto the I_2. In this case, the positive isosurface protrudes much further from the nucleus, making this area more attractive to a potential nucleophile. Experimentally, this is exactly what happens. While pyridines and amines form simple 1:1 complexes with iodine, weaker donors such as pyrazine, (12) tetramethylpyrazine, (12) quinoxaline (13) and phenazine (14) form extended chain species in which there are interactions at both ends of the iodine (Figure 2). The N-I bond length in these compounds is longer than that found in the simple adducts (2.817 to 3.092 Å) while the I-I bond distance is somewhat shorter (2.722 to 2.733 Å). It is possible that the extended chain motif is a low energy arrangement that encourages bonding to both ends of the iodine, but the effect is primarily electronic. For example, 4,4'-bipyridine, (13) 1,2-bis(pyridyl)ethylene, (15) dipyridylquinoxaline (13) and tetrapyridylpyrazine (16) are each capable of extended chain formation, yet only form molecular adducts. Conversely, 9-chloroacridine is able to form a 2:1 molecular adduct which features a bridging I_2 (Figure 3). (17)

There is yet another consequence of the orbital picture shown in Figure 1. While the terminal iodine atom is capable of acting as a Lewis acid in some instances, it is also capable of acting as a Lewis base. Thus, there are now three known instances where two nitrogen donors are bridged by an I_6 chain (2,2'-bipyridine, (18) acridine and 9-chloroacridine (17)). These are best thought of as simple donor-acceptor

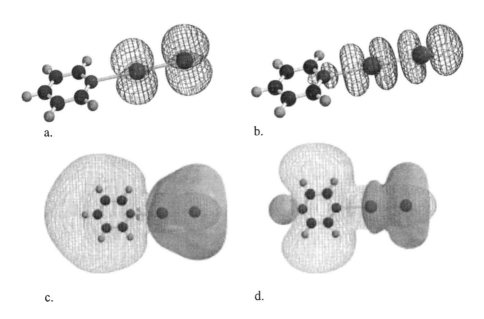

a.

b.

c.

d.

Figure 1. Results of a HF ab initio calculation on pyridine·I_2 (3-21G^* basis set), showing: (a) HOMO, (b) LUMO, (c) –10 Kcal/mol electrostatic potential isosurface (solid) and +10 Kcal/mol electrostatic potential isosurface(mesh), (d) Results of a HF ab initio calculation on pyrizine·I_2 (3-21G^* basis set) –10 Kcal/mol electrostatic potential isosurface (solid) and +10 Kcal/mol electrostatic potential isosurface(mesh).

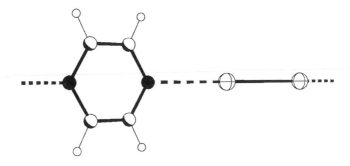

Figure 2. Structure of the extended chain complex pyrazine·I_2.

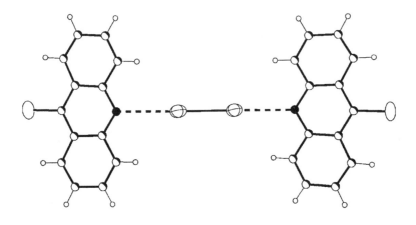

Figure 3. Structure of the molecular complex (9-chloroacridine)₂·I₂.

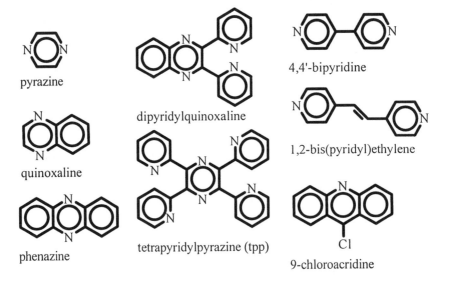

pyrazine

quinoxaline

phenazine

dipyridylquinoxaline

tetrapyridylpyrazine (tpp)

4,4'-bipyridine

1,2-bis(pyridyl)ethylene

9-chloroacridine

adducts bridged by an I_2. In the case of 2,2'-bipyridine, this leads to an extended chain, while in the two acridine complexes, dimers result (Figure 4). The factors that lead to amphoteric behavior in the N···complexed I_2 molecules are not completely understood. While subtle packing requirements may be involved, other factors may also be important. This is highlighted by the differences in the decomposition behavior between (acridine·I_2)$_2$·I_2 and (9-chloroacridine·I_2)$_2$·I_2. The former loses one equivalent of I_2, leaving a meta-stable species which is almost certainly the 1:1 molecular complex. This species then rapidly loses the remaining acceptor. The 9-chloroacridine complex loses two equivalents of I_2 to give a 2:1 complex, which is stable over a much broader temperature range. Powder X-ray diffraction shows that this complex is the bridging I_2 species shown in Figure 3 (the single crystal data was collected on a sample prepared independently by another route). (*17*)

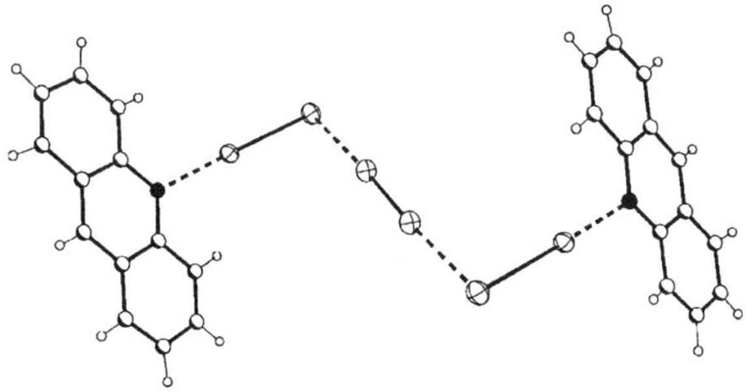

Figure 4. Structure of (acridine·I_2)$_2$·I_2.

The N···I-R Charge-Transfer Interaction

Iodine is by no means the only acceptor suitable for halogen bonding. (*19*) A variety of iodine-containing organic (and coordination) compounds have been reported to form isolatable complexes with nitrogen donors. The strength of the interaction is predictable; the greater s-character on the attached carbon and the presence of electron withdrawing substitutents increase interaction strength. (*6*) The use of organoiodide acceptors greatly increases the utility of the N···I synthon in crystal engineering applications. Iodine is a reasonably strong oxidant that is incompatible with many potential donors, but this is rarely a problem with organoiodides. As we have seen, I_2 can be quite variable in its bonding modes, which complicates crystal design. This is not an issue with organoiodides. Organoiodides do, however, come in a variety of shapes and sizes, allowing precise tuning of the acceptor volume. In addition, acceptor strength can be tuned for optimal performance by way of substituents at sites remote from the halogen bond. Finally, we will show that halogen-halogen interactions are common in these systems and are useful supramolecular synthons that complement the N···I bond.

The N⋯I bond lengths in complexes formed with organoiodides vary, depending upon the nature of the acceptor. In general, they are longer and weaker than those found in I_2 complexes. (*19*) For example, the N⋯I bond in 4,4'-bipyridine·1,4-diiodobenzene (DIB) is 3.024 (3) Å, more than 0.6 Å longer than the corresponding I_2 complex. Tetrafluoro-1,4-diiodobenzene (F_4DIB) is a significantly better acceptor, forming a complex with the same acceptor which has N⋯I bond lengths of 2.851 (3) Å. (*15*) The corresponding I-C bonds show very little distortion from those in the parent halocarbon in this class of compounds, typically increasing only by 0.1 Å.

A molecular orbital analysis of the hypothetical complex between pyridine and tetraiodoethylene (TIE) illustrates the similarity between halogen bonding involving I_2 and that involving organoiodides (Figure 5). (*11*) Again, the interaction is best described as the donation of a nitrogen lone pair into σ^*-antibonding orbital with the N⋯I-C bond angle very near to 180°.

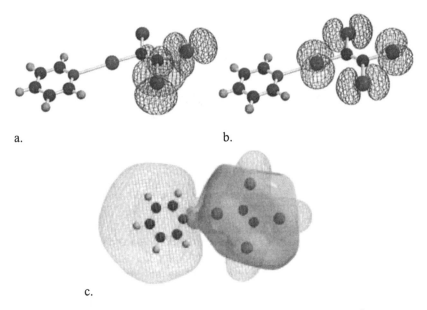

a.   b.

c.

Figure 5: Results of a HF ab initio calculation on pyridine·TIE (3-21G* basis set), showing: (a) HOMO, (b) LUMO, (c) −10 Kcal/mol electrostatic potential isosurface (solid) and +10 Kcal/mol electrostatic potential isosurface (mesh).

The highest energy orbitals in the complex can be described as iodine lone-pair orbitals, while the lowest unoccupied orbitals are essentially pure C-I σ^*-antibonding orbitals (there is also a nearly degenerate unoccupied orbital, not shown, which lies principally on the geminal iodine). As these are spatially removed from the N⋯I halogen bond and have no overlap with the acceptors π-system, there is very little effect on these orbitals when the first halogen bond is formed. This is clearly seen in Figure 5c, where the positive and negative electrostatic potential isosurfaces are overlaid. The positive isosurface extends well out from the negative "sheath". Of course, the weaker charge-transfer interaction contributes to this as well. This would suggest that all of the iodine atoms are available for halogen bonding, subject to steric interactions and other structural control vectors. This is, in fact, the case. TIE forms extended chain complexes in which 2 of the 4 iodides are complexed with diazines such as pyrazine, phenazine and quinoxaline, much like I_2. (*20*) However, similar extended chain structures are observed with the strong donors 4,4'-bipyridine, 1,4-bis(4-isoquinolyl)buta-1,3-diyne (Figure 6) (*21*) and tpp. (*22*) In the tpp complex, *each* of the four TIE iodines is involved in halogen bonding. These N···I distances are relatively short compared to the other TIE complexes, indicating that each of the interactions is strong and only minimally perturbed by the others. In addition, all of the structures (except tpp) show halogen-halogen interactions between adjacent TIE molecules (both in the extended chain plane and perpendicular to the plane), which contribute to the overall supramolecular arrangement.

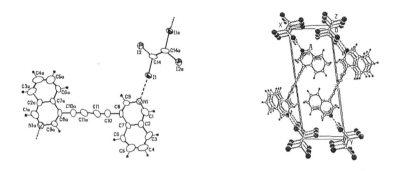

Figure 6: Structure of the molecular complex and crystal packing diagram of 1,4-bis(4-isoquinolyl)buta-1,3-diyne·TIE.

Supramolecular Organization in Halogen Bonded Complexes

One of the most striking features observed in the crystal packing of nitrogen heterocycle·iodine acceptor complexes is the segregation of donor and acceptor moieties into channels. This has been widely noted in the literature and has been the subject of some debate. It is now generally accepted that halogen-halogen (particularly iodine-iodine) attractive forces exist and they very often play a significant role in the solid state structure of halogenated compounds. (*23*) As with the N···I interaction, there is considerable directionality in these iodine-iodine contacts.

Our recent study of TIE complexes illustrates the importance of halogen self-association. (*20*) Pyrazine, (*24*) phenazine, quinoxaline, 1,4-dicyanobenzene and 2,2'-bipyridine all form complexes with TIE and they crystallize in remarkably similar arrangements (Figure 7). In each case, the donors associate with two of the four iodines, while the remaining iodines maintain I···I contacts, resulting in a sheet-like

90

structure. The reoccurring structural motif is also very similar to that of pure TIE, where some I···I contacts have been replaced with N···I contacts from the interspersed heterocycles. Of course, there are variations in halogen bond distances and angles in order to accommodate the steric requirements of the various donors. Even the asymmetric donor quinoxaline gives a similar, though highly distorted, structure. Interestingly, the complex between TIE and the very large donor 1,4-bis(4-isoquinolyl)buta-1,3-diyne (Figure 6) shows the same packing arrangement. While, it is certainly possible to construct TIE complexes which do not fit this robust packing arrangement (for example, in the case of tpp, above) it is clear that crystal engineering with organoiodides requires consideration of I···I interactions.

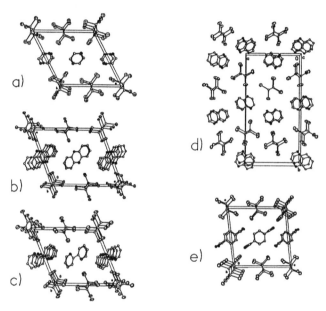

Figure 7. Complexes between TIE and: a) pyrazine, b) phenazine, c) 2,2'-bipyridine, d) quinoxaline and e) 1,4-dicayanobenzene.

Applications of Halogen Bonding

Equipped with an understanding of the relative strength and directionality of these donor-acceptor interactions, one can envision practical applications that take advantage of them. A strategic concept in the representative studies discussed below is that of a *removable* structure control vector. While the halogen bond is strong in comparison to the various forces described as "van der Waals interactions", it is weak compared to a covalent or even a hydrogen bond. Therefore, it is possible to build a crystal organized partially by halogen bonding, then remove one component of the co-crystal by thermolysis, solvolysis or some other means. This leaves the other component in a virgin state as far as elemental composition, but perhaps with an altered supramolecular organization. Thus, the halogen bond might be used to influence a chemical or physical process, but then be gently removed from the matrix, leaving a high value (by virtue of its physical state) material behind.

Halogen Bonding in Separations

Charge-transfer complexes involving iodine have been used as reagents, as model compounds for theoretical investigations and they have been observed as intermediates in chemical transformations. The potential of the N⋯X interaction as a crystal engineering vector is also beginning to be recognized, but there are few examples where it has been deliberately applied to a crystal engineering problem. One area that deserves increased attention is the resolution of enantiomers through halogen bond-facilitated co-crystallization. Early reports on the enantiomeric resolution of halo- or perfluorohaloalkanes with brucine do not explicitly discuss charge-transfer interactions, though they very likely were key to the process. (25,26) Very recently, Resnati and co-workers have described the use of (-)-sparteine hydrobromide to separate the enantiomers of 1,2-dibromohexafluoropropane. (27) In this case, N⋯Br-C interactions play a secondary role in the separation, while the primary structure control vector is the related Br⁻⋯Br-C interaction. Additional developments in this area may be expected, particularly in the case of perfluorocarbons where enantiomer resolution and absolute stereochemical assignments are challenging problems.

Halogen Bonding in Polymorph Control

We have demonstrated the use of halogen bonding for the control of polymorphism, a problem of considerable significance to a variety of industries. (28-30) There are two known polymorphs of tpp, crystallizing in either a tetragonal or a monoclinic space group, depending upon the solvent system used for crystal growth. The major difference between the two forms involves the orientation of the pyridine rings with regard to the pyrazyl nitrogen atoms. The tetragonal form (31,32) has an "endo, exo" conformation with adjacent pyridines orienting their nitrogens to the same side of the pyrazine ring plane. The monoclinic form (31,33) has an "endo, endo" configuration, with adjacent pyridyl rings lying on opposite sides of the pyrazine ring plane (Figure 8).

In both solution and the solid state, tpp reacts with iodine to form charge-transfer complexes. The reaction in solution gives a bis-I_2 adduct which was characterized by single crystal X-ray diffraction. Removal of the iodine by thermolysis gives exclusively the monoclinic polymorph. Exposure of crystals of either polymorph to iodine vapor, however, results in a common mono-I_2 adduct which, upon thermolysis, gives exclusively the *less stable* polymorph. (16,31) The identity of the product polymorph from each process was shown to be dependent upon the structure of the metastable charge-transfer complex. The bis-I_2 adduct has pyridyl ring orientations similar to that of the monoclinic polymorph. As the iodine evolves from the solid, the donor slips into a minimum energy geometry as the crystal reorganizes. Similarly, calculations suggest that the mono-I_2 adduct more closely resembles the tetragonal polymorph. Evolution of iodine at low temperatures does not permit the pyridyl rings to rotate into the geometry required by the monoclinic polymorph, instead trapping it in an local minimum geometry.

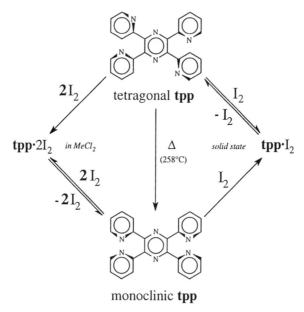

Figure 8. The interconversion of tpp polymorphs using halogen bonding.

We have examined a few other heterocyclic systems, but in each case, only the lowest energy polymorph was obtained after the evolution of the iodine acceptor. In some cases, it is likely that the structure of the charge-transfer complex is coincidentally very similar to the low energy polymorph. In others, such as the decomposition of (acridine·I_2)$_2$·I_2, the massive reorganization of the crystal required by removal of such a large percentage of the total volume probably precludes any possibility of falling into the local minimum energy well of the higher energy polymorph. The scope and limitations of this unique technique remain to be demonstrated, but the importance of controlling polymorph formation in nitrogen-containing pharmaceuticals alone is considerable.

Halogen Bonding for Controlling Solid-State Reactivity

Another area of interest are systems in which charge-transfer complexes might influence solid state chemical reactivity. This includes both inhibiting or promoting a particular reaction as well as influencing selectivities. Topochemical dimerizations of olefins and the polymerization of diacetylenes (DAs) represent two reactions which are highly sensitive to crystal packing geometries, both in terms of reactivity and, in some instances, selectivity. (*34,35*) There have been a variety of investigations aimed at controlling these processes, typically examining hydrogen bonding or π-π stacking interactions. For example, Desiraju has looked at the influence of a variety of weak interactions, including Cl···Cl interactions, on the dimerization of β-nitrostyrenes. (*36*) In these systems, the relatively weak form of halogen bonding exerted a small, but measurable influence on the crystal organization.

The photopolymerization of DA single crystals has been widely investigated and has generally been established to proceed as shown in Figure 9a. (*35*) The reaction is highly sensitive to certain geometrical features of the crystal; particularly the distance, d, between the stacked DAs and the angle between the molecular axis and the stacking axis, θ. It should be noted that θ is directly related to the degree of "slipping" along the molecular axis in a stack of DAs. In general, polymerization seems to occur when d is approximately 5 Å and θ is about 45°. (*34*) Highly flexible systems seem to have more liberal geometric requirements, but the closer the packing of the monomer is to that of the polymer, the more likely that single crystals of the polyDA can be obtained. Recently, Dougherty and co-workers have used phenyl-perfluorophenyl π-π stacking interactions to produce polyDAs with novel regiochemistry. (*37*) Their strategy was to engineer a crystal in which θ was close to 90°. Calculations suggested that this angle, combined with a very close packing distance of approximately 3.4 Å would lead to a unique *cis* assembly of the DAs as shown in Figure 9b. While crystallographic data is not available, the authors appear to have succeeded in creating this motif.

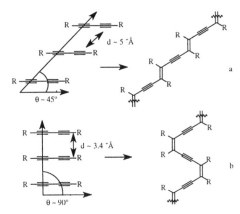

Figure 9: Geometrical requirements to form a) a trans-polydiacetylene and b) a cis-diacetylene.

We have prepared a number of 1,4-bis(N-aryl)-1,3-butadiynes in order to examine the influence of charge-transfer formation on diacetylene supramolecular organization. While there is some suggestion in the literature that removing electron density from the DA might increase the susceptibility to polymerization, (*38*) it seemed likely that the tendency for our acceptors to aggregate would inhibit the slipping along the molecular axis required to give the required θ angle. This proved to be the case, as none of these DA complexes photopolymerized. For example, the 1,4-bis(4-isoquinolyl)buta-1,3-diyne·TIE complex (Figure 6) has a stacking angle of 75°

94

and a stacking distance of 4.5 Å. (21) In addition, the formation of an extended chain structure makes an already highly rigid DA even more so. While neither the complex or the pure donor undergoes photopolymerization in this case, polymerization is actually prevented by complex formation in other instances. For example, charge-transfer complexes of the highly polymerizable compound 1,4-bis(3-quinolyl)buta-1,3-diyne are also photostable.

The ability to inhibit photopolymerization is interesting, but polyDAs are desirable materials with valuable optical properties. Charge-transfer complexes of polyDAs might find use in the construction of tunable non-linear optical materials, nanoporous solids and sensors. Thus, we are currently investigating methods for facilitating the polymerization of DA charge-transfer complexes. In an attempt to increase the flexibility of the crystal matrix, octafluoro-1,4-diiodobutane was used as the Lewis acid to form a complex with 1,4-bis(3-pyridyl)buta-1,3-diyne (Figure 10). (39) This structure has the stacking parameters d=3.61 Å, θ=71°. While clearly outside of the range required for the traditional polyDA mechanism, it is very similar to the parameters obtained by Dougherty, et. al. for their phenyl-perfluorophenyl systems. Phototolysis experiments are difficult with this complex, as the donor easily evaporates from the matrix, leaving the non-polarizable host DA behind. Despite this, preliminary thermal gravimetric analysis studies suggest at least partial polymerization of the complex.

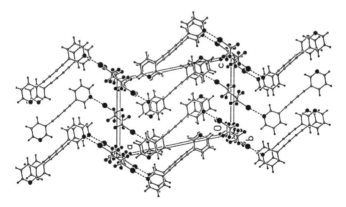

Figure 10. Structure of (1,4-bis(3-pyridyl)buta-1,3-diyne)· octafluoro-1,4-diiodobutane.

Another approach to polymerizable DA charge-transfer complexes is to construct asymmetric DAs with another structure-controlling element that encourages polymerization. Long chain urethane tail groups have been shown to encourage polymerization, though the mechanism is not completely understood. Very recently we have prepared charge-transfer complexes from a DA which features a 3-pyridyl head group and a urethane tail. (39) Complexes of this DA with TIE, tetrafluoro-1,4-diiodobenzene and octafluoro-1,4-diiodobutane, were exposed to UV radiation for 1 hour. The creamy white solids underwent dramatic color changes to deep purple, dull red and a bright red respectively. These color changes are characteristic of a highly regiospecific polymerization of the type shown in Figure 9, yet the differences indicate that the acceptor plays a role in the supramolecular arrangement of the host. (40) Additional characterization of these intriguing materials is in progress.

The application of halogen bonding to crystal engineering projects is still in its infancy. These interactions are now, however, well understood and increasing interest by several groups insures that they will play an important role in supramolecular synthesis.

Acknowledgements.

This work was funded in part by the Petroleum Research Fund and the Research Corporation.

Literature Cited

1. Perlstein, J.; Steppe, K.; Vaday, S.; Ndip, E.M.N. *J. Am. Chem. Soc.* **1996**, *118*, 8433.
2. Braga, D.; Grepioni, F. *Chem. Commun.* **1996**, 571.
3. Desiraju, G.R. *Crystal Engineering: The Design of Organic Solids.* Materials Science Monographs No. 54, Elsevier, Amsterdam, Netherlands, 1989.
4. Lachman, A. *J. Am. Chem. Soc.* **1903**, *25*, 50 and references therein.
5. Probably the most influential studies were those of Mulliken and co-workers, see for example: Mulliken, R.S. *J. Am. Chem. Soc.* **1952**, *74*, 811.
6. For a discussion of the meaning of these terms as well as a historical summary of some of the early studies in this area, see: Bent, H.A. *Chem. Rev.* **1968**, *68*, 587.
7. Dumans, J.M.; Gomel, L.; Guerin, M. in *The Chemistry of Functional Groups, Supplement D*; Patai, S,; Rappoport, Z. Eds.; Wiley, New York, 1983, 985.
8. Legon, A.C. *Chem. Eur. J.* **1998**, *4*, 1890.
9. This is the estimated energy of a C-Cl···N interaction from calculations performed on a dimer of chloro-cyanoacetylene. J.P.M. Lommerse, A.J. Stone, R. Taylor and F.H. Allen, *J. Am. Chem. Soc.* **1996**, *118* 3108. The strength of this C-X···N interaction increases on moving from chlorine to bromine to iodine, and would be expected to be stronger still for the dihalides.
10. Danten, Y.; Guillot, B.; Guissani, Y. *J. Chem. Phys.* **1992**, *96*, 3795 and references therein.
11. Hanks, T.W., *unpublished results*. All computational results discussed here were performed using the Hartree-Fock SCF method with a 3-21G* basis set. This approach consistently overestimates the N···I bond length by 0.2-0.3 Å. All other bonds and angles are in good agreement with X-ray crystallographic data. Orbital and isosurface results are qualitatively consistent with higher order calculations.
12. Bailey, R.D.; Buchanan, M.L.; Pennington, W.T. *Acta Cryst.* **1992**, *C48*, 2259.
13. Bailey, R.D.; Drake, G.W.; Grabarczyk, M.; Hanks, T.W.; Hook, L.L.; Pennington, W.T. *J. Chem. Soc., Perkins 2* **1997**, 2773.
14. Uchida, T.; Kimura, K. *Acta Cryst.* **1984**, *C40*, 139.
15. Bailey, R.D.; Hanks, T.W.; Pennington, W.T., *unpublished results.*
16. Bailey, R.D.; Grabarczyk, M.; Hanks T. W.; Pennington, W.T. *J. Chem. Soc., Perkin 2* **1997**, 2781.

17. Rimmer, E.L.; Bailey, R.D.; Pennington W.T.; Hanks, T.W., *unpublished results.*
18. Pohl, S. *Z. Naturforsch.* **1983**, *B38*, 1535.
19. Pennington, W.T.; Bailey, R.D.; Holmes, B.T.; Hook, L.L.; Watson, R.P.; Warmoth, M.; Hanks, T.W. *Trans. Am. Cryst. Assoc.; Cryst. Eng. Symp.* **1998**, *33*, 145.
20. Bailey R.D.; Hook, L.L.; Watson, R.P.; Warmoth, M.; Hanks, T.W.; Pennington, W.T. *Cryst. Eng.* In press.
21. Phelps, D.; Crihfield, A.; Hartwell, J.; Hanks, T.W.; Pennington, W.R.; Bailey, R.D. *Mol. Cryst. Liquid Cryst.*, In Press.
22. Bailey R.D.; Pennington, W.T. *unpublished results.*
23. Pedireddi, V.R.; Reddy, D.S.; Goud, B.S.; Craig, D.C.; Rae, A.D.; Desiraju, G.R. *J. Chem. Soc. Perkins Trans. 2* **1994**, 2353 and references therein.
24. Dahl, T.; Hassel, O. *Acta Chem. Scand.* **1968**, *22*, 2851.
25. Skell, P.S.; Pavlis, R.R.; Lewis, D.C.; Shea, K.S. *J. Am. Chem. Soc.* **1973**, *95*, 6735.
26. Wilen, S.H.; Bunding, K.A.; Kascheres, C.M.; Weider, M.J. *J. Am. Chem. Soc.* **1985**, *107*, 6997.
27. Farina, A.; Meille, S.V.; Messina, M.T.; Metrangolo, P.; Resnati, G.; Vecchio, G. *Angew. Chem. Int. Ed.* **1999**, *38*, 2433.
28. McCrone, W.C. In *Physics and Chemistry of the Organic Solid State, Vol. II*; Eds.: Fox, D.; Labes, M.M.; Weissberger, A.; Interscience, New York, 1963, 725.
29. Bernstein, J. *Organic Crystal Chemistry*, Eds.: Garbarczyk, J.B.; Jones, D.W. Oxford Science, Oxford, 1991.
30. Bavin, M. *Chem. & Ind.* **1989**, 527.
31. Bailey, R.D.; Grabarczyk, M.; Hanks, T.W.; Newton, E.M.; Pennington, W.T.; Electronic Conference on Trends in Organic Chemistry (ECTOC-1), Eds. Rzepa, H.S.; Goodman, J.M. (CD-ROM), (Royal Society of Chemistry publications, 1995). See also URL http://www.ch.ic.ac.uk/ectoc/papers.
32. Greaves, B.; Stoeckli-Evans, H. *Acta Cryst.* **1992**, *C48*, 2269.
33. Bock, H.; Vaupel, T.; Nather, C.; Ruppert, K.; Havlas, Z. *Angew. Chem. Int. Ed. Engl.* **1992**, *31*, 299.
34. For example, see Baughman, R.H.; Yee, K.C. *J. Poly. Sci.: Macromolecular Rev.* **1978**, *13*, 219.
35. *Polydiacetylenes;* Bloor, D.; Chance, R.R., Eds.; NATO ASI Series E, 102; Martinus Nijhoff Publishers, Boston, MA.
36. Pedireddi, V.R.; Sarma, J.A.R.P.; Desiraju, G.R. *J. Chem. Soc. Perkins 2* **1992**, 311.
37. Coates, G.W.; Dunn, A.R.; Henling, L.M.; Dougherty, D.A.; Grubbs, R.H. *Angew. Chem. Int. Ed. Engl.* **1997**, *36*, 248.
38. Several coordination compounds of diacetylenes have been reported which undergo polymerization. In addition, N-methylation of a photostable pyridyl-diacetyelene has been shown to result in a photoreactive material, see: Subramanyam, S,; Blumstein, A. *Macromolecules* **1991**, *114*, 3247.
39. Crihfield, A.; Hanks, T.W.; Pennington, W.T. *unpublished results.*
40. Polydiacetylenes are highly colored due to the extended conjugation in the backbone. They are also often thermochromic, solvatochromic and even mechanochromic. It is likely that the various acceptors cause small changes in the orientation of the polymer backbone, giving rise to the observed color.

Chapter 7

Arene–Arene Double T-Contacts: Lateral Synthons in the Engineering of Highly Anisotropic Organic Molecular Crystals

Michael Lewis, Zhengyu Wu, and Rainer Glaser*

Department of Chemistry, University of Missouri at Columbia, Columbia, MO 65211

An arene-arene double T-contact describes a pair of intermolecular arene-arene T-contacts between two molecules that each contain two spacer-connected arene rings. We have been employing such double T-contacts as lateral synthons in the crystal engineering of highly anisotropic organic crystals. In particular, this type of intermolecular interaction has been employed as a stabilizing factor to overcome the intrinsic preference for the anti-parallel alignment of dipolar molecules. Here, we review pertinent properties of benzene and of the benzene-benzene interaction and present and discuss arene-arene double T-contacts in crystals of a representative number of symmetrical and unsymmetrical acetophenone azines, $X–C_6H_4–CMe=N–N=CMe–C_6H_4–Y$, including the structures of two near-perfectly dipole parallel-aligned crystals.

Introduction

For materials to exhibit NLO activity (*1*) two physical properties are essential. First, a crystal must be asymmetric. For organic crystals, this prerequisite is relatively easy to meet because 20% of all organic molecules crystallize in chiral space groups (*2*). The second and much more challenging criterion for a crystal to be NLO active is the presence of a macroscopic dipole moment. This is not trivial, because most polar

[1]Corresponding author.

molecules tend to pack such that their dipole moments cancel each other (Scheme 1a), thus giving a crystal dipole moment of zero. Most crystals that exhibit a non-zero dipole moment do so because of an incomplete cancellation of the molecular dipole moments (Scheme 1b). In those cases, the crystal dipole moment is considerably less than the sum of the molecular dipole moments and such cases are far from ideal. The most desirable result is perfect dipole alignment (Scheme 1c) so that the dipole moment of the crystal is equal of at least approximately equal to the sum of the individual molecules.

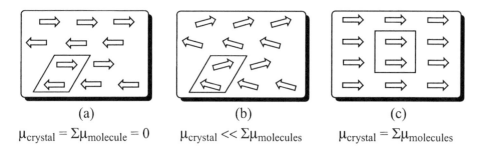

(a)	(b)	(c)
$\mu_{crystal} = \Sigma\mu_{molecule} = 0$	$\mu_{crystal} \ll \Sigma\mu_{molecules}$	$\mu_{crystal} = \Sigma\mu_{molecules}$

Scheme 1. (a) Complete cancellation, (b) incomplete or partial cancellation and (c) complete reinforcement of molecular dipole moments.

While the interest in polar organic materials has been tightly linked to applications in nonlinear optics, highly polar crystalline materials are of interest in and of themselves. Molecular orientation and parallel dipole alignment is important for many optical and electrical applications. A polar, anisotropic environment is key for nonlinear optical responses in crystals (*vide supra*) and in liquid crystals (*3, 4, 5*). Similarly, the NLO activity of dyes in polymer films depends on the parallel alignment of molecular dipole moments (*6*). Other optical phenomena that depend on molecular anisotropy include thermal conductivity (*7*), ferroelectric responses (*8*) and ferroelastic responses (*9*). From a fundamental point of view, our materials will allow for systematic studies of the effects of a highly polar environment on molecular properties. The polar alignment of the molecules will cause large electric fields within the crystals and the properties of the free molecule and of the molecule in the crystal can be expected to be quite different. It is possible to measure internal electric fields with atomic resolution (*10, 11*) and our work might enable systematic studies of internal electric fields in polar crystals.

We have designed donor-acceptor-substituted π-conjugated molecules (Scheme 2) aimed at overcoming the parallel dipole alignment challenge by incorportating structural motifs that enhance the likelihood of obtaining the scenario illustrated in Scheme 1c. The first key issue of our design concept is the minimization of the ground state dipole moment. Relatively small ground state dipole moments are required to minimize the intrinsic advantage of the dipole antiparallel-aligned arangement over the dipole parallel-aligned arrangement. This design feature is accomplished by placing an azine bridge in the center of the molecule. The azine bridge can be viewed as two polar acceptor imine groups with opposite orientations, and thus acts as a conjugation stopper (*12*) (Scheme 2). The imine-acceptor conjugation in one half of the molecule results only in a very small contribution to the overall dipole moment for that segment. The overall dipole moment is largely due to the donor-imine conjugation. Our *ab initio* studies have shown that the azine bridge is effective in preventing through-conjugation when placed between donor- and acceptor-substituted π-systems (*12*) and extensive NMR studies fully support this view (*13*). The second important design concept is the use of aromatic systems as alignment units (Scheme 2). Intermolecular arene-arene interactions (*vide infra*) are employed to compensate for electrostatic repulsions of parallel dipole alignment.

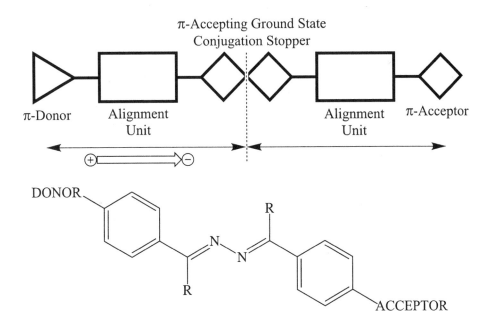

Scheme 2. Dipole design concept. The azine bridge acts as a π-accepting conjugation stopper and the phenyl rings function as lateral synthons whose intermolecular interactions help to overcome the intrinsically preferred antiparallel dipole alignment of molecular dipoles.

In this context we have been investigating the stereochemistry (*14*, *15*, *16*), electronics (*12*, *14*, *16*) and crystal packing (*14*, *16*, *17*, *18*, *19*, *20*, *21*) of symmetric and unsymmetric acetophenone azines with aims at overcoming the dipole alignment challenge. The azines that we have studied have the general structure shown in Scheme 2. All of the crystal structures we have obtained thus far have intermolecular double arene-arene interactions as the dominant packing motif thus underscoring the importance of arene-arene interactions in controlling the packing. A double arene-

arene interaction occurs when two arene rings of one molecule interact with the two arene-arene rings of an adjacent molecule. In this chapter, we will briefly review the literature on pertinent properties of benzene and of the benzene-benzene interaction and present and discuss arene-arene double T-contacts in crystals of a representative number of symmetrical (X = Y) and unsymmetrical (X ≠ Y) acetophenone azines, $X-C_6H_4-CMe=N-N=CMe-C_6H_4-Y$, including the structures of two near-perfectly dipole parallel-aligned crystals.

Benzene Structure and Quadrupole Moment

Benzene has only one unique diagonal traceless quadrupole moment tensor component, Θ_{zz} (z-axis is symmetry axis), which can be measured via electric field gradient induced birefringence (22) and diode laser spectroscopy (23). The most recent measurement is that of Sumpf (23) which gave $\Theta_{zz} = -1.0 \bullet 10^{-25}$ esu. This value is close to the frequently cited value of $\Theta_{zz} = -6.47 \pm 0.37$ a.u. reported in 1981 by Battaglia and Buckingham from birefringence data (22). Because the birefringence method is the most popular way for the determination of quadrupole moments, we use that experimental value for comparison to our *ab initio* calculations.

The first structure determination of benzene was that of Cox (24) in 1932; X-ray diffraction of a single crystal at -22°C showed a D_{6h} structure with a CC bond length of about 1.42 Å. The most precise structural work is the electron diffraction study of Strand (25) which yielded equilibrium distances r_e(CC) = 1.3929 Å and r_e(C–H) = 1.0857 Å and we used this geometry for our benzene quadrupole moment (single point) calculations. Results of our calculations of the quadrupole moment of benzene are given in Table 1 (26). All unabridged quadrupole moment tensor components are negative. The negative charge (electrons) is distributed further from the center of mass than the positive charge (nuclei) and the quadrupolarity of benzene molecule can thus be characterized as {- + -} in every direction. Q_{zz} is about 25% larger in magnitude than Q_{xx} and Q_{yy}. Compared with the experimental value –6.47 ± 0.37 au (22), our data fall into a ±10% error range. RHF/cc-pVDZ and RHF/cc-pVDZ++ give the closest values. The choice of a good basis set is the most critical issue.

Table I. Quadrupole Moment Tensor Components and Quadrupole Moment

Method/Basis Set	$Q_{xx}=Q_{yy}$[a]	Q_{zz}[a]	$<Q_{ii}>$[a]	Θ[a]	Θ[b]	%Error
RHF/6-311G(d,p)	-31.670	-40.631	-34.657	-8.962	-6.663	3.137
RHF/6-311++G(d,p)	-31.936	-41.319	-35.063	-9.383	-6.976	7.991
RHF/6-311G(2d,p)	-31.532	-40.409	-34.491	-8.877	-6.600	2.166
RHF/6-311++G(2d,p)	-31.835	-41.251	-34.974	-9.415	-7.000	8.359
RHF/cc-pVDZ	-31.775	-40.381	-34.644	-8.606	-6.398	-0.956
RHF/cc-pVDZ++	-31.671	-40.504	-34.615	-8.833	-6.567	1.654
MP2/6-311G(d,p)[c]	-32.240	-39.972	-34.817	-7.732	-5.748	-11.014
MP2/6-311++G(d,p)[c]	-32.540	-40.779	-35.286	-8.239	-6.126	-5.174

[a]Units in DÅ. [b]Units in a.u. [c]MP2 active space included all electrons.

Benzene-Benzene Interactions

The structure of benzene dimer has been studied theoretically (*27*) and experimentally (*28*) in the gas-phase, solution-phase and the solid state and all results suggest that the edge-to-face structure, or T-contact (**A** and **B** in Scheme 3), is more stable than the parallel-displaced face-to-face structure (**C**). The T-shaped structure can form numerous geometries, two of which are shown as **A** and **B** in Scheme 3 with the most stable structure being **A**.

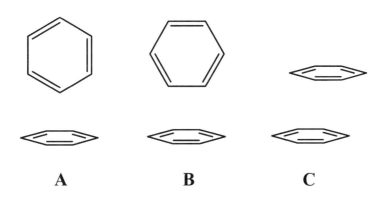

A   **B**   **C**

Scheme 3. (a) Edge-to-face contact or T-contact with one C-H pointing toward the benzene π-cloud. (b) T-contact with two C-H bonds pointing to the π-cloud. (c) Off-set face-to-face contact.

Initial *ab initio* gas-phase calculations at the MP2/MIDI-1+s,p(c) level of theory (*29*) suggested that there were four energy minima on the benzene dimer potential energy surface (PES). Three of the four structures were T-shaped with the most stable one illustrated by **A** and the fourth structure was the parallel-displaced face-to-face structure **C**. The most stable T-structure had an intermolecular distance of 5.0 Å between the molecular centers and was more stable than the parallel-displaced face-to-face structure with an intermolecular distance of 4.3 Å by approximately 0.4 kcal/mol (*29*). Later calculations using the better 6-31+G* (*30*) and DZ+2P (*31*) basis sets in conjunction with the MP2 method found that the parallel-displaced face-to-face structure was more stable by 0.9 and 0.2 kcal/mol, respectively. The intermolecular distance for the T-shaped structure was the same as the MP2/MIDI-1+s,p(c) calculations, 5.0 Å between the molecular centers, however the parallel-displaced face-to-face structure had much closer intermolecular distance of 3.85 Å. The fact that the T-shaped and parallel-displaced structures of benzene are nearly isoenergetic was reaffirmed using higher levels of theory, MP4 and CCSD(T), and better basis sets, AUG-cc-pVDZ (*32*). The CCSD(T)/AUG-cc-pVDZ calculations have intermolecular distances of 5.1 and 4.0 Å between the molecular centers for the T-shaped and parallel-displace face-to-face structures, respectively. The most recent

CCSD(T) dynamics calculations on the benzene dimer (*33*) suggest that there exists only one minimum on the PES and it is the T-structure. This study located four other stationary points, however they were either first or second order saddle points. The parallel-displaced face-to-face structure **C** was found to be a transition state for the automerization of the T-structure. The intermolecular distance between the molecular centers for the T-shaped structure in this study was 5.21 Å. The T-structure was also calculated to be more stable than the face-to-face structure in aqueous solution (*34*) and liquid benzene (*35*) via molecular dynamics and Monte Carlo simulations, respectively. The predominant structure for the calculations of liquid benzene is T-shaped with an intermolecular distance of 5.5 Å between the molecular centers. Finally, molecular mechanics calculations aimed at simulating the benzene crystal also predict the T-shaped structure to be most stable (*36*) with intermolecular distances of approximately 5.0 Å.

The large body of experimental work on benzene dimer also points to the T-shaped structure being more stable than the face-to-face aggregate. Early work employed electric deflection measurements (*37*, *38*) to show that benzene dimer possesses an electric dipole moment, thus ruling out face-to-face structure and suggesting a T-shaped one. A T-shaped structure was also hypothesized by mass-selective two-photon ionization experiments (*39*) and molecular jet optical absorption spectra (*40*). The latter study went so far as to suggest a C_{2v} symmetry for the T-shaped benzene dimer, however, subsequent time-of-flight multiphoton mass spectra (*41*) revealed that the two benzene rings were not perpendicular but rather they deviated from orthogonality by $10°$-$20°$. A binding energy of 1.6 ± 0.23 kcal/mol was determined for this arrangement using a reflection time-of-flight mass spectrometer after two-photon ionization (*42*). More recent mass-selective Raman spectroscopic studies (*43*, *44*) and mass-selective hole-burning experiments (*45*) also point to a floppy T-shaped structure, however one such mass-selective Raman study (*46*) reports the presence of both an edge-to-edge and a T-shaped structure with intermolecular distances of 3.7-4.2 and 5.0 Å, respectively. A more precise intermolecular distance of 4.96 Å was reported for the T-shaped dimer using Balle/Flygare Fourier transform microwave spectroscopy (*47*). NMR spectroscopic measurements of neat benzene (*48*) show the presence of both the T-shaped and face-to-face structures with the T-shaped structure being favored by approximately 0.7 kcal/mol. The experimental data on solid state benzene unambiguously reflects the preference for the T-shaped structure. Benzene crystallizes into two unique lattices; benzene I is orthorhombic with the space group P*bca* and benzene II is monoclinic with space group $P2_1/c$. The crystal structure of benzene I was determined by X-ray crystallography at $-3°C$ (*49*) and by neutron diffraction at -55 and $-135°C$ (*50*) and the benzene molecules pack at almost perfect right angles, $90°22'$, with intermolecular closest contacts of approximately 2.75 Å (*49*). Crystals of benzene II were grown and examined at $21°C$ and 25 kilobars (*51*). The closest intermolecular contacts in benzene II are similar to those in benzene I however the molecules are no longer orthogonal and they pack such that the planes of the two rings are at $120°$ angle with respect to each other (*51*).

Intermolecular Arene-Arene Double T-Contacts

The prevalent geometric motif of intermolecular interaction in our crystals is the arene-arene double T-contact. A double T-contact occurs when two phenyl rings of one molecule interact with the two phenyl rings of an adjacent molecule to produce two T-contacts. Arene-arene double T-contacts can achieve numerous arrangements and three of them are shown in Scheme 4. Pattern α can only occur when the space group connecting the phenyl rings is not twisted and the two phenyl rings within each molecule lie in the same plane. Patterns β and γ are two of the numerous motifs that result from a twisted spacer group with orthogonal or nearly orthogonal phenyl rings.

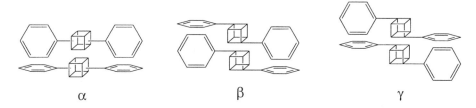

α   β   γ

Scheme 4. Three possible double T-contacts between molecules with two spacer-connected arene units.

The azine bridge we employ in our dipolar molecules is the twisted spacer. This may seem contrary to simple notions of conjugation and valence bond theory, however it is in line with our theoretical view of the azine bridge as a conjugation stopper. The twisted nature of the azine bridge is not just a result of packing effects in the crystal, but rather it is an intrinsic property that is also witnessed for the free molecules in the gas phase and in solution. Quantum mechanical *ab initio* calculations show that the optimal structure of our azines in the gas phase is a twisted structure (*12*) and extensive NMR studies supporting this view (*13*). The twisted geometry of the azine bridge in our molecules precludes the formation of double T-contacts of the α-type (Scheme 4). Instead, all of the double T-contacts that we observe in our crystals have the general motif described by types β and γ. It is important to recognize that β and γ are structural isomers or association isomers because the β aggregate can be converted into the γ aggregate by moving the top molecule to the bottom.

Double T-Contacts Between Acetophenone Azines

The crystal structures of all of our azines are highly anisotropic and they only exist in parallel or antiparallel dipole aligned arrangements. One of the driving forces behind this high degree of orientational order is the double T-contact; double T-contacts are the dominant structural motif in every structure. The β- and γ-type double T-contacts for selected symmetric acetophenone azines are shown in Figures 1 and 2 and the double T-contacts for selected unsymmetric azines are given in Figures 3 and 4. In all cases, each azine engages in *both* β- and γ-type double T-contacts.

Symmetric azines carry two identical substitutents and the three azines shown in Figures 1 and 2 are the parent compound (**H-H**) and the Cl- and I-substituted molecules (**Cl-Cl** and **I-I**). The key factor governing the formation of the double T-contact is the twisted azine spacer group. The three azines all assume a *gauche* conformation with –C=N–N=C– dihedral values, τ, of 138.7°, 134.7° and 141.8° for **H-H**, **Cl-Cl** and **I-I**, respectively. The dihedral angles relating the twist of the two phenyl rings with respect to the C_{para}-C_{imine} bond planes, ϕ_1 and ϕ_2, are 0.3° and 19.7° for **H-H**, 29.3° and 30.5° for **Cl-Cl**, and 8.0° and 8.0° for **I-I**. In all of our azines the N–N and the two C_{para}-C_{imine} bonds are almost parallel. Therefore, the sum of τ' (where $\tau' = 180 - \tau$), ϕ_1 and ϕ_2 provides an excellent approximation for the angle ω between the best planes of the benzene rings. The ω values for **H-H**, **Cl-Cl** and **I-I** are 61.3°, 105.1° and 54.2° and these angles can be visualized by looking down the N–N bonds as shown on the right side of Figures 1 and 2.

The two monomers in each β- and γ-type double T-contact are stereoisomers and they can be described by the *P/M* nomenclature for helicity (*52*). In this nomenclature the molecule has *P*-helicity if the shortest rotation around the N–N bond that results in an eclipsing of the phenyl rings is a clockwise rotation. Conversely, if the shortest rotation to effect eclipsing of the phenyl rings is counter-clockwise, then the molecule has *M*-helicity. For the dimers in Figures 1 and 2 the *P*-stereoisomer is always on top and the *M*-stereoisomer is on the bottom. The views on the right of Figures 1 and 2, looking down the N–N bonds, are illustrate the *P*- and *M*-helicities in the most compelling fashion.

The β- and γ-type double T-contacts formed in the crystal structure of **Cl-Cl** are very different than those formed in the crystal structures of **H-H** and **I-I**. The β-type double T-contacts (Figure 1) in **H-H** and **I-I** are arranged so that the CC aromatic bond that is *anti* to the N–N bond in the *P*-stereoisomer is in contact with the C-C aromatic bond that is *anti* to the N–N bond in the *M*-stereoisomer. In contrast, the β-type double T-contacts in **Cl-Cl** are arranged so that the aromatic CC bond that is *syn* to the N–N bond in the *P*-stereoisomer is in contact with the aromatic CC bond that is *anti* to the N–N bond in the *M*-stereoisomer. Likewise, a similar difference exists between the γ-type double T-contacts in the **H-H** and **I-I** crystal structures and those in the **Cl-Cl** crystal structure. The γ-type double T-contacts in **H-H** and **I-I** (Figure 2) have the CC aromatic bond that is *syn* to the N–N bond in the *P*-stereoisomer in contact with the CC aromatic bond that is *syn* to the N–N bond in the *M*-stereoisomer. The γ-type double T-contacts in the **Cl-Cl** crystal structure, however, are between the CC aromatic bond that is *anti* to the N–N bond in the *P*-stereoisomer and the CC aromatic bond that is *anti* to the N–N bond in the *M*-stereoisomer. Note that in all three structures each azine partakes in at least one β-type and one γ-type double T-contact. Therefore, both the *syn* and *anti* aromatic CC bonds of each azine partake in a double T-contact. The double T-contacts in **H-H**, **Cl-Cl** and **I-I** deviated from orthogonality by 30°, 15° and 35°, respectively, and this is similar to what is observed in experimental work on benzene dimers in the gas phase (*41*, *43*, *44*, *45*). Gas phase experiments predicted a deviation of 10° to 20° from orthogonality and the crystal structure of benzene II has T-contacts that deviate about 30° from perpendicularity.

The double T-contacts in the other symmetric acetophenone azines that we have studied have the same general motif as those illustrated in Figures 1 and 2. The double T-contacts in the crystal structure of 4-bromoacetophenone azine have the same general structure as those in the symmetric **Cl-Cl** structure. The crystal structures of the symmetric 4-fluoro- and 4-methoxy-substituted acetophenone azines have double T-contacts that resemble those in **H-H** and **I-I**.

The crystal structures of the unsymmetrical acetophenone azines that we have studied all exhibit similar double T-contacts to those in **H-H** and **I-I**. Three examples

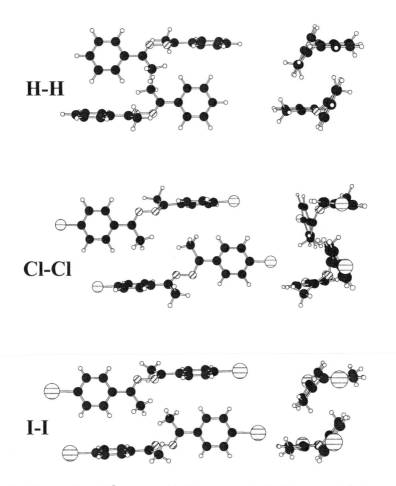

Figure 1. Illustration of β-type double T-contacts in the X-ray crystal structures of selected symmetric acetophenone azines. P-stereoisomers are on top and the M-stereoisomers are on the bottom.

of unsymmetrical donor-acceptor-substituted acetophenone azine double T-contacts are shown in Figure 3 and 4. Two of the examples have a methoxy group as the donor substituent and bromine and chlorine as the acceptors, **MeO-Br** and **MeO-Cl**, and one has an amino group as the donor and a nitro group as the acceptor, **NH$_2$-NO$_2$**. The β-type double T-contacts formed by these three crystal structures are shown in Figure 3 and the γ-type double T-contacts are illustrated in Figure 4.

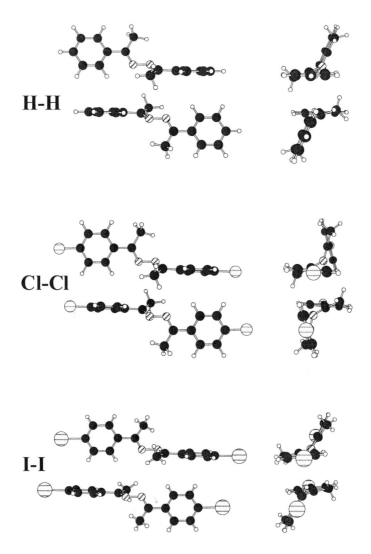

Figure 2. Illustration of γ-type double T-contacts in the X-ray crystal structures of selected symmetric acetophenone azines. P-stereoisomers are on top and the M-stereoisomers are on the bottom.

In the same fashion as the symmetric azines, the double T-contacts of the unsymmetrical azines are a result of the *gauche* conformations of the two phenyl groups within each monomer. As was the case for the symmetric azines, the –C=N–N=C– dihedral values, τ, are not always the same for the *P*- and *M*-stereoisomers. For **MeO-Br** and **MeO-Cl** the *P*- and *M*-stereoisomers do not have the exact same structure. The τ-values for **MeO-Br** are 135.8° for the *P*-isomer and 137.4° for the *M*-isomer and for **MeO-Cl** the values are 134.7° and 136.5°. The *P*- and *M*-stereoisomers in **NH$_2$-NO$_2$** are isostructural and the τ value is 135.5°. Defining the dihedral angles that relate the twist of the two phenyl rings with respect

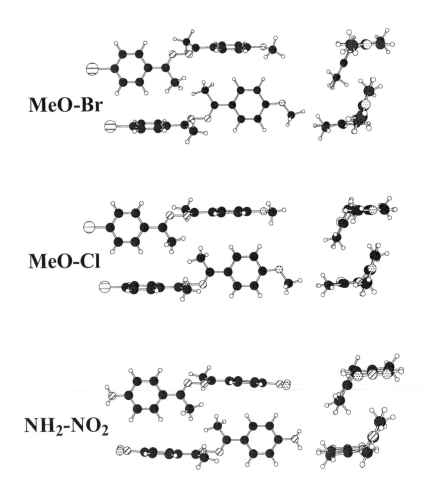

Figure 3. Illustration of β-type double T-contacts in the X-ray crystal structures of selected unsymmetric acetophenone azines. P-stereoisomers are on top and the M-stereoisomers are on the bottom.

to the C_{para}–imine bond planes, ϕ_1 and ϕ_2, becomes less arbitrary when the azine is not symmetric and we assign ϕ_1 to the phenyl ring with the donor substituent and ϕ_2 to the acceptor-substituted ring. For **MeO-Br** the ϕ_1 values are 17.4° for the *P*-isomer and 8.7° for the *M*-isomer and the ϕ_2 values are 2.6° (*P*-isomer) and 13.6° (*M*-isomer). The values for **MeO-Cl** are 15.4° (ϕ_1, *P*-isomer), 7.9° (ϕ_1, *M*-isomer), 3.3° (ϕ_2, *P*-isomer) and 14.3° (ϕ_2, *M*-isomer). The ϕ_1 and ϕ_2 values for **NH$_2$-NO$_2$** are 10.7° and 8.1°, respectively. Thus, the angle ω between the best planes through the benzene

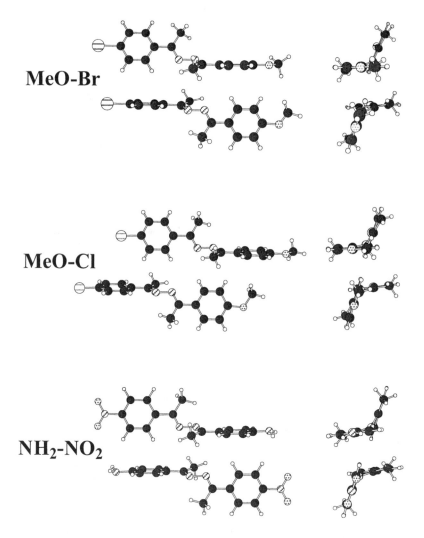

Figure 4. Illustration of the γ-type double T-contacts in the X-ray crystal structures of selected unsymmetric acetophenone azines. P-stereoisomers are on top and the M-stereoisomers are on the bottom.

rings for **MeO-Br** is 64.2° for the *P*-isomer and 64.9° for the *M*-isomer. For **MeO-Cl** the ω values are 64.0° and 65.7° for the *P*- and *M*-stereoisomers, respectively. The ω value for NH_2-NO_2 is 63.3°. These angles are best visualized via the Newman-type views down the N–N bonds shown on the right side of Figures 3 and 4.

The β- and γ-type double T-contacts formed in the crystal structures of **MeO-Br**, **MeO-Cl** and NH_2-NO_2 all have the same motif as those formed in the crystal structures of **H-H** and **I-I**. The β-type double T-contacts of **MeO-Br**, **MeO-Cl** and NH_2-NO_2 (Figure 3) are arranged so that the CC aromatic bond that is *anti* to the N–N bond in the *P*-isomer is in contact with the CC aromatic bond that is *anti* to the N–N bond in the *M*-isomer. The γ-type double T-contacts in **MeO-Br**, **MeO-Cl** and NH_2-NO_2 (Figure 4) have the CC aromatic bond that is *syn* to the N–N bond in the *P*-isomer in contact with the CC aromatic bond that is *syn* to the N–N bond in the *M*-isomer. As was the case for the symmetrical azines, the *syn* and *anti* aromatic CC bonds of each unsymmetrical azine monomer partake in at least one β-type and one γ-type double T-contact. The double T-contacts in the three unsymmetrical acetophenone azine crystal structures shown in Figures 3 and 4 all deviate from orthogonality by 20° to 25° and this further supports the experimental and theoretical preference for non-perpendicular arene-arene T-structures. The crystal structures of other donor-acceptor-substituted mixed acetophenone azines studied in our group also contain monomers with *gauche* conformations and the dimers form similar motifs as in **MeO-Br**, **MeO-Cl** and NH_2-NO_2.

Summary

The quadrupole moment of benzene is significant and we employ this property of aromatic systems in the design of parallel dipole aligned materials. The intrinsic conformational preferences and flexibility of azines allow for optimal intermolecular double T-contacts. The dipole moment in all of our mixed azines is along the N–N bond and thus, Figures 3 and 4 illustrate the near perfect parallel dipole alignment in the crystal structures of **MeO-Br** and **MeO-Cl**. The molecular dipole moments of the **MeO-Hal** azines are 3-4 D in the gas phase (*12*) and the degree of alignment achieved in their crystals exceeds 90 percent of the optimal value (*20*). In all of the other unsymmetrical azine crystals that we have studied the monomers align such that complete dipole cancellation is observed and an example of this is NH_2-NO_2 (Figures 3 and 4). The anti-parallel dipole aligned crystals will never show any nonlinear optical properties, however they are important when discussing the directing effects of aromatic rings in the crystal structures of acetophenone azines. We only observe parallel and anti-parallel dipole alignment within the double T-contacts of our crystals. Never do we witness any intermediate crystal packing moieties. This high degree of anisotropy is achieved because the arene-arene double T-contact acts as an extremely efficient lateral synthon.

References

1. Zyss, J. *Molecular Nonlinear Optics, Materials, Physics, and Devices*; Academic Press, Inc.: New York, NY, 1994.
2. Sakamoto, M. *Chem. Eur. J.* **1997**, *3*, 684-689.
3. Walba, D. M.; Korblova, E.; Shao, R.; Maclennan, J. E.; Link, D. A.; Glaser, M. A.; Clark, N. A. *Science* **2000**, *288*, 2181-2184.
4. Trzaska, S. T.; Hsu, H. F.; Swager, T. M. *J. Am. Chem. Soc.* **1999**, *121*, 4518-4519.
5. Simoni, F.; Francescangeli, O. *J. Phys. Condens. Matter* **1999**, *11*, R439-R487.
6. Atassi, Y.; Chauvin, J.; Delaire, J. A.; Delouis, J.-F.; Fanton–Maltey, I.; Nakatani, K. *Pure & Appl. Chem.* **1998**, *70*, 2157-2166.
7. Kurabayashi, K.; Goodson, K. E. *J. Appl. Phys.* **1999**, *86*, 1925-1931.
8. Saad, B.; Galstyan, T. V.; Dinescu, L.; Lemieux, R. P. *Chem. Phys.* **1999**, *245*, 395-405.
9. Luty, T.; Eckhardt, C. J. *J. Phys. Chem.* **1996**, *100*, 6793-6800.
10. Kohler, B. E.; Wohl, J. C. *J. Chem. Phys.* **1995**, *102*, 7773-7781.
11. Geissinger, P.; Kohler, B. E.; Woehl, J. C. *J. Phys. Chem.* **1995**, *99*, 16527-16529.
12. Glaser, R.; Chen. G. S. *J. Comput. Chem.* **1998**, *19*, 1130-1140.
13. Lewis, M., PhD Dissertaion, University of Missouri, **2001**.
14. Chen, G. S.; Wilbur, J. K.; Barnes, C. L.; Glaser, R. *J. Chem. Soc. Perkin Trans. 2* **1995**, 2311-2317.
15. Glaser, R.; Chen, G. S.; Anthamatten, M.; Barnes, C. L. *J. Chem. Soc. Perkin Trans. 2* **1995**, 1449-1458.
16. Glaser, R.; Chen, G. S.; Barnes, C. L. *J. Org. Chem.* **1993**, *58*, 7446-7455.
17. Lewis, M.; Barnes, C. L.; Hathaway, B. A.; Glaser, R. *Acta Cryst. C* **1999**, *55*, 975-978.
18. Lewis, M.; Barnes, C. L.; Glaser, R. *J. Chem. Crystallogr.* **1999**, *29*, 1043-1048.
19. Lewis, M.; Barnes, C. L.; Glaser, R. *Acta Cryst. C* **2000**, *56*, 393-396.
20. Lewis, M.; Barnes, C. L.; Glaser, R. *J. Chem. Crystallogr.* **2001**, *31*, in press.
21. Lewis, M.; Barnes, C. L.; Glaser, R. *Can. J. Chem.* 1998, **76**, 1371-1378.
22. Battaglia, M. R.; Buckingham, A. D.; Williams, J. H. *Chem. Phys. Lett.* **1981**, *78*, 421-423.
23. Waschull, J.; Heiner, Y.; Sumpf, B.; Kronfeldt, H. D. *J. Mol. Spectrosc.* **1998**, *190*, 140-149.
24. Cox, E. G. *Proc. Roy. Soc.* **1932**, *A135*, 491.
25. Kochikov, I. V.; Tarasov, Y. I.; Kuramshina, G. M.; Spiridonov, V. P.; Yagola, A. G.; Strand, T. G. *J. Mol. Struc.* **1998**, *445*, 243-258.
26. Wu, Z.; Glaser, R., to be published.
27. Hobza, P.; Selzle, H. L.; Schlag, E. W. *Chem. Rev.* **1994**, *94*, 1767-1785.
28. Garrett, A. W.; Zwier, T. S. *J. Chem. Phys.* **1992**, *96*, 3402-3410.
29. Hobza, P.; Selzle, H. L.; Schlag, E. W. *J. Chem. Phys.* **1990**, *93*, 5893-5897.
30. Hobza, P.; Selzle, H. L.; Schlag, E. W. *J. Phys. Chem.* **1993**, *97*, 3937-3938.
31. Hobza, P.; Selzle, H. L.; Schlag, E. W. *J. Am. Chem. Soc.* **1994**, *116*, 3500-3506.
32. Hobza, P.; Selzle, H. L.; Schlag, E. W. *J. Phys. Chem.* **1996**, *100*, 18790-18794.

33. Spirko, V.; Engkvist, O.; Soldan, P.; Selzle, H. L.; Schlag, E. W.; Hobza, P. *J. Chem. Phys.* **1999**, *111*, 572-582.
34. Chipot, C.; Jaffe, R.; Maigret, B.; Pearlman, D. A.; Kollman, P. A. *J. Am. Chem. Soc.* **1996**, *118*, 11217-11224.
35. Jorgensen, W. L.; Severance, D. L. *J. Am. Chem. Soc.* **1990**, *112*, 4768-4774.
36. Allinger, N. L.; Lii, J.-H. *J. Comp. Chem.* **1987**, *8*, 1146-1153.
37. Steed, J. M.; Dixon, T. A.; Klemperer, W. *J. Chem. Phys.* **1979**, *70*, 4940-4946.
38. Janda, K. C.; Hemminger, J. C.; Winn, J. S.; Novick, S. E.; Harris, S. J.; Klemperer, W. *J. Chem. Phys.* **1975**, *63*, 1419-1421.
39. Hopkins, J. B.; Powers, D. E.; Smalley, R. E. *J. Phys. Chem.* **1981**, *85*, 3739-3742.
40. Law, K. S.; Schauer, M.; Bernstein, E. R. *J. Chem. Phys.* **1984**, *81*, 4871-4882.
41. Bornsen, K. O.; Selzle, H. L.; Schlag, E. W. *J. Chem. Phys.* **1986**, *85*, 1726-1732.
42. Ernstberger, B.; Krause, H.; Kiermeier, A.; Neusser, H. J. *J. Chem. Phys.* **1990**, *92*, 5285-5296.
43. Venturo, V. A.; Felker, P. M. *J. Chem. Phys.* **1993**, *99*, 748-751.
44. Henson, B. F.; Hartland, G. V.; Venturo, V. A.; Felker, P. M. *J. Chem. Phys.* **1992**, *97*, 2189-2208.
45. Scherzer, W.; Kratzschmar, O.; Selzle, H. L.; Schlag, E. W. *Z. Naturforsch. A* **1992**, *47*, 1248-1252.
46. Ebata, T.; Hamakado, M.; Moriyama, S.; Morioka, Y.; Ito, M. *Chem. Phys. Lett.* **1992**, *199*, 33-41.
47. Arunan, E.; Gutowsky, H. S. *J. Chem. Phys.* **1993**, *98*, 4294-4296.
48. Laatikainen, R.; Ratilainen, J.; Sebastian, R.; Santa, H. *J. Am. Chem. Soc.* **1995**, *117*, 11006-11010.
49. Cox, E. G.; Cruickshank, D. W. J.; Smith, J. A. S. *Proc. Roy. Soc. London Ser. A* **1958**, *247*, 1-21.
50. Bacon, G. E.; Curry, N. A.; Wilson, S. A. *Proc. Roy. Soc. London Ser. A* **1964**, *279*, 98-110.
51. Piermarini, G. J.; Mighell, A. D.; Weir, C. E.; Block, S. *Science* **1969**, *165*, 1250-1255.
52. Eliel, E. L.; Wilen, S. H.; Mander. L. N. *Stereochemistry of Organic Compounds*; John Wiley & Sons Inc.: New York, 1994, p. 1120.

Chapter 8

Relationship between Molecular Structure, Polarization, and Crystal Packing in 6-Arylfulvenes

Stuart W. Staley, Matthew Lynn Peterson, and Lavinia M. Wingert

Department of Chemistry, Carnegie Mellon University, Pittsburgh, PA 15213

Comparison of the single crystal X-ray structures of two centrosymmetric isomorphous 6-(4-XC_6H_4)-6-methylfulvenes (**2**, X = OMe and **4**, X = COMe) with those of two noncentrosymmetric analogs (**1**, X = NMe_2 and **3**, X = NMeCOMe) has given insight into the intermolecular interactions that promote packing in polar vs. nonpolar crystals. It appears that a strong donor group such as NMe_2 in **1** may promote polar order in the crystal through interactions between adjacent molecular dipoles in the absence of strong competing forces such as hydrogen bonding. Crystal packing for **2** is analyzed in detail with the aid of electrostatic potential maps calculated from X-ray structure factors.

The organic solid state represents one of the major frontiers of physical organic chemistry. This circumstance derives from a number of recent developments, including advances in X-ray diffraction and solid state NMR spectroscopy, rapid progress in the fields of molecular recognition and supramolecular chemistry, and the focus of modern post-industrial societies on new tailor-made materials. Advances in electronic structure theories, such as density functional theory, have also played an important role.

The formation of organic crystals represents a quintessential example of molecular recognition. Since intermolecular forces are of the order of only a few kcal/mol, the balance between molecular conformation and intermolecular interactions is usually very subtle and difficult or impossible to predict (*1*). A major goal of chemists interested in the solid state is to *understand* and ultimately to *control* the ordering of organic molecules in the crystal lattice (*i.e.*, to "synthesize" supramolecular structures) (*2*). In this regard, a key problem concerns the relationship between molecular polarity and the propensity of molecules to crystallize in polar space groups.

Previously, it was commonly assumed that polar molecules such as those employed in nonlinear optical (NLO) studies tended to pack in centrosymmetric space groups owing to the stabilization provided by pairwise antiparallel alignment of molecular dipoles (*3-5*). However, in 1991 Whitesell, Davis *et al.* (*6*) showed by a statistical analysis using the Cambridge Structural Database (*7*) that the magnitude of the molecular dipole moment (μ) does not vary significantly between centrosymmetric and noncentrosymmetric space groups. Note that this finding should not be taken to mean that dipole-dipole interactions do not play a key role in crystal packing.

Investigators have commonly employed strong intermolecular interactions such as hydrogen bonding to influence crystal packing (*8-10*). In the present contribution we report some of our studies of crystal packing involving weaker intermolecular forces, *viz.*, weak hydrogen bonds and electrostatic interactions, including dipole-dipole interactions, between polar molecules.

6-Arylfulvenes were chosen for several reasons. a) The fulvene substructure is only weakly polar but is a highly polarizable electron acceptor (*11,12*). (The effect of electron donors is illustrated by the resonance structures in Figure 1.) b) Polarity in these molecules can be modulated by torsion about the fulvene-aryl bond, which can be strongly influenced by the steric effect of the second substituent at C_6 (*13*). c) Donor/acceptor-substituted benzenes in which the 6-fulvenyl group is the acceptor have been shown to possess large molecular hyperpolarizabilities ($\mu\beta$) by EFISH measurements (*14-16*). However, no NLO studies had been reported for the solid state prior to our initial report (*13*).

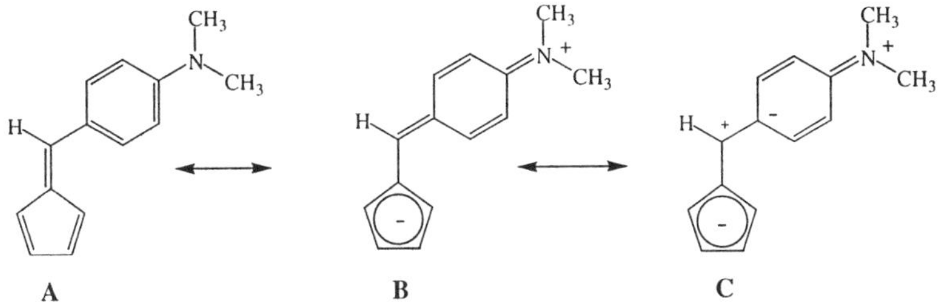

Figure 1. Resonance structures for donor-substituted 6-arylfulvenes.

Results and Discussion

Methodology for 6-Aryl-6-methylfulvenes

Our goal in this study was to examine structural and crystal packing effects from the perspective of substituent effects for a wide range of para-phenyl substituents. The aryl ring in this class of compounds is twisted 50-60° out of the plane of the fulvene ring according to HF/6-31G geometry optimizations. Hence, changes in the C_6C_7 torsional angle ($\omega(C_6C_7)$) in this region will have a maximum effect on the π overlap across this bond and therefore on the polarization of the fulvene ring. Unfortunately, study in the solid state is limited by the availability of crystalline members of this structural type. Thus, of the fourteen 6-aryl-6-methylfulvenes known to us (*12,17,18*), only four (**1-4**) have been reported to crystallize at room temperature. Fortunately, **1-4** span almost the complete range of para-phenyl substituents (X) from strong donors to strong acceptors. The Hammett σ_p values for X in **1-4** are -0.32, -0.12, 0.26 and 0.47, respectively (*19*).

1 X = NMe$_2$
2 X = OMe
3 X = NMeCOMe
4 X = COMe

Compounds **1-4** pack in three different arrangements with one donor- and one acceptor-substituted compound (**2** and **4**, respectively) having isomorphous crystal structures in space group $P2_1/c$. Our analysis of crystal packing in this series of compounds is organized as follows. First we examine the geometry-optimized structures of the isolated ("gas phase") molecules calculated at the HF/6-31G level of theory (*20,21*) and compare the effects of the para-phenyl substituents. We next analyze changes between the gas phase and the crystal for isomorphs **2** and **4** in order to determine the effect of the substituents on *isomorphous* structures. The crystal structures of nonisomorphs **1** and **3** are then analyzed in the context of both *inter*molecular (packing) and *intra*molecular (substituent) effects. Finally, we discuss the role of polarization in the packing motifs of **1-4**.

Geometry-Optimized Molecular Structures

The most important structural parameters for the current analysis ($r(C_5C_6)$, $r(C_6C_7)$, $\omega(C_5C_6)$ and $\omega(C_6C_7)$) are those that reflect polarization of the fulvene substructure. Note that the HF/6-31G-optimized value of $r(C_5C_6)$ increases (Figure 2) while that of $r(C_6C_7)$ decreases (Figure 3) with increasing donor strength (or decreasing acceptor strength) of the para substituent, indicating greater π-electron donation to the fulvene ring (see Figure 1) in the order **4 < 3 < 2 < 1**. Analogous trends are observed for plots of $\omega(C_5C_6)$ and $\omega(C_6C_7)$ versus σ_p, which show that the former increases (Figure 4) whereas the latter decreases (Figure 5) with increasing donor strength of the para substituent. We propose that $\omega(C_6C_7)$ decreases on going from acceptors to donors owing to increased π delocalization and that $\omega(C_5C_6)$ increases as a steric response to this decrease. This interpretation is supported by a

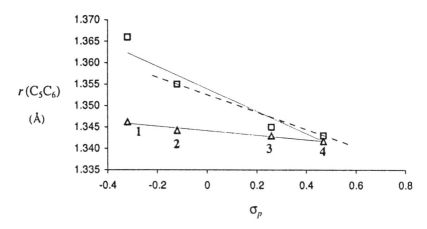

*Figure 2. Plot of r(C₅C₆) vs. σ_p for the HF/6-31G-optimized (Δ) (r^2=0.93) and X-ray (□) structures of **1-4**.*

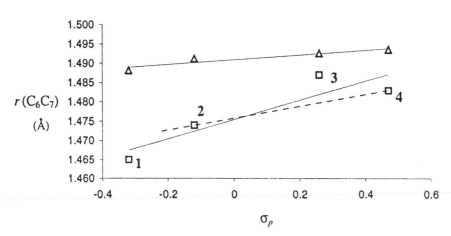

*Figure 3. Plot of r(C₆C₇) vs. σ_p for the HF/6-31G-optimized (Δ) (r^2 = 0.82) and X-ray (□) structures of **1-4**.*

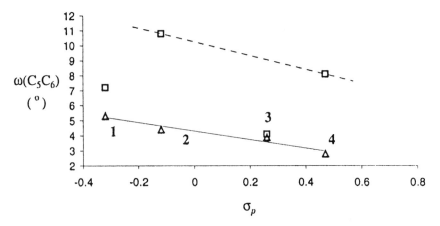

Figure 4. Plot of ω(C₅C₆) vs. σₚ for the HF/6-31G-optimized (Δ) (r² = 0.81) and X-ray (□) structures of 1-4.

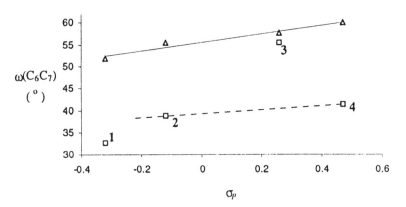

Figure 5. Plot of ω(C₆C₇) vs. σₚ for the HF/6-31G-optimized (Δ) (r² = 0.95) and X-ray (□) structures of 1-4.

plot of $\omega(C_5C_6)$ vs. $\omega(C_6C_7)$ for **1-4** and four analogs (X = NH₂, Me, CN, NO₂), for which $r^2 = 0.99$.

Effect of Crystal Lattice on Molecular Structure

Crystal Packing of Isomorphs 2 and 4

Compounds **2** and **4** are isomorphous as a consequence of both compounds forming chains of molecules connected by what could be characterized as weak hydrogen bonds between the CH bonds of the fulvene ring and the OMe or COMe oxygen atom (Figure 6) (*22,23*). The molecules in these chains interact with molecules in adjacent chains through edge-to-face interactions between the hydrogens of the aryl ring and the π system of the fulvene moiety. The contribution of the latter interaction is clearly illustrated by electrostatic potential maps for **2** (*vide infra*).

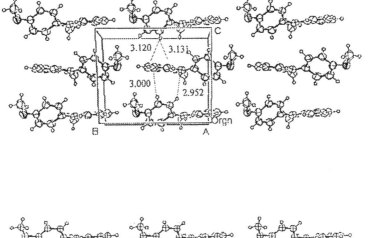

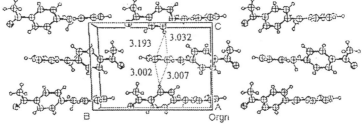

Figure 6. Packing diagrams for isomorphs 2 and 4 viewed along the a axis; top: 2; bottom: 4.

Previous investigators have concluded that packing forces dominate the *intra*molecular electronic effects of para-substituted phenyl groups with regard to torsion about single bonds in the solid state (*24,25*). Nevertheless, a central working hypothesis of this study is that the crystal lattice will exert a relatively constant effect (or at least an effect proportional to σ_p) on the molecular structures of isomorphous compounds. This idea is supported by the observation that the fulvene/aryl intermolecular closest contacts (Figure 6) and interplanar angles in **2** and **4** (51.4 and 51.1°, respectively) are nearly identical. Consequently, substituents should have similar effects on the optimized and X-ray structures. (Hydrogen bonding to the oxygen in **2** and **4** should decrease the donor ability of the former and increase the acceptor ability of the latter. However, these changes are in the same direction and represent differences of degree rather than of kind.) If this assumption is valid, then changes in molecular parameters on going from the "gas phase" to the crystal that deviate from those expected for a series of isomorphs can be related specifically to the effects of different packing arrangements.

The validity of this hypothesis is supported by a comparison of the correlation line for the optimized parameters in Figures 2-5 with the dotted line connecting the points for the corresponding X-ray parameter in **2** and **4** (the "isomorph line"). Unfortunately, we are limited to a two-point isomorph line by the paucity of data. However, in each case the slope of the isomorph line is very similar to that of the correlation line, consistent with the idea that the dotted lines are representative of structural changes that would occur with various para-phenyl substituents within a hypothetical series of isomorphs exemplified by **2** and **4**.

Crystal Packing of 3

The effect of crystal packing in **1** and **3** can now be analyzed on the basis of deviations from the isomorph lines in Figures 2-5. First, note that all of the points for **3** are almost superimposed on those for the optimized structure. This indicates that the structures of the aryl and fulvene moieties of **3** *are only slightly perturbed in the solid state.*

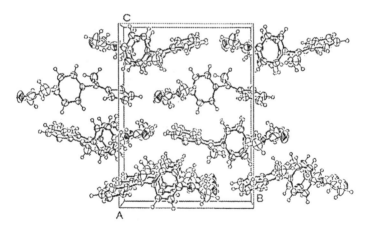

Figure 7. Packing diagram for 3 viewed along the a axis.

The packing of **3** (Figure 7, space group $P2_12_12_1$) bears a resemblance to that of **2** and **4**. In each case major motifs are chains of molecules connected by putative C-H···O hydrogen bonds (*22,23*) between the acetyl or methoxy oxygen and hydrogens on the fulvene ring and by edge-to-face interactions (interplanar angle in **3** = 93.8°) in adjacent chains. However, there is a crucial difference. In order to accommodate the NMeCOMe group, which possesses eight of the 12 closest (<3.0 Å) intermolecular contacts, the dihedral angle between the aryl ring and the substituent increases from 53.2° in the optimized structure to 90.0° in the crystal. This eliminates the possibility of any π delocalization between the nitrogen atom and the phenyl ring in the solid state, as evidenced by an increase in $r(C_{10}\text{-N})$ from 1.431 Å in the optimized structure to 1.470 Å (accompanied by a decrease of r(N-CO) from 1.371 to 1.250 Å) in the crystal. Consequently, this substituent should act as a pure σ electron-withdrawing group in the crystal. The fact that the parameters for **3** in Figures 2-5 change little on going from the gas phase to the crystal indicates *that the important edge-to-face interactions do not significantly influence these parameters.*

Crystal Packing of 1

Most of the points for **1** in Figures 2-5 lie on the opposite side of the isomorph lines from those for **3**. Thus, r(C₆C₇) and ω(C₆C₇) are smaller whereas r(C₅C₆) is larger than "predicted". These differences are all in the direction of greater π delocalization for **1** compared to that expected for an isomorph of **2** and **4**. Interestingly, the X-ray point for ω(C₅C₆) in **1** (Figure 4) lies on the same side of the isomorph line as does that for **3**, *i.e.*, ω(C₅C₆) is *smaller* than "predicted". This undoubtedly also results from increased through-resonance in the X-ray structure of **1** compared to the optimized structure owing to the more polar environment of the crystal (see next section) (*26,27*). The increase in steric interaction between the hydrogens on C₁ and C₈ that is expected when ω(C₅C₆) and ω(C₆C₇) *both* decrease is partially relieved by changes in other parameters, such as the bond angles at C₅ and C₆. This distributes intramolecular nonbonded interactions more evenly between the three groups at C₆.

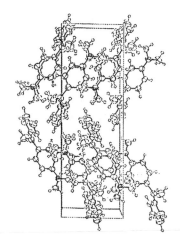

Figure 8. Packing diagram for 1 viewed along the b axis.

A packing diagram for **1** (space group $P2_12_12_1$) is given in Figure 8 (*13*). The molecules pack so that the local dipole moments of the aryl rings are aligned in a head-to-tail manner. The molecules packing in this manner form a pentamer with their four nearest neighbors, the interplanar angle between the central aryl ring and the surrounding aryl rings being 57.6° (Figure 9). The fulvene ring in **1** also packs in a head-to-tail manner with nearest-neighbor fulvene rings with an interplanar angle of 75.6°. These two types of intermolecular interactions between local dipoles act in concert to polarize the molecule in the crystal.

120

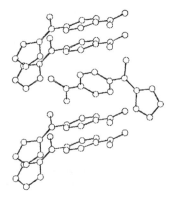

Figure 9. Pentamer motif in the crystal packing of 1. Hydrogen atoms have been omitted for clarity.

Why does **1** pack so differently from **2-4**? One possibility is that the nitrogen atom in **1** is less capable of forming chains of molecules through C-H···X hydrogen bonding compared to the oxygen atoms in **2-4**. Indeed, several recent statistical studies of hydrogen bonding have revealed no special importance for C-H···N interactions, in contrast to C-H···O interactions (*22,28*). Further, the aryl/fulvene edge-to-face interactions that are a prominent feature of the packing of **2-4** do not carry over as a component of the crystal lattice of **1**. Instead the edge-to-face interactions are between identical rings. A major component of the packing energy appears to derive from interactions between antiparallel dipoles interacting through pentamer-type structures (*13*).

Electrostatic Potential Maps for 2

Additional information regarding the nature of intermolecular interactions in these molecules is provided by electrostatic potential maps of **2**, which were obtained from X-ray data collected to sin Θ/λ = 0.7 Å^{-1} at 123 K. Anisotropic heavy atom positions, isotropic hydrogen positions, and atomic monopole functions were refined and electrostatic potential maps were derived using the VALRAY program package (*29*). Although the size of the data set relative to the number of parameters to be refined limited our calculations to a monopolar refinement of the electron distribution at each atom (*30*), this level is adequate for addressing the major issues related to crystal packing.

Two views of the electrostatic potential surrounding an individual molecule in the crystal are shown in Figures 10 and 11, along with corresponding maps derived from HF/STO-3G molecular orbital (MO) calculations on **2** with the same geometry as in the crystal (except that CH bond lengths and CCH bond angles were optimized

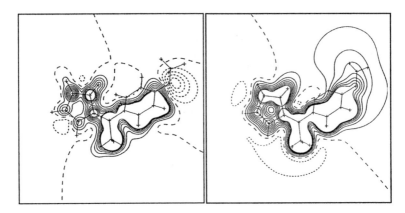

*Figure 10. Electrostatic potential maps for the X-ray geometry of **2** in the bc plane, a = 0.75; a) left: calculated at HF/STO-3G; b) right: calculated by VALRAY from the structure factors. Solid, large dashed and small dashed lines represent electropositive, neutral and electronegative potential, respectively.*

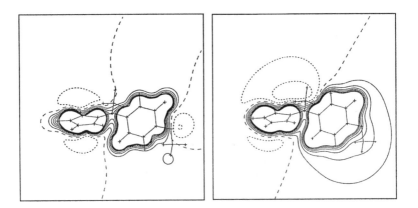

*Figure 11. Electrostatic potential maps for the X-ray geometry of **2** in the ab plane, c = 0.50; a) left: calculated at HF/STO-3G; b) right: calculated by VALRAY from the structure factors. Solid, large dashed and small dashed lines represent electropositive, neutral and electronegative potential, respectively.*

at HF/STO-3G). Comparison of the electrostatic potential maps derived from diffraction data and from MO calculations indicates that the crystal environment induces significant changes in the electronic structure of **2**. A deep well of negative electrostatic potential near the oxygen of the "gas phase" molecule Figures 10a and 11a) is replaced in the crystal by a broad region of positive electrostatic potential (Figures 10b and 11b). One possible interpretation is that this is a consequence of the above-mentioned C-H···O hydrogen bonding.

Further, the more or less equal distribution of negative electrostatic potential above and below the fulvene ring of the isolated molecule (Figure 11a) is shifted to one face and to one side of the ring in the crystal (Figure 11b). The face of the fulvene ring with the more negative electrostatic potential (*i.e.*, the top face in Figure 11a) is the face with the shorter intermolecular aryl/fulvene contacts. (A "three-dimensional" view of the negative electrostatic potential around the fulvene ring is shown in Figure 12.) Finally, Figure 13 displays contour maps of the electrostatic potential for neighboring molecules in the *ac* plane of the crystal lattice. Note the substantial overlap between electropositive regions at the periphery of the aryl ring and the electronegative region of the fulvene π system. This represents supporting evidence for the contribution of edge-to-face interactions to this crystal lattice.

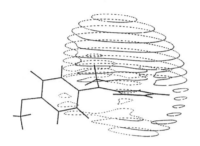

Figure 12. Three dimensional map of the negative electrostatic potential of 2 in the crystal lattice.

Polarization in the Crystal

Two factors can lead to changes in polarity on incorporation of a molecule into a crystal. These are changes in the molecular structure and interactions between neighboring molecules. The effect of the former is illustrated for **1-4** in Table 1, which shows that μ for an isolated molecule of **1** increases by 1.3 D at the HF/6-31G level of theory on going from the optimized to the X-ray geometry. In contrast, the values of μ for **2** and **3** are almost unchanged whereas that for **4** *decreases* by 1.3 D. Thus, **1** is the only member of this series which undergoes structural changes in the crystal that significantly increase polarity.

The origin of these changes is revealed by the changes in π charge listed for the fulvene and aryl rings in Table 1. The π-electron density in the fulvene ring of **1** increases the most on going to the crystal, but the π density in the aryl ring decreases the least. This indicates that the change is largely occurring by through-resonance (Figure 1, resonance form **B**). The changes in the other compounds are more localized in nature. Thus, the π-charge shift in **2** is more aryl→fulvene in character while the small changes for fulvene and large changes for aryl in **3** and **4** reflect the cutting off of π donation to aryl due to twisting of the substituent in **3** and an enhanced π-electron withdrawal by the acetyl group in the crystal structure of **4**.

The important point here is that through-resonance plays an important role in the crystal packing of **1**. High polarity, rather than favoring centrosymmetry, may actually promote the formation of polar crystals in certain circumstances. This is evident in related *p*-(dimethylamino)phenylfulvenes **5-8**. Of these, **5** (*27*), one polymorph of **6**

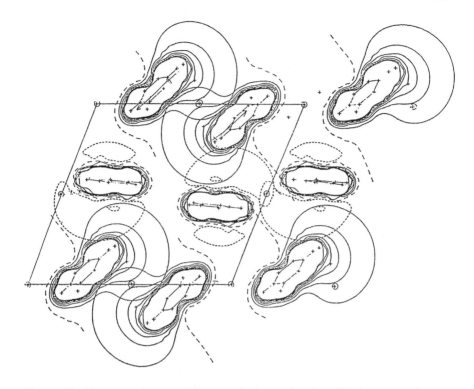

Figure 13. Electrostatic potential map of nine molecules of 2 in the ac plane, b = 0.50. Solid, large dashed and small dashed lines represent electropositive, neutral and electronegative potential, respectively.

Table I. Dipole Moments and Changes in π Charge for Optimized and X-Ray Geometries of **1-4**

| Cmpd | μ^a | | $\Delta\sum\rho_\pi (C_1\text{-}C_6)^d$ | $\Delta\sum\rho_\pi (C_7\text{-}C_{12})^d$ |
	Optimized[b]	X-ray[c]		
1	3.66	4.99	–0.0354	+0.0083
2	1.94	1.87	–0.0240	+0.0107
3	5.33	5.55	+0.0001	+0.0464
4	3.88	2.53	–0.0086	+0.0462

[a] In Debye. [b] HF/6-31G//HF/6-31G. [c] HF/6-31G//X-ray. [d] Change in π charge on going from the optimized to the X-ray geometry. The π charge is defined as the charge in the orbitals perpendicular to each ring.

124

(13) and **7** (31) pack in $P2_1/c$ or $P2_1/n$, but a second polymorph of **6** (13) and **8** pack in polar space groups ($Pca2_1$ and $P2_12_12_1$, respectively). (Second harmonic generation has been observed qualitatively in **1**, **6** ($Pca2_1$), **8**. (13) and **3**.) It is interesting that the pentamer packing motif (Figure 9) occurs in both **1** and **6** ($Pca2_1$), even though the space groups are different.

5 X = H
6 X = Ph
7 X = p -C$_6$H$_4$NMe$_2$

8

Summary

Although packing forces are undoubtedly more important than intrinsic (*i.e.*, intramolecular) substituent effects in influencing torsions about single bonds, the latter effects can be observed by comparing closely related isomorphs. It is hypothesized that isomorphous structures can play a key role in analyzing the solid state structures of nonisomorphs. This idea might give significant insight into crystal packing but requires validation.

Electrostatic potential maps could prove to be an important tool for analyzing crystal packing. This is because they give an indication of shifts in electron density in the crystal yet require smaller data sets than needed for electron density or electron density difference maps.

Finally, we believe that there might be an intriguing connection between molecular polarity and polar order in the crystal. Contrary to conventional wisdom, the most polar 6-arylfulvenes appear to be the most likely of this class of molecules to crystallize in polar space groups. Understanding the fundamental causes of this behavior represents a most worthwhile goal for future investigations.

Acknowledgments

We thank the Petroleum Research Fund, administered by the American Chemical Society, and the National Science Foundation for partial support of this research and Prof. Robert F. Stewart for extensive help in the use of VALRAY. The molecular orbital calculations were performed on a computer donated to Carnegie Mellon University by the Intel Corp.

Literature Cited

1. Gavezzotti, A. *Acc. Chem. Res.* **1994**, *27*, 309-314.
2. Desiraju, G. R. *Angew. Chem., Int. Ed. Eng.* **1995**, *34*, 2311-2327.
3. Meredith, G. R. In *Nonlinear Optical Properties of Organic and Polymeric Materials*; Williams, D. J., Ed.; ACS Symp. Ser. 233; American Chemical Society: Washington, DC, 1983; pp 27-56.
4. Nicoud, J. F.; Twieg, R. J. In *Nonlinear Optical Properties of Organic Molecules and Crystals*; Chemla, D. S.; Zyss, J., Eds.; Academic Press, Inc.: New York, NY, 1987; Vol. 1, pp 227-296.
5. Munn, R. W.; Hurst, M. *Chem. Phys.* **1990**, *147*, 35-43.
6. Whitesell, J. K.; Davis, R. E.; Saunders, L. L.; Wilson, R. J.; Feagins, J. P. *J. Am. Chem. Soc.* **1991**, *113*, 3267-3270.
7. Allen, F. H.; Kennard, O. *Chem. Design Automation News* **1993**, *8*, 31.
8. Aakeröy, C. B. *Acta Crystallogr.* **1997**, *B53*, 569-586.
9. Chin, D. N.; Zerkowski, J. A.; MacDonald, J. C.; Whitesides, G. M. In *Organized Molecular Assemblies in the Solid State*; Whitesell, J. M., Ed.; Wiley: Chicester, UK, 1999; pp 185-253.
10. Subramanian, S.; Zaworotko, M. J. *Coord. Chem. Rev.* **1994**, *137*, 357-401.
11. Replogle, E. S.; Trucks, G. W.; Staley, S. W. *J. Phys. Chem.* **1991**, *95*, 6908-6912.
12. Kresze, G.; Goetz, H. *Chem. Ber.* **1957**, *90*, 2161-2176.
13. Peterson, M. L.; Strnad, J. T.; Markotan, T. P.; Morales, C. A.; Scaltrito, D. V.; Staley S. W. *J. Org. Chem.* **1999**, *64*, 9067-9076.
14. Ikeda, H.; Kawabe, Y.; Sakai, T.; Kawasaki, K. *Chem. Phys. Lett.* **1989**, *157*, 576-578
15. Kawabe, Y.; Ikeda, H.; Sakai, T.; Kawasaki, K. *J. Mater. Chem.* **1992**, *2*, 1025-1031.
16. Kondo, K.; Goda, H.; Takemoto, K.; Aso, H.; Sasaki, T.; Kawakami, K.; Yoshida, H.; Yoshida, K. *J. Mater. Chem.* **1992**, *2*, 1097-1102
17. Sardella, D. J.; Keane, C. M.; Lemonias, P. *J. Am. Chem. Soc.* **1984**, *106*, 4962-4966.
18. Gugelchuk, M. M.; Chan, P. C.-M.; Sprules, T. J. *J. Org. Chem.* **1994**, *59*, 7723-7731.
19. Chapman, N. B.; Shorter, J. *Correlation Analysis in Chemistry*; Plenum Press: New York, NY, 1972; pp 439-533. The value for NHCOMe was used for NMeCOMe.
20. Hehre, W. J.; Ditchfield, R.; Pople, J. A. *J. Chem. Phys.* **1972**, *56*, 2257-2261.
21. GAUSSIAN94W (Revision 4.1): Frisch, M. J.; Trucks, G. W.; Schlegel, H. B.; Gill, P. M. W.; Johnson, B. G.; Robb, M. A.; Cheeseman, J. R.; Keith, T. A.; Petersson, G. A.; Montgomery, J. A.; Raghavachari, K.; Al-Laham, M. A.; Zakrzewski, V. G.; Ortiz, J. V.; Foresman, J. B.; Cioslowski, J.; Stefanov, B. B.; Nanyakkara, A.; Challacombe, M.; Peng, C. Y.; Alaya, P. Y.; Chen, W.; Wong, M. W.; Andres, J. L.; Replogle, E. S.; Gomperts, R.; Martin, R. L.; Fox, D. J.; Binkley, J. S.; DeFrees, D. J.; Baker, J.; Stewart, J. P.; Head-Gordon, M.; Gonzalez, C.; Pople, J. A. Gaussian, Inc.: Pittsburgh, PA 15213.
22. Rowland, R. S.; Taylor, R. *J. Phys. Chem.* **1996**, *100*, 7384-7391.

126

23. Gu, Y.; Kar, T.; Scheiner, S. *J. Am. Chem. Soc.* **1999**, *121*, 9411-9422.
24. L'Esperance, R. P.; Van Engen, D.; Dayal, R.; Pascal, R. A., Jr. *J. Org. Chem.* **1991**, *56*, 688-694.
25. Glaser, R.; Chen, G. S.; Anthamatten, M.; Barnes, C. L. *J. Chem. Soc., Perkin Trans. 2* **1995**, 1449-1458.
26. Spackman, M. A.; Weber, H. P.; Craven, B. M. *J. Am. Chem. Soc.* **1988**, *110*, 775-782.
27. Wingert, L. M.; Staley, S. W. *Acta Crystallogr.* **1992**, *B48*, 782-789.
28. Gavezzotti, A. *Crystallogr. Rev.* **1998**, *7*, 5-121.
29. Stewart, R. F.; Spackman, M. A. *VALRAY Users Manual*; Department of Chemistry, Carnegie Mellon University: Pittsburgh, PA, 1983.
30. van der Wal, R. J.; Stewart, R. F. *Acta Crystallogr.* **1984**, *A40*, 587-593.
31. Rau, D.; Behrens, U. *J. Organomet. Chem.* **1990**, *387*, 219-231.

Chapter 9

Optical Properties of Organic Wide Band-Gap Semiconductors under High Pressure

S. Guha[1], W. Graupner[2], S. Yang[3], M. Chandrasekhar[3], and H. R. Chandrasekhar[3]

[1]Physics Department, Marquette University, Milwaukee, WI 53201-1881
[2]Department of Physics, Virginia Polytechnic Institute and State University, Blacksburg, VA 24061-0435
[3]Department of Physics and Astronomy, University of Missouri, Columbia, MO 65211

Optical studies under hydrostatic pressure are presented using photoluminescence, Raman scattering and photo-induced absorption from two families of conjugated molecules, a polycrystalline "soft" oligomer and an amorphous "rigid" material. These studies provide insight into the geometry of the molecules, namely, planarization of the molecules under high pressure in the "soft" oligomer. Electronic spectroscopy under pressure elucidates the nature of the ground and excited electronic states of these materials.

1. INTRODUCTION

1.1 Conjugated Molecules: Organic conjugated polymers have been the focus of an enormous amount of experimental and theoretical work in the last decade due to their exciting electrical and optical properties. High electrical conductivity upon doping, high photoluminescence quantum yields up to 70 % and strong nonlinear optical response makes them suitable for a large number of potential applications and they have been commercially used since 1989 [1]. The delocalized and highly polarizable character of the π-electronic cloud makes the conjugated organic compounds attractive for photonics and optoelectronics. The combination of the semiconducting properties and high luminescence efficiency of conjugated polymers/molecules have paved the way for the development of polymer light-emitting diodes [2], light-emitting electrochemical cells [3] and solid state lasers [4,5].

Isolated conjugated molecules have a quasi one-dimensional electronic structure fundamentally different from inorganic semiconductors. The strong coupling of the electronic states results in electronic excitations that are peculiar to such low-dimensional systems. However in the solid state, the electronic properties of these materials depend significantly on the three-dimensional interactions. Recent theoretical methods of quantum-chemistry and solid-state physics in conjunction with experimental measurements have provided valuable insight into the electronic and

128

optical properties of both isolated and interacting conjugated chains (oligomers and polymers) [6,7].

1.2 Excitation/Emission Processes in Conjugated Molecules: The primary photoexcitations in isolated chains include emission from singlet excited states, triplet-triplet and polaron absorption which are formed either by electron-hole recombination or from intersystem crossing between singlet and triplet manifolds. In optical studies polarons are formed via dissociation of singlet excitons while in light emitting diodes polarons are the product of charge injection. Figure 1 shows a scheme of the electronic states and transitions in conjugated molecules. Absorption of light from the singlet ground state S_0 occurs via the creation of a singlet excited state S_1. After photoexcitation, de-excitation can occur via several mechanisms:

- Nonradiative recombination (NR) leading to a depopulation of S_1 via creation of phonons.
- Radiative recombination of S_1 leading to photoluminescence (PL).
- Population of the lowest lying triplet state T_1 via intersystem crossing (ISC) from S_1, which can be probed by triplet-triplet absorption (TT). The radiative decay of these excitations competes with nonradiative routes and takes place from the lowest excited state, in agreement with Kasha's rule [8].
- Dissociation of S_1 (=bound electron-hole-pair) into a free electron hole pair, which creates one positive and one negative polaron, giving rise to intra-polaron absorption (P).
- Population of triplet and polaron states leading to a small depopulation of the ground state, giving rise to a bleaching of the absorption (PB).

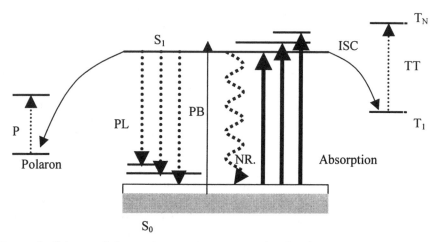

Figure 1. Scheme of electronic states in conjugated molecules.

1.3 Materials Under Investigation: The choice of materials include compounds of two generic families with distinct interchain and intrachain interactions, namely an oligo- and a polyphenyl. The oligophenyl, *para* hexaphenyl(PHP), is known to form monoclinic crystallites of space group P2₁/a [9], and is characterized by a torsional degree of freedom between the neighboring phenyl rings [10]. In the crystalline state the molecules are arranged in layers forming a herringbone type of arrangement as shown in Fig. 2 (a) [11]. The polyphenyl ladder type poly *para* phenylene is called m-LPPP due to the methyl group in the Y position as shown in Fig. 2 (b). It does not form crystallites due to the bulky side groups and shows no torsional degree of freedom between the neighboring phenyl rings owing to the methyl bridge between the rings. m-LPPP exhibits a reduction of defects coupled with high intrachain order due to planarization of phenyl rings, and excellent solubility owing to the large side groups. It has proved to be the best material within the PPP family from the point of view of stability and PL quantum yields. Up to 100 phenyl rings were polymerized with an average conjugation length of about 10 phenyl rings. Both these materials are of technological importance, high chemical purity and have been used as active materials in polymer LEDs. Both PHP and m-LPPP are known to show stimulated emission [12,13] and for m-LPPP highly directional, polarized emission with a linewidth of about 2 nm is observed from optically excited waveguides [14].

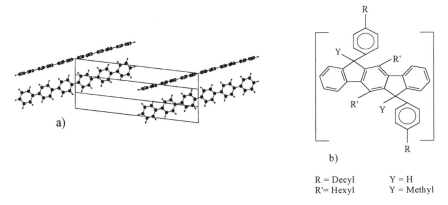

R = Decyl   Y = H
R'= Hexyl   Y = Methyl

Figure 2. Chemical structure of (a) PHP and (b) methylated laddertype poly para-phenylene (m-LPPP).

1.4 The Influence of Hydrostatic Pressure: In order to understand the influence of intermolecular interactions on the electronic properties of oligo- and polymers of *para*-phenylenes (PPP), many experiments have been based on studying the difference in PL quantum yields (PLQY) between films and solutions. For example in m-LPPP, the PLQY of solution and film are 100% and 30%, respectively [15]. PLQY of 85% in solution and that of 35% in film are observed in decyloxy PPP (DO PPP) [16]. These studies indicate that in the solid state, intermolecular interactions alter the luminescence efficiencies. We adopt an alternate method of enhancing intermolecular interaction, namely, by the application of hydrostatic pressure.

Hydrostatic pressure changes the intermolecular distances and tunes the structural and electronic properties of conjugated molecules without changing the chemical nature of the material. This technique is different from changing the intermolecular interaction via chemical substitution, which typically introduces torsion to the polymer backbone [17]. The basic impact of pressure is mainly structural resulting in changes in the molecular geometry of materials. Such structural changes also alter the electronic properties: they lead to an increase of the overlap between adjacent electronic orbitals. In terms of the band picture of solids, the relative shifts affect the electronic transitions resulting in some very interesting changes in the optical and electronic properties. For example in conjugated molecules, pressure shifts the empty π^* orbitals with respect to the occupied π orbitals (S_0-S_1 transition in Fig. 1).

High pressure research is largely based on the pioneering work of P.W. Bridgman [18], who developed many of the important techniques and made a variety of measurements on the macroscopic properties of condensed systems. Over the years Bridgman's pressure cell has undergone many modifications; a widely used pressure cell at present is the Merrill-Bassett type diamond anvil cell [19] with preindented stainless steel gaskets [20]. An extensive review of the effect of high pressure on the electronic properties of molecules and solids can found in Ref. (21). In this chapter we explore the pressure induced changes in the conjugated molecules as seen in their vibrational and electronic properties via Raman, photoluminescence (PL), absorption and photo-induced absorption (PIA) studies. These studies provide insight into the fundamentals of intermolecular interactions which affect the luminescence properties of these molecules. These investigations are necessary for improving the PLQY of polymer LEDs for their commercial applications.

High pressure Raman spectroscopy is a very useful tool for examining molecular interactions in crystalline and amorphous solids. Since the vibrational frequencies of a harmonic solid are independent of compression, pressure-induced changes in the Raman spectrum is a measure of the anharmonicity (Grüneisen parameter) in the potential energy surface of the particular vibrational coordinate. For bulk solids the Grüneisen parameter (γ) is usually volume-independent. However, in molecular solids (both organic and inorganic) it is observed that the above law breaks down resulting in a volume dependent γ [22,23]. This arises due to the difference between the local volume compression relative to that of the bulk. For example, PHP which is a non-planar molecule shows a dramatic change upon planarization, i.e., decreasing torsional angle between neighboring phenyl rings with the application of pressure. The planarization of a molecule can be observed from changes in the ratio of the Raman intensities of the C-C stretch mode at 1280 cm^{-1} to the C-H bend mode at 1220 cm^{-1}. Such effects provide an avenue for understanding the arene-arene interactions which have been recently studied in organometallic compounds. [24,25]

2. STRUCTURAL PROPERTIES

2.1 Non-planar Versus Planar Molecules: The luminescence and electronic properties of poly *para*-phenylene (PPP) type molecules depend strongly on the molecular arrangement of the phenyl rings and intermolecular interactions. Films of PHP grown on a sapphire substrate show highly anisotropic polarized emission indicating a preferred growth of crystallites, where the chains have a specific orientation with respect to the substrate [26]. Though the electronic properties of PHP and m-LPPP are similar, structurally they are very different. PHP is a non-planar and polycrystalline material in contrast to the planar, amorphous m-LPPP.

There have been numerous studies of the crystal structure of the oligophenyls of PPP, starting with the determination of the biphenyl structure in the later nineteen twenties. X-ray data and molecular simulations for the oligophenyls of PPP show that increasing the oligophenyl length enhances the crystallographic packing resulting in a decrease of the torsional angles (δ) between the neighboring phenyl rings [9]. Hydrostatic pressure is an alternate way of enhancing crystallographic packing and obtaining insights into the torsional angles between neighboring phenyl rings. Raman scattering under high pressure provides an excellent tool for understanding the nature of these torsional angles between neighboring phenyl rings in non-planar PHP.

2.2 Raman Spectrum of PHP: The Raman spectrum of the oligophenyls is mainly characterized by four intense modes of A_g symmetry. The mode displacement pattern for three of these modes is shown as an inset of Fig. 3 (a). A theoretical approach by Rumi and Zerbi [27] shows that Raman intensities are a test bench for probing the change in the potential well between neighboring phenyl rings when oligo-p-phenylenes change from a non-planar to a planar geometry. In particular, the ratio of intensity of the C-C stretch mode at 1280 cm^{-1} to the C-H in-plane bending mode at 1220 cm^{-1} (I_{1280}/I_{1220}) provides a test for the torsional angle between two neighboring phenyl rings, since simulations show that a higher number of conjugated phenyl rings results in a lower torsional angle between the phenyl rings. For a biphenyl molecule where $\delta \approx \pm 50°$, $I_{1280}/I_{1220} \sim 25$. Beyond six phenyls, $\delta \approx \pm 20°$ and $I_{1280}/I_{1220} \sim 1$.

The temperature dependence of PHP at atmospheric pressure shows that I_{1280}/I_{1220} decreases with increasing temperature (10 to 300 K). This is indicative of a planar conformation at room temperatures changing to a non-planar conformation at lower temperatures. From this temperature dependence, the activation energy required to change from a non-planar to a planar configuration ($\Delta E_{np \to p}$) is obtained as ~0.036 eV. This is close to the theoretical value of $\Delta E_{np \to p}$=0.06 eV for a PPP chain [28].

2.3 Pressure Effects: While temperature affects the population of the higher-lying (more planar) states, pressure increases the proximity of the molecules. The most striking feature of the Raman spectrum of PHP is the decrease in the intensity of the 1280 cm^{-1} peak relative to the 1220 cm^{-1} peak, (I_{1280}/I_{1220}) at high pressures as shown in Fig. 3. Figure 3(b) shows the ratio of the integrated intensities, I_{1280}/I_{1220} as a function of pressure. At 1 bar the ratio of the intensities is 0.84 and at higher pressures beyond 15 kbar, the average value of $I_{1280}/I_{1220} \sim 0.3$. The solid line is a guide to the eye. This decrease in intensity of the inter-ring C-C stretch mode at higher pressures is due to the planarization of the molecule. Indeed, at 20 kbar, no change is observed in the intensity ratio between 10 and 300 K.

The Raman modes as a function of pressure are shown as a inset in Fig. 3(b). The frequencies of all modes increase linearly with pressure beyond 15 kbar and are fit to $\omega(P)=\omega(0)+(d\omega/dP) \times P$, where P is in kbar. At lower pressures the slopes seem to be

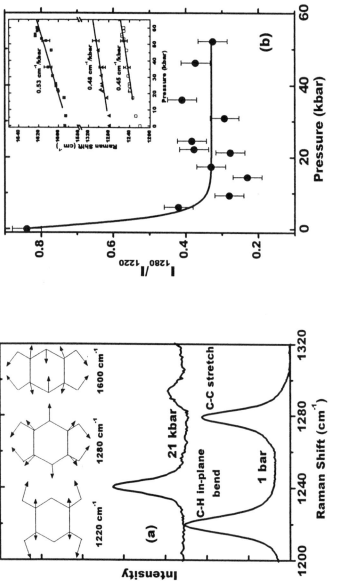

Figure 3. (a) Raman spectra of PHP at 1 bar and 21 kbar. The inset shows the mode displacement pattern for the 1220 cm⁻¹, 1280 cm⁻¹ and the 1600 cm⁻¹ modes. (b) Ratio of the intensity of the 1280 cm⁻¹ mode to the 1220 cm⁻¹ mode as a function of pressure at T = 300 K, in PHP. The solid line is a guide to the eye. The inset shows the Raman frequencies of the 1220 cm⁻¹, 1280 cm⁻¹ and the 1600 cm⁻¹ as a function of pressure.

somewhat different because of the non-planar geometry of the molecule. All the three Raman modes have similar pressure coefficients.

Pressure-induced vibrational frequency shifts in a bulk crystal are often characterized by

$$\frac{\omega_i(P)}{\omega_i(0)} = \left[\frac{V(0)}{V(p)}\right]^{\gamma_i} \qquad (1)$$

where $\omega_i(P)$ is the wavenumber of the ith mode at the applied pressure P, V(P) is the volume of the bulk solid at pressure P and γ_i is a vibrational mode Grüneisen parameter [29]. Since the bulk modulus and the volume dependence under pressure are not known for PHP, we estimate the Grüneisen parameter from Eq. 1. by using the change in volume under pressure for p-terphenyl [30]. Figure 4 shows a plot of $\log[\omega(P)/\omega(0)]$ versus the change in volume under pressure for the three Raman modes in PHP, beyond 15 kbar. Using a linear fit for all the three modes, it can be clearly seen that none of the fits goes through the origin, as one would have expected if Eq. 1 was valid. Linear regression analysis for all three lines in Fig. 4 (where the intercepts are non-zero) yield a value of ~0.09 for the slope. The deviation from the above equation for the three modes implies that they have different compressibilities associated with them. In order to simulate pressure effects, recent calculations of absorption in PPP show that changing the perpendicular distance between chains has a much stronger effect on both broadening and energy positions as compared with changing the in-plane interchain distance [31]. Therefore, it is not very surprising that the volume ratio V(0)/V(P) is different for the three Raman modes.

2.4 Discussion of Planarization: Both the temperature- and pressure dependent Raman studies indicate that the functional dependence of the potential energy of two neighboring phenyl rings versus torsional angle is "W"-shaped as shown in Fig. 5. Increasing the temperature promotes the molecule to a higher energy state, namely, the more planar configuration. This description however, does not require a change of the shape in the potential energy curve. Upon increasing pressure the potential energy curve does change: it becomes narrower and starts losing the "W" shape changing towards a "U" as shown schematically by the dashed and the dotted lines in Fig. 5. This in turn means that the energetic difference $\Delta E_{np \to p}$ between the non-planar and the planar conformation of the molecule decreases with increasing pressure. Using the known average bulk modulus of terphenyl (67 kbar) in the range of up to 10 kbar and the unit cell parameters for PHP, the average energy stored per PHP molecule at a pressure of 10 kbar is of the order of 0.1 eV. This is much higher than the activation energy (experimentally we obtain $\Delta E_{np \to p}$=0.036 eV) required for planarization. Therefore at 10 kbar all the molecules are completely planarized.

In order to obtain a quantitative estimate of the changes in the Raman spectrum due to planarization we have used a restricted Hartree-Fock method to calculate the Raman spectrum of biphenyl both in planar form (D_{2h} symmetry) and with the phenyl

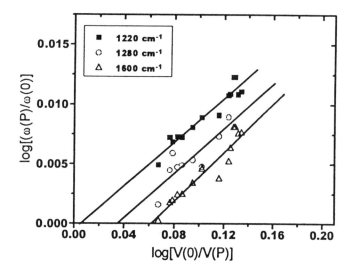

Figure 4. A log–log plot of the wavenumber ratio ω(P)/ω(P) Vs. the volume ratio V(0)/V(P) for the 1220 cm^{-1}, 1280 cm^{-1} and the 1600 cm^{-1} modes in PHP, beyond 15 kbar. The volume ratio is obtained from the compressibility of p-terphenyl.

rings tilted at 52° (D$_2$ symmetry), which is the minimum energy condition for the molecule [32]. The calculation for the non-planar biphenyl yields I$_{1280}$/I$_{1220}$=14.5 and for the planar geometry I$_{1280}$/I$_{1220}$ decreases to 3.2. In comparing these theoretical results of the biphenyl molecule to the experimental results on PHP, one has to compare the ratio of I$_{1280}$/I$_{1220}$ for the non-planar to the planar geometry. Experimentally the non-planar geometry corresponds to the 1 bar case and the planar geometry corresponds to higher pressures. The experimentally obtained ratio of the intensities of the non-planar geometry to the planar geometry is 2.8 (=0.84/0.3) for PHP which is in close agreement to the calculated value of 4.5 (=14.5/3.2) for a biphenyl molecule.

3. ELECTRONIC PROPERTIES- Steady-State Photoexcitation

3.1 PL Emission Under Pressure: Both m-LPPP and PHP show similar emission properties, even though the structural properties are quite different. This is due to the

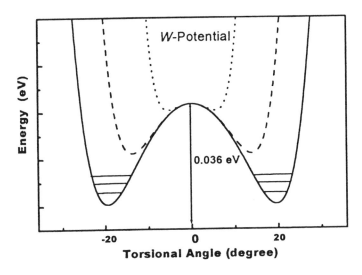

Figure 5. Schematic of the potential energy curve between two neighboring phenyl ring versus torsional angle in PHP. The solid line is the potential at 1 bar whereas the inne dashed and the dotted line represent the potential energy curve at higher pressures.

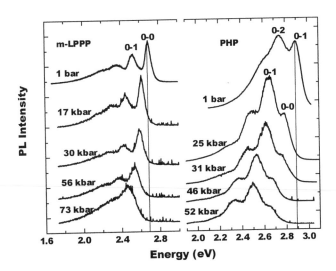

Figure 6: PL spectra of a m-LPPP film and PHP at various pressures. These measurements are at 300 K.

fact that emission takes place from the longest conjugated segments and that the excited electronic state is planar also for PHP. However, under pressure we find that emission properties behave differently, indicating the differences in intermolecular interaction in a planar versus nonplanar molecule. A vibronic progression is seen in the PL emission of both materials (Fig. 6). This is indicative of the coupling of the C-C=C-C stretch vibration of the conjugated backbone. The vibronic peaks result from the overlap of the wavefunctions of the electronic ground state $|\psi_{gi}>$ and excited state $|\psi_{ej}>$, where i and j refer to the vibrational energy levels. The emissive transition that is highest in energy is called the 0-0 transition which takes place between the zeroth vibronic level in the excited state to the zeroth vibronic level in the ground state. The 0-1 transition therefore involves the creation of one phonon. The PHP sample in Figure 6 is a powder and due to self-absorption effects the 0-0 transition is not observed at 1 bar.

3.2 Red-shift and PL Broadening under Pressure: With increasing pressures conjugated molecules/polymers show a red-shift in the PL energies as shown in Fig. 6. Recent calculations show that this red-shift under pressure can be mainly attributed to a decreasing length of the backbone; changing the distance perpendicular to the backbone mainly affects the width of the optical transitions [31]. The red-shift of transition energies has also been attributed to a "gas-to crystal" shift [33]. In these molecules the HOMO and LUMO levels correspond to the π and π^* bands, respectively. The excited state is more polarizable; an increased interaction under pressure manifests itself as lowering the energy of the excited state relative to the ground state, resulting in a net red-shift of the energy gap.

It is observed that PHP has a higher pressure coefficient for both the 0-0 and 0-1 transitions compared with m-LPPP (see Fig. 7a and 7b). The PL energy positions in PHP (Fig. 7b) show a strong pressure dependence below 15 kbar signaling planarization of the molecule, consistent with the Raman data. The experimental results agree well with the first principles calculation of a PPP chain, which predicts a decrease in the band-gap energy by almost 1 eV when the torsional angle between the phenyl rings changes from 50° to 0° [28]. Such effects are not observed in m-LPPP owing to the planar nature of the material (Fig. 7a).

The widths of the PL vibronics as a function of pressure are quite different in m-LPPP and PHP (insets of Fig. 7a and Fig. 7b). The m-LPPP sample shows a broadening of the PL vibronics in contrast to the PHP sample. Broadening of the PL vibronics with increasing pressures implies that the compressibility of the ground and excited electronic states are different. Such an analysis has been used in Ref. (34) to obtain the ratio of the force constants of the ground and excited states and the horizontal displacement of the ground state with respect to the excited state in the configuration coordinate.

PL broadening as a function of pressure also sheds light on the nature of the intermolecular interactions. The lattice is more compressible perpendicular to the chain axis due to the weaker bonds. In PHP the stronger effect is the planarization of

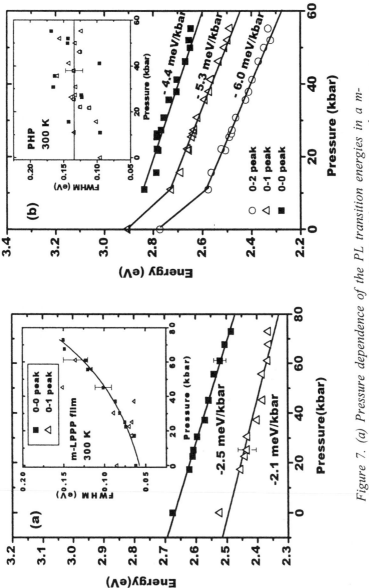

Figure 7. (a) Pressure dependence of the PL transition energies in a m-LPPP film at 300 K. The inset shows the FWHM of the transition peaks as a function of pressure. (b) Pressure dependence of the PL transition energies in PHP powder at 300 K. The inset shows the FWHM of the 0-0 and 0-1 PL peaks as a function of pressure.

the phenyl rings which also contributes to increasing the compressibility in the a-b direction. Broadening of the PL vibronics in m-LPPP is an indication that there is a stronger interaction perpendicular to the chain axis compared to PHP. These interactions can be further understood by studying the excited states as a function of pressure using photomodulation techniques, which is described in the next section.

4. ELECTRONIC PROPERTIES-Excited State Spectroscopy

Phase sensitive photomodulation techniques can be used to obtain the spectra and the lifetime of various transitions like the T-T absorption, intra polaron absorption and photobleaching (see Figure 1). The technique of transient or quasi steady state photo-induced absorption allows population of excited states and an opportunity to probe the optical transitions from these states. The main advantage of this technique is that it probes both emissive and absorptive states. The signal from these states is enhanced by the presence of defects in the conjugated polymers.

4.1 Defects in Conjugated Molecules: Polarons are charged species encountered in conjugated systems. Any perturbation of the regular periodic one dimensional arrangement of building blocks of a polymer can be seen as a defect separating two conjugated segments. This includes a torsion of two neighboring units around a single bond, a bent arrangement of the polymer such as an ortho or meta substitution on an otherwise *para*-substituted polymer. Furthermore an interruption of the alternating sequence of single and double bonds, any chain end or even interaction with another neighboring chain is a defect. Such defects giving rise to a polaron formation are illustrated in Figure 8. All these defects have one feature in common: they separate two conjugated segments from each other by an energy barrier and can therefore stabilize the permanent formation of a charge transfer state such as a polaron pair formed by the dissociation of a singlet exciton [35]. The polarons can either overcome these barriers if thermally excited or recombine: the PIA due to them disappears [36]. It is observed for nearly defect free conjugated molecules that the polaron absorption peaks are not seen in quasi steady state spectroscopy due to the lack of defect induced stabilization [37].

m-LPPP by far has the least number of defects of all the PPP type molecules (average conjugation length ~50 phenyl rings). Hence the polaron signal is very low in the pure sample. In order to detect PIA due to polarons in m-LPPP one has to photo-oxidize the sample; for example by irradiating it with the UV line (351.1 nm) of an Ar-ion laser in air. The experimental details for obtaining photomodulation spectra in m-LPPP can be found in Ref. (31).

4.2 Photomodulation Spectra Under Pressure: Figure 9 shows the photomodulation spectra of photo-oxidized m-LPPP at various pressures. The thick solid line

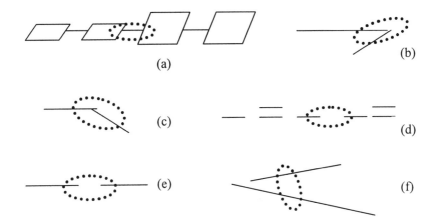

Figure 8. "Defects" which stabilize polarons. (a) torsion of two neighboring units around a single bond, a bent arrangement of the polymer such as an (b) ortho or (c) meta substitution, (d) an interruption of the alternating sequence of single and double bonds, (e) any chain end or (f) interaction with another neighboring chain. The polaron pair is symbolized by the dotted ellipses.

represents the 1 bar data. All the measurements were done at 77 K. The TT absorption occurs at 1.3 eV. Around 1.9 eV the polaronic transition is observed with its vibronic replica at 2.1 eV and the photobleaching (PB) occurs at 2.7 eV. The TT and PB are observed in the pure m-LPPP film (not photo-oxidized). PIA as a function of pressure shows that TT, PB and polaron absorption all show a red-shift with increasing pressures. The rates of the shift of TT and PB for the pure and photo-oxidized films are slightly different. Since PB is related to absorption, the shift rates of PB (-2.4 meV/kbar) and 0-0 PL peak (Fig. 7b) are very similar.

The TT-absorption shifts at a smaller rate than the PB for both the pure and photo-oxidized films. Theoretical calculations show that the singlet-singlet transition and TT-transition shift to lower energy when the chain length of the molecules becomes shorter. Also the shift rate of the TT-transition (-1.26 meV/kbar) is smaller than the shift rate of singlet-singlet transition (-2.5 meV/kbar) when the chain length of the molecule changes since the triplet states are more localized [38]. It is observed that the polaron absorption peak disappears at higher pressures indicating that pressure destabilizes the defects (see Fig. 9). Since smaller lifetimes correspond to less stable states we conclude that increasing intermolecular interaction destabilizes localized states such as triplets and polarons, as theoretically predicted [39].

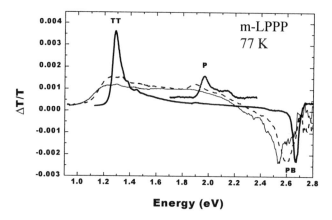

Figure 9. Photomodulation spectra of photo-oxidized m-LPPP for 38 (dashed) and 66 (thin solid) kbar. The thick solid lines represent the 1 bar data. The chopper frequency was at 145 Hz and the sample temperature at 77 K.

Along with a red-shift of the various transitions, broadening is also observed for the TT, PB and polaron transitions under pressure. Recent calculations of the dielectric function by changes in the intra- and intermolecular geometry of the molecule simulate pressure effects [31]. Changing only the length of the polymer chain results in a red-shift of the imaginary part of the dielectric function $Im(\varepsilon)$ (which is directly related with absorption) without any broadening. However, a reduction in the in-plane distance and the perpendicular distance results mainly in a broadening of the absorption peak. The increased intermolecular interactions result in a stronger overlap of the localized valence and conduction bands. Consequently, the observed red-shift and broadening of the various transitions in m-LPPP under high pressure can be understood by combining both mechanisms described above. Pressure produces reduced interchain distances that cause broadening as well as shortened inter-ring bonds that shift the transitions to the red.

5. CONCLUSIONS

We have presented a detailed study of the impact of pressure on the optical properties of conjugated molecules and polymers. These studies elucidate the changes in vibrational and electronic properties of these materials in the presence of enhanced intermolecular interactions. Raman scattering studies provide a direct means of studying the structural changes, namely, planarization of the molecule under application of pressure. Steady-state and excited-state optical spectroscopy enables the study of emission from singlet excitons, bleaching of the singlet exciton absorption and induced absorption from triplet excitons and polarons. Since hydrostatic pressure changes the molecular geometry, the study of triplet and polaron absorption and emission from singlet excitons under pressure provides a means of understanding the changes in the electronic properties of conjugated molecules.

ACKNOWLEDGEMENTS: We would like to thank our collaborators C. Ambrosch-Draxl, R. Glaser, G. Leising, P. Puschnig, R. Resel, each of whom played important roles in the development of this field. We would also like to thank Q. Cai and Chris Martin for their invaluable help in the laboratory. We are indebted to U. Scherf and K. Müllen for the donation of the m-LPPP powder. This work was supported by the University of Missouri Research Board and the Austrian National Bank project 6608.

References

1. Roth, S.; Graupner, W.; McNeillis, P. *Acta Physica Polonica* **1995**, 87, 699; Miller; J.S. *Adv. Mater.* **1993** 5, 587.
2. Burroughes, J.H.; Bradley, D.D.C.; Brown, A.R.; Marks, R.N.; Friend, R.H.; Burn, P.L.; Holmes, A.B. *Nature* **1990**, *347*, 539.
3. Pei, Q.; Yu, C.; Zhang, Y.; Heeger, A.J. *Science* **1995**, *269*, 1086.
4. Tessler, N.; Denton, G.J.; Friend, R.H. *Nature* **1996**, *382*, 695
5. Stagira, S.; Zavelani-Rossi, M.; Nisoli, M.; DeSilvestri, S.; Lanzani, G.; Zenz, G.; Mataloni, P.; Leising, G. *Appl. Phys. Lett.* **1998**, *73*, 2860.
6. Beljonne, D.; Shuai, Z.; Friend, R.H.; Bredas, J.L. *J. Chem Phys.* **1995**, *102*, 2042.
7. Cornil, J.; Santos, dos D.A. , Crispin, X.; Silbey, R.; Bredas, J.L. *J. Am. Chem. Soc.* **1998**, *120*, 1289.
8. Pope, M.; Swenberg, C.E. *Electronic Processes in Organic Crystals*; Oxford University Press: New York, 1982.
9. Baker, K.N.; Fratini, A.V.; Resch, T.; Knachel, H.C.; Adams, W.W.; Socci, E.P.; Farmer, B. L. *Polymer* **1993**, *34*, 1571.
10. Socci, E. P.; Farmer, B.L.; Adams, W.W. *J. Polymer Sci. B* **1993**, *31*, 1975.
11. Guha, S.; Graupner, W.; Resel, R.; Chandrasekhar, M.; Chandrasekhar, H.R.; Leising, G. Mat. Res. Soc. Symp. Proc. **1998** *488*, 867.
12. Graupner, W.; Leising, G.; Lanzani, G.; Nisoli, M.; De Silvestri, S.; Scherf, U. *Phys. Rev. Lett.* **1996**, *76*, 847 .
13. Piaggi, A.; Lanzani, G.; Bongiovanni, G.; Mura, A.; Graupner, W.; Meghdadi, F.; Leising, G. *Phys. Rev. B* **1997**, *56*, 10133.
14. Zenz, C.; Graupner, W.; Tasch, S.; Leising, G.; Müllen, K.; Scherf, U. *Appl. Phys. Lett.* **1997**, *71*, 2566.

142

15. Tasch, S.; Niko, A.; Leising, G.; and Scherf, U. *Appl. Phys. Lett.* **1996**, *68*, 1090.
16. Yang, Y.; Pei, Q.; Heeger, A. J. *Appl. Phys. Lett.* **1996**, *79*, 934.
17. Loi, M. A.; List, E. J. W.; Gadermaier, C.; Graupner, W.; Leising, G.; Bongiovanni, G.; Mura, A.; Pireaux, J.-J. *Synthetic Metals* (in print).
18. Bridgman, P.W. *Physics of high Pressure*; G. Bell and Sons, Ltd.: London, 1949.
19. Jayaraman, A. *Rev. of Mod. Phys.* **1983**, *55*, 65.
20. Dunstan, D.J. *Rev. Sci. Instrum.* **1989**, *60*, 3789.
21. Drickamer, H.G. and Frank, C.W. *Electronic Transitions and the High Pressure Chemistry and Physics of Solids*, Chapman and Hall, Ltd., 1973.
22. Zallen, R. Phys. Rev. B **1974**,*9*, 4485.
23. Zallen R.; Slade, M.L. Phys. Rev. B **1978**, *18*, 5775.
24. Niemeyer, M. *Organometallics* **1998**, *17*, 4649.
25. Glaser, R.; Haney, P.E.; Barnes, C. L. *Inorg. Chem.* **1996**, *35*, 1758.
26. Zojer, Z.; Knupfer, M.; Resel, R.; Meghdadi, F.; Leising, G.; Fink, J. Phys. Rev. B **1997**, *56*, 10138.
27. Rumi, M.; Zerbi, G. *Chem. Phys.* **1999**, *242*, 123.
28. Ambrosch-Draxl, C.; Majewski, J.A.; Vogl, P.; Leising, G. *Phys. Rev B* **1995**, *51*, 9668.
29. Born, M. and Huang, K. *Dynamical theory of Crystal lattices*; Oxford University Press: Oxford, 1954; pp38-40
30. Landolt-Börnstein Physikalisch-Chemische Tabellen, Springer-Verlag, 1961
31. Yang, S.; Graupner, W.; Guha, S.; Puschnig, P.; Martin, C.; Chandrasekhar, H.R.; Chandrasekhar, M.; Leising, G.; Ambrosch-Draxl, C. *SPIE proceedings* **1999**, *3797*, 26.
32. Guha, S.; Graupner, W.; Resel, R.; Chandrasekhar, M.; Chandrasekhar, H.R.; Glaser, R.; Leising. G. Phys. Rev. Lett. **1999**, *82*, 3625.
33. Lacey, R. J.; Batchelder, D. N.; Pitt, G.D. J. Phys. C: Solid State Phys. **1984**, *17*, 4529.
34. Guha, S.; Graupner, W.; Yang, S.; Chandrasekhar, M.; Chandrasekhar, H.R.; Leising. G.; Phys. Stat. Sol. (b) **1999**, *211*, 177.
35. Graupner W.; Cerullo, G.; Nisoli, M.; De Silvestri, S.; Lanzani, G.; List, E.J.W.; Leising, G.; Scherf, U. *Phys. Rev. Lett.* **1998**, *81*, 3259.
36. Graupner, W.; Leditzky, G.; Leising, G.; Scherf, U. Phys. Rev.B, **1996**, *54*, 7610.
37. Graupner, W.; Eder, S.; Mauri, M.; Leising, G.; Scherf, U. *Synth. Met.* **1995** *69*, 419.
38. Beljonne, D.; Cornil, J.; Friend, R.H.; Janssen, R. A. J.;. Brédas, J. L. *J. Am. Chem. Soc.* **1996**, *118*, 6453. Cornil, J.; Beljonne, D.; Brédas, J.L. *J. Chem. Phys.* **1995**, *103*, 842.
39. Vogl, P.; Campbell, D. K. *Phys. Rev. Lett.* **1989** *62*, 2012.

Chapter 10

Optical Probes of Crystal Growth Mechanisms: Intrasectoral Zoning

Richard W. Gurney[1], Miki Kurimoto[1], J. Anand Subramony[1,2], Loyd D. Bastin[1], and Bart Kahr[1,*]

[1]Department of Chemistry, University of Washington, Seattle, WA 98195-1700
[2]Current address: Applied Materials, 3195 Kifer Road, M/S 2955, Santa Clara, CA 95051

Impurities present in crystallizing solutions may become segregated within regions of the resulting crystals that correspond to specific slopes of growth hillocks. When the impurities are dyes or luminophores, chemical zoning of this sort results in distinct patterns of color or light that serve to identify growth active surface structures and crystal growth mechanisms. Moreover, the optical probes reveal the specificity of non-covalent interactions between molecules and anisotropic hillocks. These phenomena are illustrated herein for three crystals, potassium dihydrogen phosphate, potassium sulfate, and α-lactose monohydrate, grown in the presence of an azo dye, a luminescent benzene derivative, and a naturally occurring anthraquinone, respectively.

Introduction

Optical probes are a mainstay of the biochemical scientist eager to illuminate the specificity of non-covalent interactions (*1,2*). As crystal growth from solution is also governed by the specificity of non-covalent interactions, we evaluate here to what extent dyes and luminophores can be used to reveal aspects of non-covalent assembly during crystal growth.

There is of course an obvious distinction between the use of optical probes for the study of biochemistry and crystal growth: space. It is easy to imagine an optical label dangling from a protein of interest into the cytoplasm as it makes its way about a cell. Most crystals, on the other hand, are close packed and can not easily accommodate guests of arbitrary size. Mitscherlich's Principle of Isomorphism, a long-standing constraint on the preparation of solid solutions, asserts that mixed crystals shall be limited to components having similar sizes and shapes (*3,4*). The successful use of the Principle of Isomorphism in single crystal matrix isolation is well illustrated by the host/guest hydrocarbon pairs in Figure 1. In the examples of naphthalene in durene (*5,6*), diphenylmethylene in benzophenone (*7,8,9*), and pentacene in *p*-terphenyl (*10,11,12,13*) exquisite matches of hosts and guests enabled scientists to address particular scientific questions. McClure has also illustrated many examples of the single crystal matrix isolation of inorganic ions (*14*).

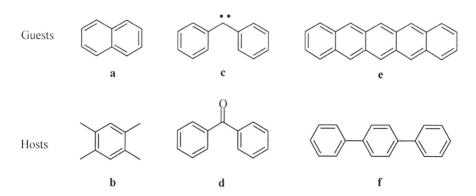

Guests

a   c   e

Hosts

b   d   f

Figure 1. Analyte guest and crystal host pairs in classic examples of single crystal matrix isolation: (a) naphthalene, (c) diphenylmethylene, and (e) pentacene in (b) durene, (d) benzophenone, and (f) p-terphenyl, respectively.

In the aforementioned experiments it is typically assumed that the guest is uniformly distributed throughout the host; however, this need not be the case. If the crystal growth conditions are not carefully controlled or the relative concentration of host and guest varies, the crystal is subject to concentric or temporal zoning in which composition changes during growth, Figure 2a. Favorable and unfavorable conditions for the incorporation of an impurity may oscillate resulting in concentric polygons in cross section, Figure 2b. As growth circles record a tree's history, concentric zoning records a crystal's maturation and morphological history (*15,16,17*). In the fields of mineralogy and petrology the concentric zoning of trace metals or elemental ions within minerals has been studied in great detail (*18,19*).

Heterogeneous incorporation of impurities among crystal growth sectors, volumes of a crystal that have grown in a particular direction through a specific face, is termed intersectoral zoning, Figure 2c. Here, the segregation is spatial as opposed to temporal (*20,21,22*). This is a general feature of impure crystals in which symmetry distinct facets express different affinities for impurities. While intersectoral zoning has been described in minerals with trace elements (*18,19*) and in synthetic crystals marked by dyes (*23,24,25,26*), a molecular level of understanding was first presented by Addadi, Lahav, Leiserowitz, and their colleagues in their extensive studies of solid solutions (*27,28,29*). Their 'Tailor-made Auxiliaries' of crystal growth demonstrated that nonsymmetry related facets must incorporate impurities differently, often resulting in a lowering of the host's symmetry (*30,31*).

Impurities may also inhomogeneously deposit within a single growth sector depending on the crystal's surface topography. Surfaces of crystals grown in the lower supersaturation regime often propagate through a dislocation which produces growth spirals or hillocks: shallow stepped pyramids with single or multiple dislocations at the apex (*32,33*). Polygonization of hillocks partitions the parent face into vicinal regions, each having slightly different inclinations. Impurity partitioning within vicinal regions, intrasectoral zoning, results from the selective interactions of impurities with particular stepped hillock slopes, Figure 2e. Intrasectoral zoning therefore provides more detailed information about recognition mechanisms because the active growth surfaces at the time of incorporation are more fully defined. The identification of intrasectoral zoning patterns enabled the determination of the mechanism of trace element partitioning in the minerals calcite [$Ca(CO_3)$], topaz [$Al_2SiO_4(F,OH)_2$] and apatite [$CaF_2(Ca_3(PO_4)_2)_3$] (*18,19,34,35,36,37,38*).

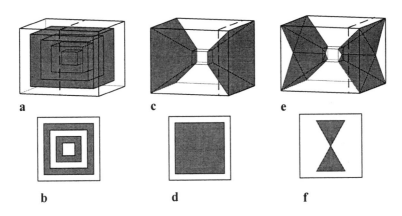

Figure 2. Idealizations of compositional zoning in an orthorhombic crystal grown from seed: (a,b) concentric, (c,d) intersectoral, and (e,f) intrasectoral zoning. (b,d,f) represent cross sections, indicated by broken lines in (a,c,e), respectively.

In recent years we have explored in detail the intersectoral zoning of dyes in simple, transparent host crystals. We showed how structural studies of dyed K_2SO_4 crystals (*39*) led to the design of new mixed crystals with prescribed properties such as lasing (*40*). Dyed acetate crystals were used as models for the study of matrix-assisted room temperature phosphorescence (*41*) and optical probes inside ferroelectric crystals such as Rochelle salt were used to monitor phase transitions (*42*). We reinvestigated molecular crystal hosts for dyes (*43,44*) and also studied biomolecules with intrinsic chromophores (*45*). Yet, despite the hundreds of examples of intersectoral zoning of dyes and luminophores we had observed, not a single unambiguous case of intrasectoral zoning was encountered.

Zaitseva and coworkers perfected KH_2PO_4 (KDP, space group *I*-42*d*) (*46*) crystal growth conditions as a prerequisite to the development of the National Ignition Facility. In their hands, amaranth (**1**), Figure 3, an anionic dye that we had shown to have an exclusive affinity for the {101} surfaces of KDP (*47,48,49*), was both inter- and intrasectorally zoned (*50*). This observation required introduction of the dye

146

during late growth thereby coloring only a thin surface layer so that patterns were not confounded by moving dislocation cores, Figure 4a.

Figure 4b highlights the (101) face of the KDP/**1** crystal in Figure 4a. The stripes result from **1** partitioning within the **A** and **B** regions of the polygonized hillocks prevalent on the pyramidal faces, Figure 4d. A differential interference contrast (DIC)

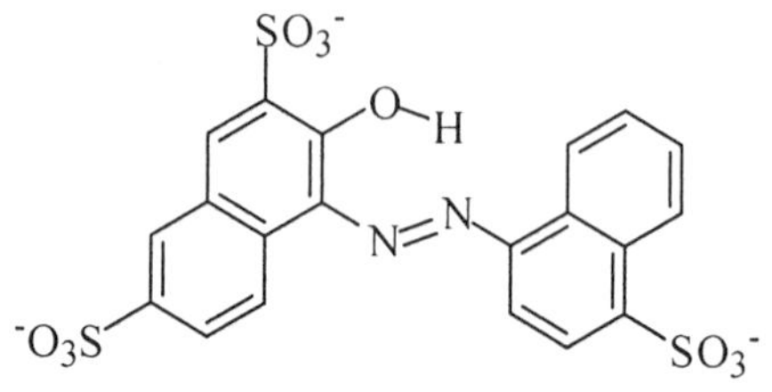

Figure 3. Amaranth (1).

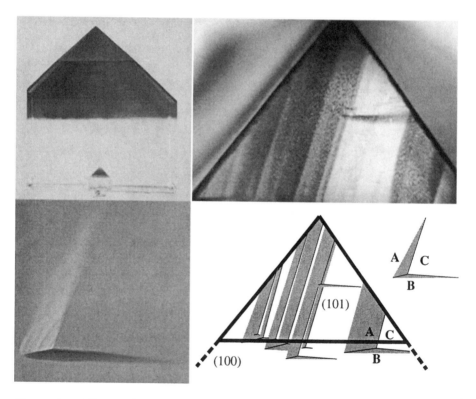

Figure 4. a) Top, Left: Photograph of KDP/1. b) Top, Right: The heterogeneous incorporation in the (101) pyramidal growth sector is apparent in this closer view (linear dimension 3.25cm). c) Bottom, Left: DIC micrograph of a pure KDP hillock (linear dimension 0.15mm). d) Bottom, Right: Schematic representation of (b).

micrograph (*51*) of a representative {101} hillock is shown in Figure 4c. As illustrated in Figure 4d, the majority of hillock centers terminated at the [010] intersection. The likely positions of the centers are illustrated had they not been overgrown. The horizontal band in the center right of Figure 4b presumably corresponds to a **B** slope (*52*).

As the identification of intrasectoral zoning patterns was instrumental in determining mixed crystal growth mechanisms of minerals with trace elements as well as the partitioning of **1** within the pyramidal growth sectors of KDP, we sought other examples among our dyed crystals. Below, we demonstrate how the analogous use of a fluorescent probe for K_2SO_4 crystals and a dye with α-lactose monohydrate crystals can be used to identify active growth surfaces.

Salt Crystals: Sulfonated Anilines and Potassium Sulfate

Even though, to the best of our knowledge, growth hillocks on K_2SO_4 crystals have never before been observed, Buckley commented on the "rounded and uneven" nature of the {021} surfaces (*25*), which Tutton had earlier referred to as "striated and distorted (*53,54*)." When crystals of K_2SO_4, Figure 5, were grown from a solution containing mM quantities of aniline-2-sulfonate (**2**), Figure 6, five distinct growth sectors {001}, {011}, {021}, {110} and {111} became luminescent, and each displayed a unique optical signature (*55*). In addition to this high degree of intersectoral zoning, the {021} growth sectors displayed a pronounced blue banding characteristic of intrasectoral zoning where **2** was found in the **B** but not the **A** regions, Figure 7. The partitioning of **2** results from the selective recognition of a specific slope of the hillocks on the K_2SO_4 (021) face, as emphasized in the comparative DIC and fluorescence micrographs in Figure 8.

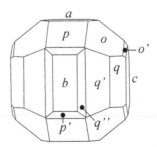

Figure 5. K_2SO_4 crystals are typically bound by {021}(q'), {110}(p), {010}(b), {111}(o) and {011}(q) faces and occasionally display {001}(c), {031}(q''), {100}(a), {112}(o'), and {130}(p') facets. Indices are based on optical goniometry (a = 5.772 Å, b = 10.072 Å, c =7.483 Å, space group Pmcn) (25,56).

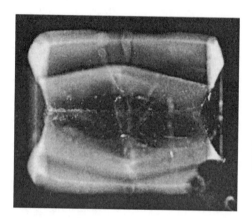

Figure 6. Aniline-2-sulfonate (2).

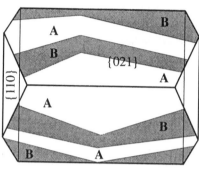

Figure 7. (a) Left: Luminescence ($\lambda_{fluor.}$ = 388 nm and $\lambda_{phosp.}$ = 454 nm) of a $K_2SO_4/2$ mixed crystal illuminated with ultraviolet light (λ = 254 nm). Lateral dimension is 1.0 cm. Dark spot in the lower right hand corner of photo is epoxy used for mounting during goniometry. Emission from 2 does not correspond to growth sector boundaries but to vicinal regions on the {021} faces. (b) Right: Idealized representation of crystal shown on the left.

Figure 8. (a) Left: DIC micrograph of intrasectorally zoned $K_2SO_4/2$ mixed crystal shown in Figure 7. The two "comet-like" objects are regions having slopes similar to the "top" portion of the image. (b) Right: Luminescence micrograph of the region shown on left. Luminescence of 2 is visibly confined to the steeper sloped region of the hillocks. The horizontal axis is parallel to the [100] direction, the top of the image is closer to (010) than (001).

The hillocks on the surface of $K_2SO_4/\mathbf{2}$ mixed crystals are ideally represented in Figure 9. The step train propagating toward (010) has the largest advancement velocity, υ_A, and has straight steps. As the steps begin to curve, Figure 10a, a boundary is formed between the **A** and **B** vicinal faces: compare Figures 8, 9 and 10a. The step flats shown in the slice in Figure 9 index as (021) on either side of the apex, while the step risers in the **A** and **B** regions index as (031) and (011) faces, respectively.

It is evident from DIC micrographs of the {021} faces of pure and mixed K_2SO_4 crystals that **2** does not induce the formation of the hillocks but affects their overall morphology, Figure 10. The doped crystals are often dominated by a small number of macroscopic hillocks, while the pure K_2SO_4 crystal surfaces are generally covered with a greater number of smaller hillocks.

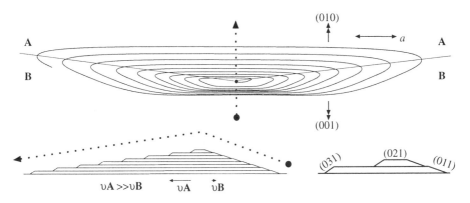

Figure 9. Idealized representation of K_2SO_4 {021} hillock.

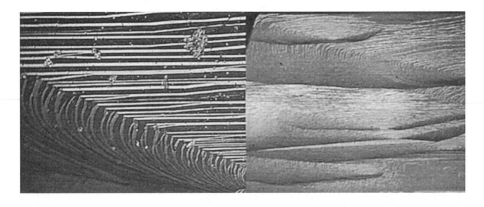

Figure 10. DIC micrographs. (a) Left: $K_2SO_4/\mathbf{2}$ mixed crystal (lateral dimension 2.5 mm). (b) Right: pure K_2SO_4 crystal (lateral dimension 2.5 mm).

Sugar Crystals: Carminic Acid and α-Lactose Monohydrate

Recently, we have shown that α-lactose monohydrate (LM) crystals incorporate biopolymers such as green fluorescent protein (GFP) in specific growth sectors, namely (010). The protein gains substantial kinetic stability since the large amplitude vibrations that precede denaturation are dampened in the crystal (57). In order to better understand the partitioning mechanism of guests within LM crystals, we sought simpler compounds that displayed analogous behavior.

α-Lactose monohydrate (LM) is a simple disaccharide made from galactose and glucose, Figure 11. An aqueous grown LM crystal (space group $P2_1$) (58) is shaped like a hatchet with (010), {0-11}, {100}, {110}, {1-10}, {1-50} and (0-10) faces. LM crystals nucleate at the apex of the hatchet and grow unidirectionally toward +*b* through a spiral growth mechanism with dislocations propagating along [010] (59), as confirmed by the recent observation of hillocks on the (010) face by atomic force microscopy (AFM) (60,61).

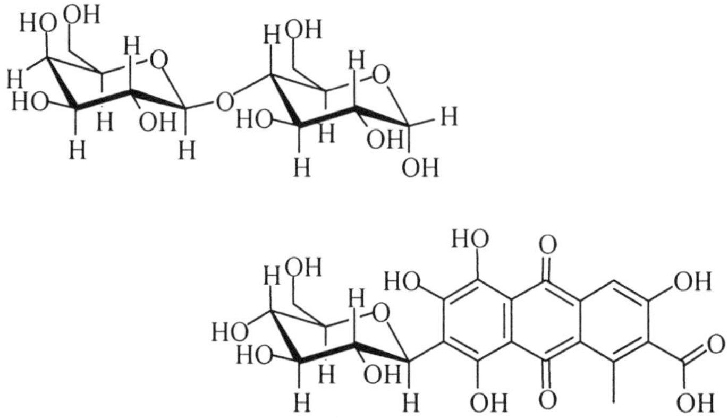

Figure 11. α-Lactose (4-O-β-D-galactopyranosyl-α-D-glucopyranose) (top) and carminic acid (3) (7-β-D-glucopyranosyl-9,10-dihydro-3,5,6,8-tetrahydroxyl-1-methyl-9,10-dioxo-2-anthracenecarboxylic acid) (bottom).

DIC microscopy of a pure LM crystal reveals a typical single polygonized hillock that partitions the (010) surface into four vicinal faces pairwise related by two-fold symmetry, Figure 12a. An idealized representation of the hillock based on AFM and DIC results is pictured in Figure 12b. The elementary growth steps have an average step height equal to the *b* dimension and have a greater advancement velocity in the **B** and **B'** vicinal faces as compared to the **A** and **A'** vicinal faces.

Carminic acid (3), Figure 11, a natural red dyestuff produced by the female Chochineal beetle (*Dactylopius coccus*) (62), recognizes the (010) face of LM exclusively, as does GFP. If the {0-11} face of a LM/3 crystal is observed obliquely **3** is seen to be partitioned within two distinct bands, suggestive of intrasectoral zoning, Figure 13. This partitioning was quickly interpreted when brightfield absorbance and corresponding DIC micrographs of the (010) face of the LM/3 crystal were compared, Figure 14. It is observed that **3** selectively interacts with the **B** and **B'** vicinal faces of the hillock, Figure 14a. Furthermore, it is conceivable that incorporation of **3** is concurrent with the onset of the dislocation because the color emanates from a point other than the nucleation sight, Figure 13. Even though the center of the hillock cannot be located in the DIC micrograph due to etching, the dye reveals its location, Figure 14.

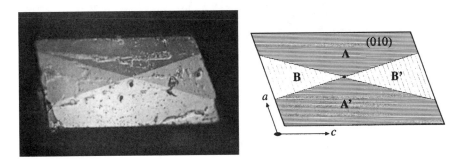

Figure 12 (a) Left: DIC micrograph of a pure LM crystal (010) surface. (b) Right: Schematic representation of the hillock on the (010) face illustrating the four vicinal regions A, A', B, and B' pairwise related by two fold rotational symmetry. Lateral direction of crystal is 0.6 mm.

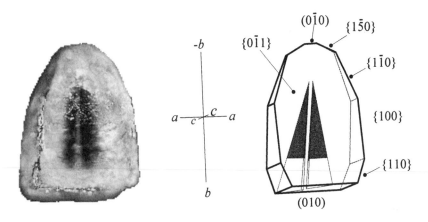

Figure 13. (a) Left: A brightfield reflected light micrograph of a LM/3 mixed crystal (1.2 (h) × 0.8 (w) × 0.5 (d) mm³. The image was electronically processed to remove glare from the edges. Absorption of 3 (λₘₐₓ = 500 nm) does not correspond to growth sector boundaries. (b) Right: Idealized representation of crystal in (a).

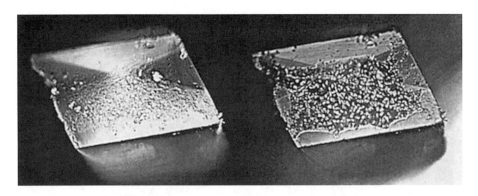

Figure 14. Micrographs of LM/3 mixed crystal (010) surface: (a) Left: brightfield reflected light showing absorbance. b) Right: and reflected light with DIC. Lateral dimension of crystal is 0.6 mm.

Conclusion

Recent studies, particularly by scanning probe microscopies, have revealed the role of steps, kinks, and other emergent structures in directing the adsorption of guests or impurities during crystal growth. Ward's studies of ledge directed epitaxy (*63,64*), and Hollingsworth's demonstration of spiral growth nucleated in incommensurate urea inclusion compounds (*65*) stand out as exemplary illustrations of this sort. Here, we have shown that colored or light emitting molecules can under certain conditions also image growth active surface structures. In so doing, molecular probes reveal to the eye the macroscopic features of crystal growth mechanisms. At the same time, the selectivity of molecular guests for vicinal slopes of a given growth spiral stand out as previously underexplored examples of molecular recognition.

Given the specificity of the recognition processes illustrated herein, it is not surprising that the optical probes, once trapped inside of the crystals, are highly oriented. We do not have space in this account to discuss the orientation of **1**, **2**, and **3** in their respective hosts, but we have studied the orientation of these and other guests in crystals by measurements of linear dichroism(*39,44,47*), fluorescence anisotropy, and in the case of guests with long-lived triplet states, single crystal EPR (*41,55*). The formation of "oriented gases" of chromophores in this way is useful for the interpretation of anisotropic physical properties, especially when the property to be measured is the electronic structure of a dye, because molecules must be oriented and isolated from one another to avoid coupling and preserve the monomolecular characteristic of interest. Thus, optical probes, "Tailor-Made Auxiliaries (*31*)" with optical signatures easily administered to crystallization solutions, preserve dynamic information about crystal growth mechanisms, the subject emphasized in this report – while the anisotropy of the guests spectral properties preserve information about the microscopic interactions between guests and hillock slopes.

Experimental

All chemicals were reagent grade, purchased from Aldrich Chemical Co. except for potassium phosphate monobasic (ProChem) and potassium sulfate (Mallinkrodt). Deionized water for crystallizations was distilled using a Corning Mega-Pure System MP-1.

Typical growth solutions for K_2SO_4 crystallizations contained > 25 mg guest per gram of host. Guest concentrations in crystals were determined from the absorbance of dissolved crystals. Typical mixed crystals were grown in 5 to 7 days by slow evaporation from dishes inside an undisturbed cabinet. For KDP crystallizations see reference 50.

Ion-free α-lactose was prepared by passing the saturated lactose solution through both cation (Amberlite IR120) and anion (Amberlite IRA401) exchange resins. Typical growth solutions for LM crystallizations contained 1 mg of guest per gram of host. Typical mixed crystals were grown in 3 to 4 days by slow evaporation from a 24 multi-well plate at room temperature inside an undisturbed cabinet.

Crystals were indexed optically using a STOE Darmstadt optical goniometer and a Leica DMLM reflected light microscope with spindle stage. Crystals were also indexed with x-rays using a Nonius KappaCCD Diffractometer. To preserve surface structures, crystals were pulled through a layer of hexanes sitting atop the growth solution, washed with hexanes, and carefully dried with Kimwipes (*32*).

Linear Dichroism measurements were performed on a SpectraCode Multipoint-Absorbance-Imaging (MAI-20) Microscope which consists of an Olympus BX 50 polarizing microscope connected to a Spectra Pro-300i triple grating monochromether (Acton Research Corporation). A fiber optic cable containing a linear stack of 20 fibers connects the microscope to the monochromether and Princeton Instruments CCD detector. The instrument is controlled using the program KestralSpec.

Ultraviolet absorption spectra were recorded with a Hitachi U-2000 spectrophotometer controlled with SpectraCalc software (Galactic Industries). Differential interference contrast images were obtained with a Leica DMLM reflected light microscope and captured with a Diagnostic Instruments SPOT digital camera.

Acknowledgements

We thank the National Science Foundation (CHE-9457374, CHE-9727372), the University of Washington Royalty Research Fund, the donors of the American Chemical Society-Petroleum Research Fund (30688-AC6), the National Institutes of Health (GM58102-01), and the University of Washington Center for Nanotechnology for financial support. We also thank Scott Lovell and Dong Qin for their assistance. We extend special thanks to Natalia Zaitseva for gifts of KDP crystals and for advice on crystal growth techniques.

154

Literature Cited

1. Slavic, J. *Fluorescent Probes in Cellular and Molecular Biology;* CRC Press: Ann Arbor, MI, 1994.
2. Haugland, R. P. *Handbook of Fluorescent Probes and Research Chemicals,* 6th ed.; Molecular Probes, Inc.: Eugene, OR, 1996.
3. Mitcherlich, E. *Abh. Akad. Wiss. Berlin* **1818–1819,** 427–437.
4. Freund, I. *The Study of Chemical Composition. An Account of Its Method and Historical Development;* Dover: New York, 1968.
5. Hutchison, C. A.; Magnum, B. W. *J. Chem. Phys.* **1958,** *29,* 952–953.
6. Hutchison, C. A.; Magnum, B. W. *J. Chem. Phys.* **1961,** *34,* 908–922.
7. Sixl, H.; Mathes, R.; Schaupp, A.; Ulrich, K.; Huber, R. *Chem. Phys.* **1986,** *107,* 105–121.
8. Anderson, R. J. M.; Kohler, B. E. *J. Chem. Phys.* **1975,** *63,* 5081–5086.
9. Cheng, C.; Lin, T. -S.; Sloop, D. J. *Chem. Phys. Lett.* **1976,** *44,* 576–581.
10. Moerner, W. E.; Orrit, M. *Science (Washington, D.C.)* **1999,** *283,* 1670–1676.
11. Weiss, S. *Science (Washington, D.C.)* **1999,** *283,* 1676–1683.
12. Moerner, W. E. *Acc. Chem. Res.* **1996,** *29,* 563–571.
13. Skinner, J. L.; Moerner, W. E. *J. Phys. Chem.* **1996,** *100,* 13251–13262.
14. McClure, D. S. *Electronic Spectra of Molecules and Ions in Crystals,* Academic Press: New York, 1959.
15. Reeder, R. J.; Paquette, J. *Sediment. Geol.* **1989,** *65,* 239–247.
16. Reeder, R. J.; Fagioli, R. O.; Meyers, W. J. *Earth-Sci. Rev.* **1990,** *29,* 39–46.
17. Ortoleva, P. J. *Earth-Sci. Rev.* **1990,** *29,* 3–8.
18. Paquette, J.; Ward, W. B.; Reeder, R. J. In *Carbonate Microfabrics;* Rezak, R.; Lavoie, D. L., Eds.; Springer: New York, 1993, pp 243–252.
19. Rakovan, J.; Reeder, R. J. *Am. Mineral.* **1994,** *79,* 892–903.
20. Reeder, R. J.; Grams, J. C. *Geochim. Cosmochim. Acta* **1987,** *51,* 187–194.
21. Brice, J. C. *The Growth of Crystals from Liquids;* Series of Monographs on Selected Topics in Solid-State Physics; Elsevier: New York, 1973; Vol. 12.
22. Kröger, F. A. *Preparation, Purification, Crystal Growth, and Phase Theory, 2nd ed.;* The Chemistry of Imperfect Crystals; Elsevier: New York, 1974; Vol. 1.
23. Lehmann, O. *Annalen der Physik und Chemie* **1894,** *51,* 47–76.
24. Neuhaus, A. *Zeit. Krist.* **1943,** *105,* 161–219.
25. Buckley, H. E. *Crystal Growth;* John Wiley & Sons, Inc.: New York, London, 1950.
26. France, W. G. *Colloid Symposium Annual* **1930,** *7,* 57–58.
27. Vaida, M.; Weissbuch, I.; Lahav, M.; Leiserowitz, L. *Isr. J. Chem.* **1992,** *32,* 15–21.
28. Weissbuch, I.; Popovitz-Biro, R.; Lahav, M.; Leiserowitz, L. *Acta Crystallogr., Sect. B* **1995,** *B51,* 115–148.
29. Aizenberg, J.; Hanson, J.; Koetzle, T. F.; Leiserowitz, L.; Weiner, S.; Addadi, L. *Chem. Eur. J.* **1995,** *1,* 414–422.
30. Vaida, M.; Shimon, L. J. W.; Weisinger-Lewin, Y.; Frolow, F.; Lahav, M.; Leiserowitz, L.; McMullan, R. K. *Science (Washington, D.C.)* **1988,** *241,* 1475–1479.
31. Weissbuch, I.; Lahav, M.; Leiserowitz, L. In *Molecular Modeling Applications in*

Crystallization; Myerson, A.S., Ed.; Cambride University Press: New York, 1999; pp 166–227.

32. Enckevort, W. J. P. v. *Prog. Cryst. Growth Charact.* **1984,** *9,* 1–50.

33. Sunagawa, I. In *Materials Science of the Earth's Interior;* Sunagawa, I., Ed.; Terra Scientific Publishing Co.: Tokyo, Japan, 1984; pp 63–105.

34. Paquette, J.; Reeder, R. J. *Geology* **1990,** *18,* 1244–1247.

35. Northrup, P. A.; Reeder, R. J. *Am. Mineral.* **1994,** *79,* 1167–1175.

36. Paquette, J.; Reeder, R. J. *Geochim. Cosmochim. Acta* **1995,** *59,* 735–749.

37. Rakovan, J.; Reeder, R. J. *Geochim. Cosmochim. Acta* **1996,** *60,* 4435–4445.

38. Rakovan, J.; McDaniel, D. K.; Reeder, R. J. *Earth Planet. Sci. Lett.* **1997,** *146,* 329–336.

39. Kelley, M. P.; Janssens, B.; Kahr, B.; Vetter, W. M. *J. Am. Chem. Soc.* **1994,** *116,* 5519–5520.

40. Rifani, M.; Yin, Y. -Y.; Elliott, D. S.; Jay, M. J.; Jang, S. -H.; Kelley, M. P.; Bastin, L.; Kahr, B. *J. Am. Chem. Soc.* **1995,** *117,* 7572–7573.

41. Mitchell, C.A.; Gurney, R.W.; Jang, S. -H.; Kahr, B. *J. Am. Chem. Soc.* **1998,** *120,* 9726–9727.

42. Sedarous, S.; Subramony, J. A.; Kahr, B. *Ferroelectrics* **1997,** *191,* 302-306.

43. Mitchell, C; Lovell, S.; Thomas, K.; Savickas, P.; Kahr, B. *Angew. Chem. Int. Ed. Engl.* **1996,** *35,* 1021-1023.

44. Lovell, S.; Subramony, P.; Kahr, B. *J. Am. Chem. Soc.* **1999,** *121,* 7020-7025.

45. Chmielewski, J. A.; Lewis, J.; Lovell, S.; Zutshi, R.; Savickas, P.; Mitchell, C.; Subramony, J. A. Kahr, B. *J. Am. Chem. Soc.* **1997,** *119,* 10565-10566.

46. Rashkovitch, L. *The KDP-Family of Single Crystals;* Dover: New York, 1993.

47. Kahr, B.; Jang, S. H.; Subramony, J. A.; Kelley, M. P.; Bastin, L. *Adv. Mater.* **1996,** *8,* 941–944.

48. Subramony, J. A.; Jang, S. H.; Kahr, B. *Ferroelectrics* **1997,** *191,* 292–300.

49. Kahr, B.; Lovell, S.; Subramony, J. A. *Chirality* **1998,** *10,* 66–77.

50. Zaitseva, N.; Carman, L.; Smolsky, I.; Torres, R.; Yan, M. *J. Cryst. Growth* **1999,** *204,* 512–524.

51. Pluta, M. In *Specialized Methods: Advanced Light Microscopy;* Elsevier: New York, 1989; Vol 2, pp 146–196.

52. Subramony, J. A. Ph.D. Dissertation, Purdue University, 1999.

53. Tutton, A. E. *J. Chem. Soc.* **1894,** 628–717.

54 Vogels, L. J. P.; Verheijen, M. A.; Bennema, P. *J. Cryst. Growth* **1991,** *110,* 604–616.

55. Gurney, R.W.; Mitchell, C.A.; Ham, S.; Bastin, L.D.; Kahr, B. *J. Phys. Chem.* In Press.

56. Ojima, K.; Nishihata, Y.; Sawada, A. *Acta Crystallogr., Sect. B* **1995,** *B51,* 287–293.

57. Kurimoto, M.; Subramony, P.; Gurney, R. W.; Lovell, S., Chmielewski, J.; Kahr, B. *J. Am. Chem. Soc.* **1999,** *121,* 6952–6953.

58. Fries, D. C.; Rao, S. T.; Sundaralingam, M. *Acta Crystallogr., Sect. B* **1971,** *B27,* 994–1005.

59. Visser, R. A. *Neth. Milk Dairy J.* **1982,** *36,* 167–193.

60. Dincer, T. D.; Parkinson, G. M.; Rohl, A. L.; Ogden, M. I. In *Crystal. Growth Org. Mater. 4, International Workshop 4th;* Ulrich, J., Ed.; Shaker Verlag: Aachen, Germany, 1997; pp 25–32.

61 Dincer, T. D.; Parkinson, G. M.; Rohl, A. L.; Ogden, M. I. *J. Cryst. Growth* **1999,** *205,* 368–374.

62. Baranyovitz, F. L. C.; *Endeavour* **1978,** *2,* 85–92.

63 Carter, P. W.; Ward, M. D. *J. Am. Chem. Soc.* **1993,** *115,* 11521–11535.

64 Bonafede, S. J.; Ward, M. D. *J. Am. Chem. Soc.* **1995,** *117,* 7853–7861.

65 Hollingsworth, M. D., Brown, M. E., Hiller, A. C.; Santarsiero, B. D.; Chaney, J. D. *Science (Washington D.C.)* **1996,** *273,* 1355–1358.

Liquid and Crystals

Chapter 11

Fabrication of Organic Microcrystals and Their Optical Properties

H. Oikawa, H. Kasai, and H. Nakanishi

Institute for Chemical Reaction Science, Tohoku University, Katahira 2-1-1, Aoba-ku, Sendai 980-8577, Japan

Organic microcrystals occupy the mesoscopic phase between a single molecule and bulk crystal, and they are expected to exhibit peculiar optical- and electronic-properties, depending on crystal size. The reprecipitation method was available for fabrication of organic microcrystals. The crystal size was in the range of about ten nanometer to several hundred nanometer. The excitonic absorption peak positions were shifted to short-wavelength region with decreasing crystal size. This phenomenon is not explainable by quantum confinement effect, and it is now speculated to be due to a certain coupled interaction between exciton and lattice vibration in thermally softened microcrystal lattice.

Nano-particles and super-fine particles in inorganics and metals have been investigated extensively from the viewpoints of both fundamental science and applications (*1-4*). Microcrystals occupy the mesoscopic phase between a single molecule and bulk crystal (*3-6*). In particular, it was worth noting that several reports supporting the enhancement of nonlinear optical (NLO) properties on the basis of quantum confinement effect have recently been published in semi-conductor nano-particles with sizes below 10 nm (*7-14*). These nano-particles were fabricated either by the deposition methods in a molten glass-matrix or by the vacuum-evaporation processes (*15*). On the other hand, organic compounds have essentially an abundance of physicochemical properties (*16*), in comparison with inorganic materials. However, little attention had been paid so far to fabrications of organic microcrystals, when our studies on organic microcrystals were started (*17,18*).

We have demonstrated that the "reprecipitation method" is useful and convenient to prepare some kinds of organic microcrystals (*19*): Polydiacetylene (PDA) derivatives (*20-22*), low-molecular weight aromatic compounds such as perylene and

C_{60} (23-26), and organic functional chromophores of pseudo-isocyanine, merocyanine and phthalocyanine (27, 28). In any case, the crystal size was commonly in the range of several tens of nanometers to sub-micrometer (20, 23, 28). Some interesting phenomena have been confirmed, e.g., the enlargement in conversion of solid-state polymerization in diacetylene monomer microcrystals (29), the shift of the excitonic absorption peak position to the short-wavelength region with decreasing crystal size (20-22), and the appearance of the emission peak from free-exciton energy level in perylene microcrystals with decreasing crystal size and the subsequent shift of the emission peak position to the high energy region (23-25).

In the present chapter, we will provide an interpretation of the reprecipitation method, and then discuss the fabrications of fibrous PDA microcrystals as well as ordinary PDA ones, microcrystallization processes to control the crystal size and shape, and linear optical properties dependence on crystal size.

Reprecipitation Method

Figure 1 shows the scheme of the reprecipitation method (20-27). A target compound is first dissolved in an alcohol or in acetone so that its concentration is about 10^{-3} M. Next, a few micro-liter of the diluted solution should be injected rapidly into a vigorously stirred poor solvent (10 mL), using a microsyringe. It follows that the target compound is reprecipitated and microcrystallized in a poor solvent. Finally, one can obtain organic microcrystals dispersed in the dispersion medium. If the target compound would be solid-state polymerizable diacetylene monomer as shown in Figure 1, then the monomer microcrystals dispersed are further polymerized by UV-irradiation, and then the corresponding PDA microcrystals are formed (30-32). Water is commonly used as a poor solvent, while hydrocarbon such as n-hexane, cyclohexane and decalin are employed as a poor solvent in the case of water-soluble ionic chromophores (27). We can control the crystal size and shape by changing some factors in the reprecipitation process: Concentration of an injected solution, temperature of the poor solvent, and an added surfactant (33).

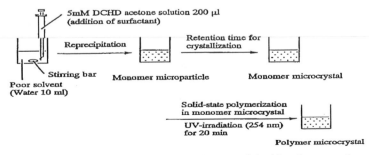

Figure 1 Reprecipitation method schematically exemplified for the case of a diacetylene such as DCHD [1,6-di(N-carbazolyl)-2,4-hexadiyne] and the corresponding solid-state polymerized DCHD, poly(DCHD).

However, no suitable water-miscible good solvents are found in some cases such as slightly soluble titanyl-phthalocyanine. The supercritical fluid crystallization (SCFC) technique was attempted in this case (28,34). We have succeeded in controlling the crystal size and the crystal forms by changing the temperature of supercritical acetone fluid and the composition of acetone-water mixture used as a cooling solvent as shown in Figure 2. In particular, γ-form of titanyl-phthalocyanine (micro)crystals is noted in the field of xerography (35).

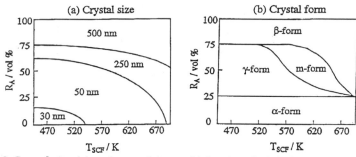

Figure 2 Crystal size (a) and crystal forms (b) for titanyl-phthalocyanine microcrystals prepared by SCFC technique. T_{SCF} represents temperature of supercritical acetone, and R_A is the volume ratio of acetone in acetone-water mixture used as a cooling solvent. Reproduced with permission from Ref. 34. Copyright 1999.

Various Types of Organic Microcrystals and Microcrystallization Processes

Figure 3 shows the typical SEM (scanning electron microscopy) photographs of poly(DCHD) [poly(1,6-di(N-carbazolyl)-2,4-hexadiyne)] microcrystals (20,21). The crystal size, which was also determined by DLS (dynamic light scattering) technique, was evidently influenced by the water temperature. These obtained microcrystals are suggested to be a single crystal in principle from HRTEM (high resolution transmission electron microscopy) observation (36). In addition, poly(DCHD) microcrystals prepared in the presence of a surfactant such as SDS [sodium dodecylsulfate] at the elevated temperature of 60°C have grown as fibers with retention time after reprecipitation as shown in Figure 4 (37). The contour length of fibrous microcrystals is more than 1 μm, and the diameter was about 50 nm. Hence, this diameter was not so different from those of initially formed amorphous-like DCHD particles as described below (20,21,33).

As already mentioned, the crystal size and shape of microcrystals are changeable by the reprecipitation conditions. In other words, it is important to investigate the microcrystallization processes for the purpose of controlling the crystal size and shape. Here, we have focused on poly(DCHD) and perylene microcrystals mainly by the measuremenrs with SEM and SLS (static light scattering) measurements.

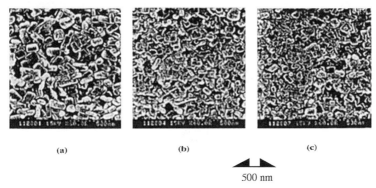

Figure 3 SEM photographs of poly(DCHD) microcrystals fabricated by the reprecipitation method. The values of d and WT mean the average crystal sizes and water temerature, respectively: (a), d = 150 nm, WT = 0 °C; (b) d = 100 nm, WT = 20 0 °C; (c) d = 70 nm, WT = 50 °C.

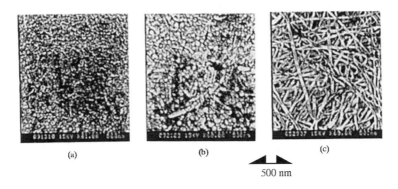

Figure 4 SEM photographs of DCHD microparticles and fibrous microcrystals with various retention times of (a) 0 minute, (b) 7 minutes, and (c) 20 minutes at 333 K in the presence of SDS. Reproduced with permission from Ref. 37.
Copyright 2001 John Wiley & Sons.

Figure 5 depicts the SEM photographs of DCHD particles and microcrystals with retention time after injecting DCHD-acetone solution into water (20, 21). In Figure 5, the shape was converted from sphere-like to cubic-like, but the average size seems to be almost same. During this period the excitonic absorbance based upon π-conjugated poly(DCHD) chains, measured at λ_{max} = 650 nm, increased gradually and saturated with retention time. The low absorbance at the initial stage means that solid-state polymerization did not proceed enough. Thus, sphere-like DCHD particles at the initial stage are said not to be solid-state polymerizable microcrystal but to be amorphous-like particle (20, 21, 38). In fact, we could not observe any X-ray diffraction pattern peaks from DCHD particles formed at the initial stage (21).

On the other hand, we have investigated the microcrystallization process of perylene by SLS measurements (33). At constant temperature, the scattered light intensity I_s increased gradually with retention time, and then saturated. The saturated

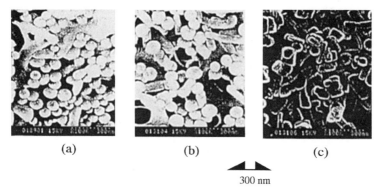

(a)   (b)   (c)

300 nm

Figure 5 SEM photographs of DCHD microparticles and microcrystals with various retention times of (a) 0 minute, (b) 5 minutes, and (c) 10 minutes. Reproduced with permission from Ref. 20. Copyright 1996.

I_s were almost proportional to the amount of the injected perylene solutions, although the crystal sizes were almost the same and about 200 nm in any cases. It follows that the saturated I_s correspond to the number of perylene microcrystals. Next, at the constant amount of the injected solution, the increments of I_s were multiplied exponentially with increasing temperature at above 40°C, and then also saturated with retention time at the given temperature. The saturated I_s was almost the same at any temperature above 40°C, and the crystal sizes were also about 200 nm. On the contrary, the saturated I_s became lower below 40°C, and the crystal size was reduced to about 120 nm in this case.

According to these data, the microcrystallization processes of DCHD and perylene were speculated to occur as illustrated in Figure 6 (*20-22,33*). In any cases, just after reprecipitation, fine droplets are first formed in an aqueous liquid. In the case of DCHD, after removing solvent into the surrounding water, the amorphous and supersaturated DCHD particles are formed, and then nucleation and crystal growth may occur in the individual amorphous particles. On the other hand, the cluster-like fine particles are considered to be once produced in the case of perylene, and then nucleation and crystal growth proceeds through thermal collision between these clusters. Therefore, the initial size of the droplets formed should be minimized to control the crystal size of poly(DCHD). We have tried to reduce the size of the droplets by both decreasing concentration of the injected solution and adding SDS. As a result, we could obtain the smallest poly(DCHD) microcrystals with a size of about 15 nm in our research (*21*).

In addition, the microcrystallization process of fibrous poly(DCHD) microcrystals is speculated in the following (*37*). In the presence of added SDS, amorphous-like DCHD particles seem to be stable thermodynamically even at the elevated temperature. Meanwhile, a part of these particles is microcrystallized, and then amorphous-like particles and microcrystals would co-exist. Next, already-formed microcrystals may act as a kind of substrate, and amorphous particles may be bound epitaxially through thermal collision on the particular crystal plane. The amorphous-

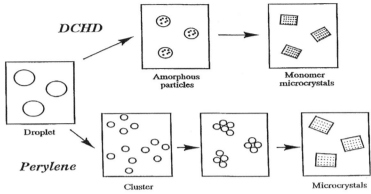

Figure 6 Proposed scheme of microcrystallization processes for DCHD and perylene in the reprecipitation method.

like DCHD particles likely is not completely random but ordered in pseudo-crystalline state. In fact, the excitonic absorbance from solid-state polymerization was observed to be low but finite even at the initial retention time. The added SDS and high temperature may contribute to stabilize the co-existence between amorphous-like DCHD particles and microcrystals, and to promote epitaxial microcrystal growth.

Crystal Sizes Dependence of Linear Optical Properties in Organic Microcrystals

Figure 7(a) shows VIS spectra of poly(DCHD) microcrystals dispersed in an aqueous liquid. The excitonic absorption peak positions, λ_{max}, of π-conjugated polymer chains were shifted evidently to the short-wavelength region with decreasing crystal size (*20-22*). The relationship between λ_{max} and crystal size is plotted in Figure 7(b). On the other hand, the value of λ_{max} (= 670 nm) in fibrous poly(DCHD) microcrystals was almost similar to that of the corresponding bulk poly(DCHD) crystals (*37*). This fact suggests that π-conjugated polymer chains are extended along the long axis of the fibrous microcrystal (*38*). The size effects on linear optical properties were also observed in perylene microcrystals (*23-25,39*).

These blue-shift phenomena of λ_{max} with decreasing crystal size are apparently similar to the behaviors reported in semi-conductor nano-particles with sizes below 10 nm. However, the crystal sizes in the present organic microcrystals are about ten times greater than those of the semi-conductor nano-particles. We believe these experimental results are a peculiar size effect in organic microcrystals, and the mechanism cannot be explained by the so-called quantum confinement effect (*7-14*). To further promote our discussion on these phenomena, the temperature dependence of λ_{max} for poly(DCHD) microcrystals with three crystal sizes was measured as shown in Figure 8 (*40*). In every case, λ_{max} was red-shifted with lowering temperature, and these three plot lines are almost parallel within experimental errors. We can regard the intercepts as the intrinsic λ_{max} at each crystal size. In addition, the half-width of the

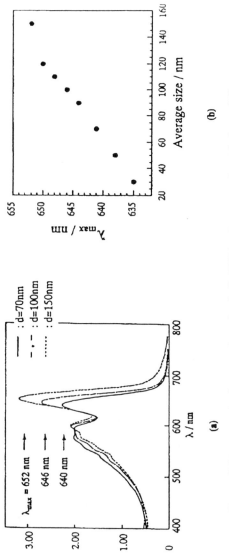

Figure 7 Linear optical properties of poly(DCHD) microcrystals with three different crystal sizes dispersed in an aqueous liquid: (a), visible absorption spectra; (b), dependence of excitonic absorption peak positions λ_{max} (nm) on crystal sizes. Reproduced with permission from Ref. 40. Copyright 1998 Gordon and Breach.

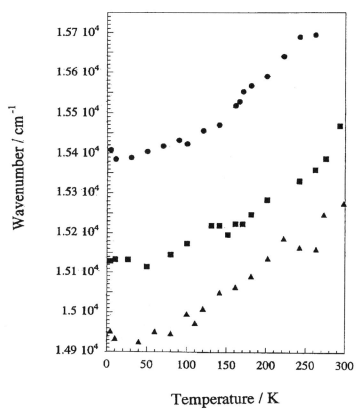

Figure 8 Excitonic absorption peak positions v_{max} (cm^{-1}) dependence on temperature for poly(DCHD) microcrystals with three different crystal sizes: ●, 50 nm; ■, 100 nm; ▲, 1 μm above. Reproduced with permission from Ref. 40. Copyright 1998 Gordon and Breach.

excitonic absorption peak was changed from *ca.* 770 cm^{-1} to *ca.* 600 cm^{-1} with decreasing temperature in the case of poly(DCHD) microcrystals with 100 nm in crystal size, whereas the half-width changed from *ca.* 1050 cm^{-1} to *ca.* 750 cm^{-1} as well in the case of 50 nm crystals.

Table 1 summarized the size effects on linear optical properties in poly(DCHD) microcrystals. Let us consider some factors to clarify the present size effects. In conclusion, the quantum size effect, light scattering effect from microcrystals dispersion, and some surface effects between microcrystals and the surrounding dispersion medium should be rejected or, at least, minor factors. As a reasonable discussion, a certain coupled interaction between exciton and lattice vibration in thermally "softened" crystal lattice in microcrystals is now speculated to bring about instabilization in the lowest exciton level and/or high occupation in exciton band (*40, 41*). This is the possible qualitative explanation at the present time. Theoretical analyses are now in progress as well.

Table I Crystal Sizes Dependence of Linear Optical Properties for Poly(DCHD) Microcrystals

Crystal Size	Small	⇔	Large
λ_{max}	High energy shift	⇔	Low energy shift
$\Delta v_{1/2}$	Broadening	⇔	Narrowing
Temperature	High	⇔	Low
Crystal Lattice	High frequency vibration	⇔	Low frequency vibration

Concluding Remarks

We have established the reprecipitation method, including supercritical fluid crystallization technique, to fabricate well-defined organic microcrystals, *i.e.*, to control crystal size, shape, and crystal forms. Next, linear optical properties of organic microcrystals were found to be dependent on crystal size, which was qualitatively explained by a certain coupled interaction between exciton and lattice vibration in soften microcrystal lattice, rather than the so-called quantum confinement effect.

References

1. Depasse, J.; Watilon, A.; *J. Colloid Interface Sci.* **1970**, *33*, 430.
2. Van de Hulst, H. C. *Light Scattering by Small Particles;* Dover Publications Inc.: New York, 1981.

3. Murray, C. B.; Norris, D. J.; Bawendi, M. G.; *J. Am. Chem. Soc.* **1993**, *115*, 8706.

4. Buffat, P.; Borel, J. P.; *Phys. Rev.* **1976**, *A13*, 2287.

5. Iijima, S.; Ichihashi, T.; *Jpn. J. Appl. Phys.* **1985**, *24*, L125.

6. Iwama, S.; Hayakawa, K.; Arizumi, T.; *J. Cryst. Growth* **1984**, *66*, 189.

7. Ekinov, A. I.; Efros, Al. L.; *Sov. Phys. Semicond.* **1982**, *16*, 772.

8. Roussignol, P.; Ricard, D.; Flytzains, C.; *Appl. Phys.* **1990**, *B51*, 437.

9. Hanamura, E.; *Solid State Commun.* **1973**, *12*, 451.

10. Kubo, R.; *J. Phys. Soc. Jpn.* **1962**, *17*, 975.

11. Hanamura, E.; *Solid State Commun.* **1987**, *62*, 465.

12. Hanamura, E.; *Phys. Rev.* **1988**, *B38*, 1228.

13. Brus, L. E.; *J. Chem. Phys.* **1984**, *80*, 4403.

14. Hache, F.; Richard, D.; Flytzains, C.; *J. Opt. Soc. Am.* **1986**, *B3*, 1647.

15. Nakamura, A.; *Hyomen (Surface)* **1992**, *30*, 330 (in Japanese).

16. Bosshard, Ch.; Sutter, K.; Prêtre, Ph.; Hulliger, J.; Flörsheimer, M.; Kaaatz, P.; Günter, P.; *Organic Nonlinear Optical Materials*, In *Adv. Nonlinear Opt., 1*, Gordon and Breach Pub.: Amstertdam, 1995.

17. Toyotama, H.; *Kinozairyo (Functional Materials)*, **1987**, *6*, 44 (in Japanese).

18. Yase, K.; Inoue, T.; Okada, M.; Funada, T.; Hirano, J.; *Hyomen Kagaku (Surface Science)* **1989**, *8*, 434 (in Japanese).

19. Kasai, H.; Nalwa, H. S.; Oikawa, H.; Okada, S.; Matsuda, H.; Minami, N.; Kakuta, A.; Ono, K.; Mukoh, A.; Nakanishi, H.; *Jpn. J. Appl. Phys.* **1992**, *31*, L1132.

20. Katagi, H.; Kasai, H.; Okada, S.; Oikawa, H.; Matsuda, H.; Liu, Z.-F.; Nakanishi, H.; *Jpn. J. Appl. Phys.* **1996**, 35, L1364.

21. Katagi, H.; Kasai, H.; Kamatani, H.; Okada, S.; Oikawa, H.; Matsuda, H.; Nakanishi, H.; *J. Macromol. Sci. Pure & Appl. Chem.* **1997**, *A34*, 2013.

22. Matsuda, H.; Yamada, S.; Van Keuren, E.; Katagi, H.; Kasai, H.; Kamatani, H.; Okada, S.; Oikawa, H.; Nakanishi, H.; Smith, E. C.; Kar, A. K.; Wherrett, S.; *SPIE Proc.* **1997**, *2998*, 241.

23. Kasai, H.; Kamatani, H.; Okada, S.; Oikawa, H.; Matsuda, H.; Nakanishi, H.; *Jpn. J. Appl. Phys.* **1996**, *35*, L1221.

24. Kasai, H.; Yoshikawa, Y.; Seko, T.; Okada, S.; Oikawa, H.; Matsuda, H.; Watanabe, A.; Ito, O.; Toyotama, H.; Nakanishi, H.; *Mol. Cryst. Liq. Cryst.* **1997**, *294*, 173.

25. Kasai, H.; Oikawa, H.; Okada, S.; Nakanishi, H.; *Bull. Chem. Soc. Jpn.* **1998**, *71*, 2597.

26. Fujitsuka, M.; Kasai, H.; Masuhara, A.; Okada, S.; Oikawa, H.; Nakanishi, H.; Watanabe, A.; Ito, O.; *Chem. Lett.* **1997**, 1211.

27. Kamatani, H.; Kasai, H.; Okada, S.; Matsuda, H.; Oikawa, H.; Minami, N.; Kakuta, A.; Ono, K.; Mukoh, A.; Nakanishi, H.; *Mol. Ctyst. Liq. Cryst.* **1994**, *252*, 233.

28. Komai, Y.; Kasai, H.; Hirakoso, H.; Hakuta, Y.; Okada, S.; Oikawa, H.; Adsciri, T.; Inomata, H.; Arai, K.; Nakanishi, H.; *Mol. Cryst. Liq. Cryst.* **1998**, *322*, 167.

29. Iida, R.; Kamatani, H.; Kasai, H.; Okada, S.; Oikawa, H.; Matsuda, H.; Kakuta, A.; Nakanishi, H.; *Mol. Cryst. Liq. Cryst.* **1995**, *267,* 95.

30. Yee, K. C.; Chance, R. R.; *J. Polym. Sci.* **1978**, *16*, 431.

31. Enkelmann, V.; Leyrer, R.; Schleier, G.; Wegner, G.; *J. Mater. Sci.* **1980**, *15*, 168.

32. Enkelmann, V.; Werz, G.; Muller, M. A.; Schmidt, M.; Wegner, G.; *Mol. Cryst. Liq. Cryst.* **1984**, *105*, 11.

33. Kasai, H.; Oikawa, H.; Okada S.; Nakanishi, H.; *Bull. Chem. Soc. Jpn.* **1998**, *71,* 2597.

34. Komai, Y.; Kasai, H.; Hirakoso, H.; Hakuta, Y.; Katagi, H.; Okada, S.; Oikawa, H.; Adschiri, T.; Inomata, H.; Arai, K.; Nakanishi, H.; *Jpn. J. Appl. Phys.* **1999**, *38*, L81.

35. Watanabe, K.; Kinoshita, A.; Hirasa, N.; Itami, A.; *Konica Tech. Rept.* **1990**, *3,* 108.

36. Yase, K.; Hanada, T.; Kasai, T.; Okada, S.; Nakanishi, H.; *J. Electro. Microscopy in press.*

37. Oikawa, H.; Oshikiri, T.; Kasai, H.; Okada, S.; Tripathy, S. K.; Nakanishi, H.; *Polym. Adv. Tech. in press.*

38. Cantow, H. J., Ed.; *Polydiacetylene,* In *Adv. Polym. Sci.,* 63, Springer-Verlag, Berlin: 1984.

39. Kasai, H.; Kamatani, H.; Yoshikawa, Y.; Okada, S.; Oikawa, H.; Nakanishi,H.; *Chem. Lett.* **1997**, 1182.

40. Katagi, H.; Oikawa, H.; Okada, S.; Kasai, H.; Watanabe, A.; Ito, O.; Nozue, Y.; Nakanishi, H.; *Mol. Cryst. Liq. Cryst.* **1998**, *314*, 285.

41. Oikawa, H.; Nakanishi, H.; *Rev. Laser Eng.* **1997**, *25*, 765 (in Japanese).

Chapter 12

Some Applications of Organic Microcrystals

H. Oikawa, H. Kasai, and H. Nakanishi

Institute for Chemical Reaction Science, Tohoku University, Katahira 2-1-1, Aoba-ku, Sendai 980-8577, Japan

As described in the preceding chapter, the most striking characteristic of the reprecipitation method is that organic microcrystals prepared are stable when dispersed in a medium, owing to high ζ-potential. Hence, the optically transparent and layered thin films could be fabricated by electrostatically adsorbing poly(DCHD) microcrystals with negative ζ-potential on an inert polycation film, and the $\chi^{(3)}(\omega)$ values were enhanced mainly by concentration effect. In addition, SHG-active DAST microcrystals, having huge dipole moment, dispersed in hydrocarbon medium were oriented by applying remarkably lower electrostatic field, compared with DC field in ordinary liquid crystal molecules. The present microcrystal dispersions can be regarded as a novel "Liquid and Crystals" system.

Introduction

As discussed in the preceding chapter, organic microcrystals prepared by the reprecipitation technique are dispersed in a stable fashion, owing to high ζ-potential (1). This fact is technologically important for some applications. In the present chapter, we will demonstrate two detailed examples by utilizing this character of dispersed organic microcrystals. One is the fabrication of layered organic microcrystal thin film through electrostatic adsorption on a polyeletrolyte, and their enhanced NLO properties, and the other is the electric-field-induced orientation of polar microcrystals with large dipole moment in the dispersion liquid. The latter can be regarded as a novel "Liquid and Crystals" system. In other words, organic microcrystals are expected as novel functional materials with high optical quality for electronics and photonics devices.

Fabrication of Layered Organic Microcrystals Thin Films, and Their Enhanced NLO Properties

The ζ-potential of poly(DCHD) microcrystals was taken as -40 mV (2). Actually, the optical high quality microcrystal thin films were fabricated by electrostatically adsorbing microcrystals with negative ζ-potential on polycation, using layer-by-layer technique (3). As a binder polymer between microcrystals, polycation, PDAC [poly(diallyldimethylammonium chloride)] was used. Figure 1 demonstrates the layer-by-layer procedures. First, the clean slide glass is immersed into the polycation aqueous solution so as to be coated with PDAC thin film. After slightly rinsing, this slide glass should be immersed into poly(DCHD) microcrystals dispersion liquid, and then slightly rinsed again. This cyclic process should be performed as many times as needed.

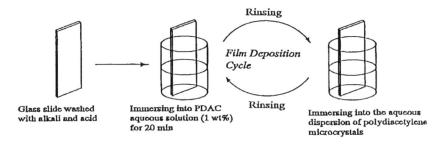

Rinsing

Film Deposition Cycle

Rinsing

Glass slide washed with alkali and acid

Immersing into PDAC aqueous solution (1 wt%) for 20 min

Immersing into the aqueous dispersion of polydiacetylene microcrystals

Figure 1 Scheme of layer-by-layer technique for fabrication of organic microcrystal thin film. Reproduced with permission from Ref. 25. Copyright 2001 Blackwell.

Figure 2 shows the SEM photographs of poly(DCHD) microcrystals electrostatically adsorbed on PDAC thin film at only one cycle (4). The isolated single poly(DCHD) microcrystal was found to be trapped within 20 minutes of the immersing time into microcrystals dispersion liquid. The number of the trapped microcrystals increased with immersing time. Single molecule spectroscopy such as scanning near-field optical microscope (SNOM) (5) should be carried out in the near future.

Figure 3 indicates the SEM photographs of poly(DCHD) microcrystals trapped with cycle times at a constant immersing time (1 hour) (4). The microcrystals trapped were densely condensed and covered completely everywhere in SEM photograph at more than 10 cycle times. The excitonic absorbance at $\lambda_{max} = 650$ nm (Refer to Figure 7 in the preceding chapter.) was multiplied almost proportionally with increasing cycle times and without changing λ_{max}, which suggests that any interactions between layers in the microcrystal thin films may not exist optically. This fact may simplify the explanation for enhanced NLO properties as discussed later. In addition, this layer-by-layer technique seems to be much convenient to control the thickness of thin film and the concentration of microcrystals, rather than conventional spin-coating method.

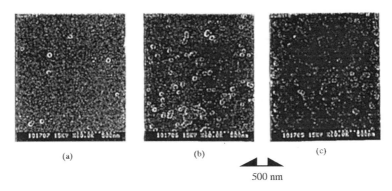

500 nm

Figure 2 SEM photographs of poly(DCHD) microcrystals thin. The immersing times are (a) 20 minutes, (b) 40 minutes, and (c) 60 minutes at only one cycle. Reproduced with permission from Ref. 4.

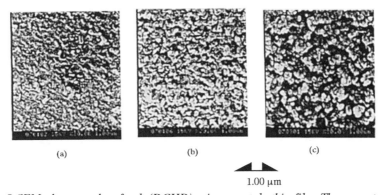

1.00 μm

Figure 3 SEM photographs of poly(DCHD) microcrystals thin film. The repeated-numbers of cycles are (a) one cycle, (b) five cycles, and (c) twelve cycles. The immersing time per one cycle was one hour. Reproduced with permission from Ref. 4.

Next, evaluations of the third-order NLO properties, *i.e.*, $\chi^{(3)}(\omega)$, of the layered poly(DCHD) microcrystal thin films have been attempted using the so-called Z-scan method (6, 7). When the laser beem with Gaussian profile are incident upon a sample with nonlinear refractive index n_2, the sample may act as a lens. As a result, the laser beam intensity at the aperture may be variable, depending on the sample position, *i.e.*, Z-position. We can estimate the corresponding values of $\chi^{(3)}(\omega)$ from the measured n_2.

Table 1 shows the values of n_2 and $\chi^{(3)}(\omega)$ for the twelve-layered poly(DCHD) microcrystal thin film. Compared with the previous poly(DCHD) microcrystals (2.4 wt%) loaded in gelatin film (8), the present values were multiplied by a factor of about 1000, due to a concentration effect. Instead of poly(DCHD) bulk crystal, one can employ the layered microcrystal thin film with high optical quality and low scattering loss as a material for NLO devices.

Table I Nonlinear Refractive Index n_2 and the Third-Order Susceptibility $\chi^{(3)}(\omega)$ for Twelve-Layered Poly(DCHD) Microcrystals Thin Film Evaluated by the Z-Scan Method

Sample	Pumping Wavelength/nm	n_2/cm^2GW^{-1}	$Re\,\chi^{(3)}(\omega)/esu$
Twelve-layered microcrystals thin film[a]	670	-4.2	-1.8×10^{-7}
Microcrystals/gelatin spin-coated film[b]	670	-1.4×10^{-2}	-8.8×10^{-10}

a) This work:: The microcrystals thin film was about 200 nm thick, and the averaged crystal size was 150 nm.

b) Ref. 22 in the preceding chapter.

In addition, as shown in Figure 4 (*4*), poly(DCHD) microcrystals were adsorbed selectively and electrostatically on photo-patterned LB polycation film: *ran*-copoly[(*N*-dodecylacrylamide)$_{0.77}$ / (*N*-(11-acryloylundecyl)-4-vinylpyridinium bromide)$_{0.03}$ / (*N*-(11-hydroxyundecyl)-4-vinylpyridinium bromide)$_{0.04}$ / (4-vinylpyridine)$_{0.16}$] (*9*). The copolymer was previously coated onto a hydrophobic-treated silicone substrate, exposed by UV light through a photo-mask for half an hour, and then developed in chloroform. This treated silicone substrate was immersed into poly(DCHD) microcrystals dispersion liquid in the same manner. This result suggests the possibility for fabrication of well-defined super-lattice structure between microcrystals and other materials.

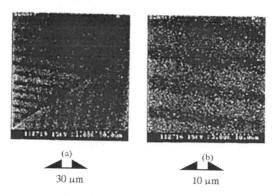

(a)
30 μm

(b)
10 μm

Figure 4 SEM photographs of photo-patterned poly(DCHD) microcrystals thin film prepared by selectively and electrostatically adsorbing poly(DCHD) microcrystals on photo-patterned polycation LB film. Reproduced with permission from Ref. 4.

Electric-Field-Induced Orientation of Polar DAST Microcrystal with Large Dipole Moment

Organic microcrystal-dispersion system prepared in the reprecipitation method usually behaves as a random and isotropic phase. If the organic microcrystals could respond sensitively to applied external fields such as electric and/or magnetic fields, polarized light, and flow field, the dispersion system would be converted from a random to an oriented and anisotropic phase, which may lead to changes of optical properties like transmittance and refractive index in the dispersion systems.

For example, to realize electric-field-induced orientation of organic microcrystals, there are two required factors. One is that organic microcrystals have to have dipole moment, and the other requires the use of a dispersion medium with low dielectric constant. The dipole moment is necessary to obtain a response to the applied electric field, and the low dielectric constant should be required so as to apply effectively an electric field to microcrystals. We have chosen the system of DAST [4'-dimethylamino-N-methylstilbazolium p-toluenesulfonate] microcrystals dispersed in decalin ($10,11$). DAST is well-known to be functional ionic chromophore as SHG-active material ($12-14$). DAST microcrystals were fabricated by reprecipitating 100 μL of DAST-ethanol solution (5 mM), containing the surfactant DTMAC [dodecyltrimethylammonium chloride], into decalin (10 mL) stirred at room temperature. This procedure is called the "inverse-reprecipitation method" (1).

Figure 5 shows the SEM photograph of the prepared DAST microcrystals. The microcrystals are plate-like and about 500 nm, which size was almost similar to that measured by DLS method. This fact implies that DAST microcrystals are dispersed in a stable fashion as the primary particle and without aggregated. DAST microcrystals obtained were confirmed to be SHG-active by the powder-test (12). In addition, XRD patterns from DAST microcrystals were in fair agreement with those of SHG-active DAST bulk crystals (13).

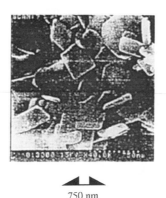

750 nm

Figure 5 SEM photograph of DAST microcrystals prepared by the inverse-reprecipitation method. Reproduced with permission from Ref. 10. Copyright 1999.

When the absorbance of DAST microcrystals in dispersion liquid is measured under an applied electric-field, there are two kinds of configurations. One is the perpendicular configuration between applied electric-field and probe light for measurement of VIS absorption spectrum, and the other is the parallel one. The absorbance was measured at λ_{max} = 550 nm (15) in VIS absorption spectrum of DAST microcrystals dispersed in decalin, and DC electric-field was applied to be 150 V/cm.

Figures 6(a) and 6(b) show the changes of absorbance with and without applied electric-field at the perpendicular and parallel configurations, respectively. In any cases, the response rate seems to be within 2 second after the electric field is switched on, and the changed absorbance was relaxed again within 5 second after switch-off. These responses in absorbance were entirely reversible, and were attributable to electric-field-induced orientation of DAST microcrystals.

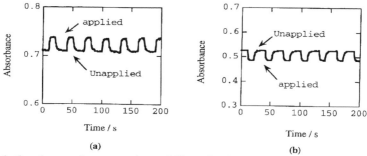

(a)   (b)

Figure 6 Absorbance changes at λ_{max} =550 nm for DAST microcrystals in dispersion liquid with and without applied electric-field (150 V/cm): (a), perpendicular configuration [E $\perp$ h v]; (b), parallel configuration [E // hv]. Reproduced with permission from Ref. 10. Copyright 1999.

Figure 7 indicates the dependence of relative changes in absorbance on applied electric-field in the case of parallel configuration. The relative changes were almost proportional to applied electric-field, and reversible until *ca.* 300 V/cm. Above 300 V/cm, the absorbance changes were irreversible, owing to electrophoresis of DAST microcrystals. Anyway, it should be kept in mind that these behaviors could be realized at the fields of less than 300 V/cm, a value that is much smaller than those (*ca.* 10^4 to 10^5 V/cm) for alignment of ordinary LC molecules (16). In other word, it is suggested that DAST microcrystals may have large dipole moments.

So, we have attempted to evaluate the dipole moment of one DAST microcrystal. First, the space group of DAST single bulk crystal is Cc (13, 17), which means that the dipole moment of DAST-cation and -anion pair is located in the a-c plane. In the following, this cation and anion pair is called "DAST molecule". Next, the dipole moments of DAST molecules row cut-out along each of a-, b-, and c-axes were calculated as listed in Table 2 (18). Using MOPAC packaged software (19), the calculations were performed on the basis of the molecular coordinates determined by single crystal XRD analysis of DAST without optimizing molecular structure. As a result, the dipole moments roughly increased proportionally with the number of

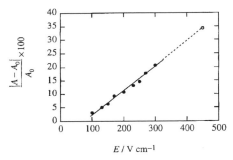

Figure 7 Dependence of relative absorbance changes on applied electric field in the case of E // hν. A_0 is the absorbance at λ_{max} =550 nm without applied electric-field. Reproduced with permission from Ref. 10. Copyright 1999.

Table II Dipole Moments of DAST Molecules Calculated along Crystal Axes

Numer of Pairs	a-axis		b-axis		c-axis	
	Total/D	One Pair/D	Total/D	One Pair/D	Total/D	One Pair/D
1	35.5	35.5	35.5	35.5	35.5	35.5
2	75.4	36.7	70.3	35.1	62.8	31.4
3	116.6	38.9	105.6	35.2	92.7	30.9
4	158.2	39.6	141.2	35.3	121.0	30.3

Lattice parameters: $C_{23}H_{26}N_2SO_3$, M_w = 410.54, Monoclinic, Space group Cc, a = 10.365(3) Å, b = 11.322(4) Å, c = 17.893(4) Å, α = 90°, β = 92.24(2)°, γ = 90°, V = 2098.2(11) Å^3, Z = 4, R = 0.033 (12,16). Reproduced with permission from Ref. 11. Copyright 2001 Elsevier.

DAST molecules cut-out in any axis. As a first approximation, we can assume that the total dipole moment per one DAST microcrystal is proportional to the number of DAST molecules. When DAST microcrystal is now 100 nm × 100 nm × 50 nm in size, the total dipole moment per one DAST microcrystal is estimated to be on the order of 10^7 D, assuming that the dipole moment of one DAST molecule is 30 D. Furthermore, the angle of dipole moment to the a-axis in the a-c plane was confirmed to be 35° as shown in Figure 8 (11).

Now, the required minimum applied electric-field was estimated from the Langevin function (20) by using the above-calculated dipole moment of one DAST microcrystal (μ = 2.8 × 10^7 D). As a result, the calculated curve from Langevin function became asymptotic to be unity at about 20 V/cm, which is about one order of

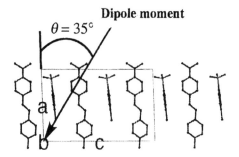

Figure 8 Direction of dipole moment in the unit cell (the projection along b-axis) of DAST crystal structure. Reproduced with permission from Ref. 11. Copyright 2001 Elsevier.

magnitude less than the experimental values in Figures 6 and 7.

On the other hand, the maximum relative change of absorbance in parallel configuration was determined theoretically as follows. The absorption coefficients in the oriented and random states $\varepsilon_{oriented}$, and ε_{random}, are represented by eqs (1) and (2) [21]:

$$\varepsilon_{oriented} \propto \frac{\int_0^{2\pi} |\mu_{mn}|^2 \sin^2 \theta \cos^2 \phi \, d\phi}{\int_0^{2\pi} d\phi} = \frac{1}{2} |\mu_{mn}|^2 \sin^2 \theta \tag{1}$$

$$\varepsilon_{random} \propto |\mu_{j=x,y,z}|^2 = \frac{1}{3} |\mu_{mn}|^2 \tag{2}$$

Where, μ_{mn} is the transition moment of DAST crystals. Therefore, the maximum relative change of absorbance, R_{abs}, is given by

$$R_{abs} = \frac{|\varepsilon_{oriented} - \varepsilon_{random}|}{\varepsilon_{random}} = \frac{3}{2} \sin^2 \theta - 1 \tag{3}$$

By assuming that the transition moment is parallel to the polar axis of the DAST crystal, that is, $\theta = 35°$ in eq. 3, the value of R_{abs} is calculated to be 0.51. This is roughly twice the experimental value of the relative changes in absorbance at 300 V/cm in Figure 7.

These discrepancies between experimental results and theoretical predictions are considered to be due to several factors: For example, ignorance of local-field (20, 22), relaxation and viscous effect for orientered DAST microcrystals (23), ζ-potential, and overestimated dipole moment of one DAST microcrystal in the theoretical discussion. On the other hand, it should be necessary to optimize experimentally the crystal size and the cell structure so as to apply the electric-filed to DAST microcrystals efficiently, and to reduce the electrostatic shielding effects by added surfactant.

Concluding Remarks

In the present chapter, some trial applications have been demonstrated, and these are the enhancement of the third-order susceptibility of layered poly(DCHD) microcrystal thin films and the electric-field-induced orientation of DAST microcrystals with large dipole moments. In comparison with organic bulk crystals as NLO materials, the microcrystal-dispersion systems can be regarded as a new concept of "Liquid and Crystals" with high optical quality and low scattering loss, and are expected to be novel optical devices in electronics and photonics fields (24,25). Remarkably, the research on organic microcrystals seems to affect not only the fields of electronics and photonics but also to the areas of catalysis and medical supplies as well as food industry.

Acknowledgement

The authors would like to express their gratitude to Drs. H. Matsuda and K. Yase, of National Institute of Materials and Chemical Research, MITI; Profs. H. Masuhara and T. Itoh, of Department of Applied Physics, Osaka University; Prof. T. Kobayashi, of Department of Applied Physics, University of Tokyo; Prof. S. K. Tripathy, of Department of Chemistry and Physics, University of Massachusetts Lowell, US; Prof. Z.-L. Liu, of College of Chemistry, Peking University, China; Prof. B. S. Wherrett, of Department of Physics, Heriot-Watt Iniversity, UK; Profs. K. Arai, H. Inomata, T. Adschiri, and Dr. Y. Hakuta of Department of Biochemistry and Chemical Engineering, Tohoku University; Profs. O. Terasaki and Y. Nozue, of Department of Physics, Tohoku University; Profs. O. Ito, T. Miyashita, and S. Okada, Drs. M. Fujitsuka and J. Aoki, and Ph.D. students in Nakanishi's Laboratory, of Institute for Chemical Reaction Science, Tohoku University, for collaborative research cited in both the preceding and the present chapters.

References

1. Nakanishi, H.; Kasai, H.; *Am. Chem. Soc. Sypm. Series*, **1997**, *672*, 183.
2. Wiersema, P. H.; Loeb, A. L.; Overbeek, J. Th. G.; *J. Colloid Interface Sci.* **1966**, *22*, 78.

178

3. Tripathy, S. K.; Katagi, H.; H.; Kasai, Balasubramanian, S.; Oshikiri, T.; Kumar, T.; Oikawa, H.; Okada, S.; Nakanishi, H.; *Jpn. J. Appl. Phys.* **1998**, *37*, L343.
4. Ref. 37 in the preceding chapter.
5. Dürig, U.; Pohl, D. W.; Rohner, F.; *J. Appl. Phys.* **1986**, *59*, 3318.
6. Sheik-Bahae, M.; Said, A. A.; Van Stryland, E. W.; *Opt. Lett.* **1989**, *14*, 955.
7. Sheik-Bahae, M.; Said, A. A.; Wei, T. H.; Hagan, D. J.; Van Stryland, E. W.; *IEEE. J. Quantum Electron.* **1990**, *26*, 760.
8. Ref. 22 in the preceding chapter.
6. Oshikiri, T.; Aoki, J.; Katagi, H.; Kasai, H.; Oikawa, H.; Okada, S.; Miyashita, T.; Nakanishi, H.; *manuscript in preparation.*
10. Fujita, S.; Kasai, H.; Okada, S.; Oikawa, H.; Fukuda, T;. Matsuda, H.; Tripathy S. K.; Nakanishi, H.; *Jpn. J. Appl. Phys.* **1999**, *38*, L659.
11. Oikawa, H.; Fujita, S.; Kasai, H.; Okada, S.; Tripathy, S. K.; Nakanishi, H.; *Colloids & Surface A, in press.*
12. Nakanishi, H.;. Matsuda, H.; Okada, S.; Kato,M.; *MRS Int. Mtg. Adv. Mater.* **1989**, *1*, 97.
13. Marder, S. R.; Perry, J. W.; Schaefer, W. P.; *Science* **1989**, *245*, 626.
14. Duan, X.-M.; Konami, H.; Okada, S. ; Oikawa, H.; Matsuda, H.; Nakanishi, H.; *J. Phys. Chem.* **1996**, *100*, 17780.
15. Marder, S. R.; Perry, J. W.; Yakymyshyn, C. P.; *Chem. Mater.* **1994**, *6*, 1137.
16. Khoo, I.-C.; *LIQUID CRYSTALS, Physical Properties and Nonlinear Optical Phenomena*, John Wiley & Sons, Inc.: New York 1995.
17. Marder, S. R.; Perry, J. W.; Schaefer, W. P.; *J. Mater. Chem.* **1992**, *2*, 985.
18. Turi, L.; Dannenberg, J. J.; *J. Phys. Chem.* **1989**, *10*, 9638.
19. Stewart, J. J. P.; *J. Comput. Chem.* **1989**, *10*, 209.
20. Debye, P. J. W.; *TOPICS IN CHEMICAL PHYSICS*, Elsevier Publishing Company, Amsterdam, 1962.
21. Fraser, R. D. B.; *J. Chem. Phys.* **1953**, *21*, 1511.
22. Moore, W.; *Basic Physical Chemistr"*, Prentice-Hall Inc.: New Jersey, 1983.
23. Marks, A. M.; *Appl. Opt.*, **1969**, *8*, 1397.
24. Saitoh, M.; Mori, T.; Ishikawa, R.; Tamura, H.; *Proc., SID* **1982**, *24*, 249.
25. Kasai, H.; Oikawa, H.; Nakanishi, H.; *The Organic Microcrystals*,In *IUPAC, Chemistry for the 21th Century*, Masuhara, H., Ed.; Blackwell Science: *in press*.

Liquid Crystals

Chapter 13

Molecular Design of Highly Fluorinated Liquid Crystals

Frédéric Guittard and Serge Géribaldi

Laboratoire de Chimie des Matériaux Organiques et Métalliques, Université de Nice, Parc Valrose, F-06108 Nice cedex 2, France (Email: guittard@unice.fr)

The aim of this work is to present recent studies of highly fluorinated liquid crystals with focus on the molecular design and stabilization of the liquid crystal phases. We point out that the typical (rod-like, or disk-like) molecular shape gives rise to liquid crystal behavior but is not a necessary condition for the existence of liquid crystalline mesophase. The incorporation of incompatible parts (flexible-rigid, lipophilic / hydrophilic or lipophilic...), i.e., the presence of amphiphilic structure, leads to a liquid crystalline behavior which can be exacerbated from molecular design. For instance, the highly fluorinated tail, due to its intrinsic physical properties, gives future prospects in this field. Its impact on the mesomorphic properties is compared to the hydrocarbon homologues of these series.

Introduction

Since the discovery of fluorine, the use of organofluorinated compounds has generated a great deal of research. In fact, the replacement of one or several hydrogen atoms by fluorine atoms confers to the resulting material unusual properties which make them useful for a variety of applications : surface coating, fire extinguishers, agrochemicals and many other fields with high added values. These newer fields have provided a research area which is devoted to the studies of compounds containing mono or polyfluorinated moieties to demonstrate their liquid crystal character. The introduction of sufficiently incompatible moieties (1) allows us to induce a liquid crystalline behavior. The introduction of fluorine and more specifically a highly fluorinated tail could be one of these incompatible species. First, we will describe the intrinsic physical properties of fluorine which lead to supramolecular properties, and we will justify the interest of fluorinated liquid crystals. Then we will present a recent example of molecular shape in each liquid crystalline type (lyotropic and thermotropic), and the impact of such substitution in connection with hydrocarbon homologous behavior. In each part references to previous works are included. The aim of this paper is to focus our attention on highly fluorinated low molecular weight liquid crystals which include a number of fluoromethylene units up to one (polymers (2,3) or copolymers (4) will not be discussed).

Fluorine Hypothesis

When we talk about fluorinated materials, it is important to consider not only the kind of groups but also the number of fluorine atoms in the molecule. In fact, the introduction of one fluorine atom or a trifluoromethyl group in low molecular weight compounds, opposite a perfluorinated tail, does not allow us to reach the same properties. The first strategy consists of using the strong electronegative property of fluorine, which confers a high dipole moment and permits modification of the electrostatic interaction often required in bioactive compounds. The resulting CF bond possesses a high thermal stability and leads to alteration of the acidity of an active site. However, due to the van der Waals radius, the replacement of a hydrogen by a fluorine atom can not be considered as an isosteric substitution. As a result, the fluorine atom is sterically closer to an oxygen rather than to a hydrogen atom (5,6).

The increase in the number of fluorine atoms within the same molecule leading to a partially fluorinated or perfluorinated tail allows us to obtain other characteristic properties. A shortness of the CF bonds linked to the strongly electronegative fluorine leads to a contraction of the carbon backbone which, associated with the weak Pauli radius and large van der Waals radius of the fluorine atom, gives a perfluorinated tail showing a large volume and a peculiar rigidity. This electronic sheath leads to a unique characteristic of fluorinated tails, which are chemically inert to strong acids, oxidizing agents, and concentrated alkalis. Hence highly fluorinated-based agents can be used in drastic media in which hydrocarbon homologues would decompose. The great interest for industrial applications of compounds having a perfluorinated tail is the result of two main physical properties :

(i)- a weak inter-tail interaction which leads to a low viscosity, necessary for the decrease of frictional coefficient and for obtaining lubricating materials (7-9). Two consequences of these weak interactions are the low surface energy and the low vapor pressure. The low value of the free surface energy compared to those obtained within hydrocarbon series provides a high surface activity of surfactants, even within thermotropic calamitic liquid crystals (10,11), with a fluoroalkyl tail. In some cases it is possible to decrease the surface tension of an aqueous solution of surfactant to 15 mN/m, or less in the case ultra low surface energy (12,13). The surface tension properties are obtained at lower concentrations (cmc) than those of hydrocarbon homologues.

(ii)- the lipophilicity of these tails. As a result, materials incorporating fluorocarbon at their surface are very poorly wettable by common solvents including aqueous media which confer soiling resistant properties to skins, coats, plastics, etc... incorporating fluorinated moiety. Furthermore, in spite of the weak inter-tails interaction, their strong incompatibility tends toward microphase separation (mainly lamellar structures) which favorably affects the anisotropic properties such as liquid crystalline properties.

The synergy of these properties is utilized, as in lubricants, cosmetics, cleaners for hard surfaces, and in surfactant formulation as extinguishing media, especially for quenching fires of petroleum and derivatives. Lastly, an interesting property of the perfluorinated compounds is their strong ability to dissolve gases such as oxygen, which allow us to consider them as an interesting alternative to blood (14,15).

In conclusion, the introduction of fluorine atoms and more specifically of a highly fluorinated tail instead of a hydrocarbon one, drastically modifies the physical properties and confers to the resulting material peculiar properties. The introduction of a fluorinated tail into a molecule does not necesserily lead, in general, to an improvement in desired activity. Indeed, fluorinated substitution can often lead to loss of activity or selectivity. However, since fluorine-containing molecules are generally more expensive (due to the highly aggressive, potentially explosive and expensive starting materials) than non-fluorinated homologous ones, a clear advantage needs to be gained to justify the inclusion of fluorine in commercial applications (16).

The molecular design allows the careful choice of the type of moiety which will be the most suitable for the amplification of the desirable qualities of the perfluorinated tail. The introduction of such tails allows for consideration of a more systematic approach in the liquid crystal field based on the molecular design of amphiphilic structure.

Apart from of the academic interest, why introduce fluorine atoms in liquid crystal materials?

The introduction and the sensible choice of the fluorine atom position within liquid crystal sytems allow us to reach nematic and smectic materials which present a considerable technological interest (17) for display (18) and other (19) applications.

- The efficiency of the nematogenic devices is often exacerbated from the introduction of fluorine onto the rigid core (20), so-called fluoro-substituents. In fact, the properties required are those of materials used in the electronic industry : optical and chemical stability, wide mesomorphic temperature range, low melting point, low viscosity and low conductivity. The low conductivities and viscosities of fluorine-based compounds explain the great interest for their use for the preparation of nematogen designs. Furthermore, the judicious choice of position of fluorosubstituents permits exhibition of appropriate dielectric anisotropies for commercial applications (18,21,22).

• On the other hand, the ferroelectric smectogens can be reached also by fluorosubstituents or by the introduction of a highly fluorinated functionality. In fact, the development of ferroelectric liquid crystal materials requires, among so many other (23) low viscosities, wide mesophases and low melting points. The fluorine approach is an interesting hypothesis for answering these requirements .

The use of fluorine atoms in liquid crystal materials can prove useful in future prospect as an interesting alternative to avoid faults or instabilities recorded in hydrocarbon series notably in the anchoring phenomena. Some of the most important qualities for a potential use as liquid crystal materials is the wide temperature range of the mesomorphism, its enantiotropy and reproductibility during the phase transition phenomenon, and its chemical stability. Due to their intrinsic physical properties, the perfluorinated moieties are chemically stable and in suitable cases, can improve the stability of the resulting mesomorphic phases.

The introduction of fluorine can be carried out on different positions (24) as a function of the required properties. The replacement of hydrogen atoms with fluorine atoms in the so-called fluoro-substituent has been described in several papers (25). Little attention has been paid to the influence of perfluorinated tails on liquid crystal behavior (26).

Lyotropic Liquid Crystals

The mesomorphic phases exhibited by fluorinated surfactants have been the subject of numerous studies, and many structures are now known within fluoroalkyl series exhibiting a lyotropic (27,28) or thermotropic (29) behavior. The phase behavior of fluoroalkyl and hydrocarbon homologs are similar. The structures investigated allow confirmation that fluorinated derivatives could exhibit cubic phases (30-32), potentially interesting for several applications and whose existence was not clearly established from previous works (33-35). The comparison hydrocarbon/fluorocarbon structures has been carried out on compound 1 as Y-shaped nonionic oligooxyethylene surfactants (32).

$$\mathbf{1} \quad F(CF_2)_nCH_2CH_2 N \overset{\displaystyle \diagup (C_2H_4O)_p\ CH_3}{\diagdown (C_2H_4O)_p\ CH_3}$$
$$n=4,6,8,10;\ p=2,3$$

In contrast to other Y-shaped hydrocarbon oligooxyethylene surfactants which form only cubic I_1 and H_1 phase, the phase behavior of surfactant 1 / water systems shows a remarkable change in the liquid crystal phase polymorphism (from H_1 to L_α) with n and p. Compared to the hydrocarbon homologs, the results confirm the influence of the greater rigidity and higher hydrophobicity of the fluoroalkyl tail, its smaller conformational freedom and its packing effects on the phase behavior.

Thermotropic Liquid Crystals

Discotic Liquids Crystals

This important class of liquid crystals has received less attention within highly fluorinated series (36). Mono substituted triphenylene, **2**, with a partially fluorinated tail, exhibits a decrease of the clearing temperature compared to symmetrical fluorinated hexasubstituted triphenylene, **3**, where decomposition appears before the clearing point. The difference can again be accounted for by the higher hydrophobicity and by the larger cross-sectional area of the fluorinated tail.

$Z = C_5H_{11}$
$Y = (CH_2)_{11}(CF_2)_8F, (CH_2)_4(CF_2)_4F$
$Y' = (CH_2)_2(CF_2)_nF; n = 6,8,10$

Calamitic Liquid Crystals

Multiblock Molecules

The mesomorphic behavior of a succession of linear aliphatic chains with different compatibility, i.e., fluorocarbon/hydrocarbon tails, could appear surprizing considering the classical idea that a calamitic liquid crystal must contain a succession of aromatic or non-aromatic rings (37). It is important to recall that the presence of incompatible moieties (lipophilic and fluorocarbon effects for example) is enough to lead to microsegregation favorable to induce the LC phase. As a consequence, the introduction of elements with sufficient incompatibility within the same molecule or in mixture leads to induction of a liquid crystal behavior. The expected behavior is the microsegregation which leads to the mesomorphic anisotropic behavior. The limit is the decrease in solubility and eventually the phase separation in the case of mixture. This fact was highlighted recently by a succession of blocks of fluoromethylene units and methylene units, so-called di (38) or tri-blocks (39-41), and more recently functionalized di-blocks (42) such as in **4**.

$$4 \quad F(CF_2)_n - (CH_2)_p \, Br$$

(n/p) : (8/2,4,6,10); (10/10); (12/10)

This concept allows us to achieve a molecular organized system such as Langmuir Blodgett films (43).

Twin Molecules

Twin molecules consist of symmetric bimesogenic parts linked by a spacer. Compounds incorporating fluoromethylene units in the spacer do not exhibit liquid crystalline properties as compared to those with a hydrocarbon spacer, 5 (44). On the other hand, the introduction of a highly fluorinated tail at the end position greatly changes the thermotropic behavior with the appearance of a LC phase, and also dramatically increases the melting point (up to 350°C).

$F(CF_2)_8(CH_2)_2 O$ ... $- (CH_2)_7 - $ ... $- O(CH_2)_2(CF_2)_8F$

5

The loss of liquid crystal character when fluoromethylene units are used in the spacer is not due to the molecular shape (twin) but rather to the relative rigidity of the twin center. In fact, other studies that have been carried out on twin molecules on the basis of an organosiloxane unit with a bimesogenic part including a chiral center, show a high spontaneous polarization, a high tilt angle, and exhibit an antiferroelectric phase (45).

Swallow Tail Molecules

Other molecular shapes have been studied, such as highly fluorinated swallow-tailed compounds 6 and 7. The presence of two chains with different compatibilities, one hydrocarbon and the other fluorinated, linked to the rigid core through an acetamide bond in 6, has a great influence on the liquid crystalline properties. When p>0, the hydrogen bonds, which do not favor, in this case, the appearance of liquid crystal character are absent. An increase in the length of the hydrocarbon moiety affects the thermal stability of the mesophase which becomes the monotropic from p >3 (46).

6 $\quad F(CF_2)_nC_2H_4 \atop H(CH_2)_p$ NC ... OCH_3

n=4,6,8; p=1,2,3,4,8,12

From compounds 7, only the substance with one methylene unit (p=1) between the carboxylic group in the swallow-tail, has a short nematic phase above the smectic phase. The fluorinated tail leads to increase in the clearing points and to stabilization of the smectic phases, compared to the non-fluorinated swallow-tailed compounds. This high incompatibility of hydrocarbon (p) and fluorocarbon (n) segments leads to dimer pairs with the dipoles aligned in one direction but without inducing ferroelectric behavior of the bulk material (47-49).

$$H(CH_2)_8\,O-\langle\!\!\!\bigcirc\!\!\!\rangle-C\!\!\!\overset{O}{\underset{O}{}}-\langle\!\!\!\bigcirc\!\!\!\rangle-C\!\!\!\overset{O}{\underset{O}{}}-\langle\!\!\!\bigcirc\!\!\!\rangle-CH=C\!\!\!\overset{CO_2(CH_2)_p(CF_2)_nF}{\underset{CO_2(CH_2)_p(CF_2)_nF}{}}$$

7

Branched Molecules

Within hydrocarbon series, the nematic phase is often obtained by the introduction of a lateral substituent into a calamitic structure. In the case of fluorinated series, **8**, the introduction of a lateral substituent (n-alkanoyloxy-methylene) has no effect on the nucleation of the nematic phase. The high incompatibility between fluorinated part and hydrocarbon moiety leads to the microsegregation and controls the order of the layer. However, the presence of lateral substituents allows decrease in the S_C-S_A transition and stabilizes, at the same time, the S_A mesophase (50). This fact again highlights the high smectogenics behavior of the fluoroalkyl chain which is the detriment of the other phase.

8 $F(CF_2)_7(CH_2)_8\,O-\langle\!\!\!\bigcirc\!\!\!\rangle-C\!\!\!\overset{O}{\underset{O}{}}-\langle\!\!\!\bigcirc\!\!\!\rangle\!\!\!\overset{\overset{O}{\|}}{\underset{\underset{O}{\|}}{}}\!\!\!\begin{matrix}C-(CH_2)_{n-1}H\\[6pt]C-\langle\!\!\!\bigcirc\!\!\!\rangle-O(CH_2)_8(CF_2)_7F\end{matrix}$

n=2-13

Within these branched compounds (51), the immiscibility of fluorinated/hydrocarbon tails allows the creation of polymers which exhibit smectic C and A mesophases instead of nematic behavior obtained within hydrocarbon series (52,53).

Polycyclic Molecules

From the description mainly used for classical rod-like liquid crystals, we present in this part a class of compounds which do not exhibit side chain or side function either in the rigid core or in the aliphatic chain and also possess a succession of rings. Numerous studies have been carried out on this common molecular shape (54-63). We will give two examples allowing the direct comparison of hydrocarbon / fluorocarbon homologues derived from a biphenyl core. This core was used because of the great interest in achieving room temperature liquid crystals (64).

- What is the impact of the substitution of a hydrocarbon tail for a fluorinated one? The comparison of the thermal behavior of two series which differ only by the nature of the aliphatic chain, fluorinated **9,10** or hydrocarbon **11,12** (65-67), shows liquid crystalline behavior only for **9,10**. The hydrocarbon monocatenar structure does not have any liquid crystal character regardless of the nature of the spacer (68,69). In the past, many studies (70-86) have been devoted to the understanding of the influence on the mesomorphic properties of the replacement of hydrogen atoms by fluorine atoms in an aliphatic chain. The introduction of the perfluorinated chain, notably partially fluorinated, can be considered as an interesting alternative to obtain smectic phases based on the weak miscibility and stiffness of the fluorinated tail.

$$X(CX_2)_n - CH_2CH_2 - spacer \langle \rangle \langle \rangle$$

$$spacer = -SC(O)-; -NH=CH-$$

9,10	X = F; n = 4,6,8	Cr $\longrightarrow$ LC $\longrightarrow$ I
11,12	X = H; n = 4,6,8	Cr $\longrightarrow$ I

- The evaluation of the mesomorphic behavior of 4-biphenyl derivatives, incorporating different spacers bonding the mesogenic core to the fluorinated tail, allows us to point out the great influence of the molecular design of this spacer for the appearance and the stability of the mesophase. **9a, 10a, 15, 19** exhibit an enantiotropic behavior with an additional smectic E phase in some cases; **16** leads to a very short range of smectic phase and **13, 14, 17, 18** suppress liquid crystal character. The data collected from the biphenyl derivatives lead first to confirmation of the enhancement of the smectic properties owing to the perfluorinated tail, and also demonstrated the impact of the spacer on these properties that can increase the smectogenic behavior of the fluorinated tail. It is clear from the above results (87,88) that the presence of a 2-perfluoroalkylethyl chain helps to generate the smectogenic character. This behavior can be revealed or enhanced from molecular design of the spacer.

$$F(CF_2)_6 - CH_2CH_2 - spacer \langle \rangle \langle \rangle$$

	spacer	transition temperatures on heating
9a	-N=CH-	Cr 57 SA 105 I
10a	-SC(O)-	Cr 51.6 SA 152 I
13	-NHC(O)-	Cr 175 I
14	-N(CH3)C(O)-	Cr 87 SA 90 I
15	-C(O)O-	Cr 80.5 SA 113.2 I
16	-OC(O)-	Cr 70.5 SA 72 I
17	-SCH2O-	Cr 71.6 I
18	-SCH2-	Cr 59.7 I
19	-O-	Cr 81.4 E 96.1 SA 105.5 I

The use of fluorinated liquid crystals and the considerable research efforts are due to the discovery of ferroelectricity in the chiral smectic C (S_{C*}) phase by R. Meyer in 1975, and the proposal of electro-optical devices using ferroelectric liquid crystals by Clark and Lagerwall in 1980. It was initially found that these materials require at least a smectic phase, two aromatic rings in the core and two terminal chains at the end, one of which contains a chiral group. Due to amphiphilic molecular architecture, we will see that ferroelectric properties can be obtained without a chiral center.

The previous studies highlighted the smectogenic character of the perfluorinated tail. Thus, association of a fluoroalkyl chain with a chiral unit via a mesogen as a connector, appears to offer more attractive alternative. With this purpose, several series of compounds containing a semifluorinated tail, two or three aromatic rings, and an aliphatic chiral tail on the other extremity, have been prepared (89-95). It is worth noting that due to their technological interest, many other works are often described through patents (see for example 96-99)

It is also worth noting that within the same series, the length of the linear tail could induce a modification of the phase sequence. For example, structures **20**. exhibit the ferroelectric S_{C*} phase and the paraelectric S_{A*} phase. Nevertheless, the phase-sequence (see Figure 1), in such structures, depends on the number of fluoromethylene and methylene units (100).

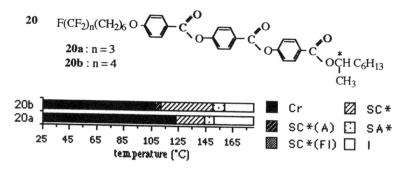

Figure 1. Phase transition temperatures for compounds 20 (100)

The helical pitch was found to be nearly temperature-independent. The introduction in the fluoroalkyl tail of an additional fluoromethylene unit (n=4 instead of 3) leads to a rather wide antiferroelectric phase sequence. However, the tilt angle and the spontaneous polarization of these compounds are of the same order.

The study of polyphilic (hydrocarbon, fluorinated, aromatic moieties) compounds leads to an understanding of the segregation of these chemical fragments for the appearance of mesogenicity (101-103). The molecule with a central hydrocarbon moiety exhibits only a S_A phase with interdigitation of the aliphatic and aromatic parts. The high immiscibility between each moiety leads to segregation and can be compared as biblock compounds: fluorocarbon tail on one hand and hydrocarbon moieties (both aliphatic and aromatic) on the other hand. Achiral polyphilic molecules have been synthesized (104), including biaryl derivatives substituted either by two perfluorinated tails at the end positions **21** (105-107) or by one perfluorinated tail and a trifluoromethyl group (108). The overall compounds exhibit unconventional ferroelectric liquid crystal properties.

It is worth noting that the chiral center is not a necessary feature for having ferroelectric properties (109) because the molecular shape can also cause this effect, as in the case of banana shaped molecules.

Monocyclics Liquid Crystals

The incompatibility of fluorocarbons with both saturated and aromatic hydrocarbons leads to a microsegregation favorable for expressing LC behavior even in monocyclic derivatives, which cannot be expected within the hydrocarbon series. In fact in recent years, it has been observed that the introduction of fluorinated tail could decrease the number of aromatic rings required for mesogenicity, and leads to monophenyl derivatives, with no hydrogen bonding (110), that exhibit unexpected LC properties. This supports the fundamental interest of the fluorinated tail for easy access to a less expensive raw precursor.

The development of liquid crystals having a single benzene ring as mesogenic core is of importance and current interest. Single benzene ring compounds would be expected to have low molecular anisotropy and would not readily display liquid crystalline properties. The presence of two non-miscible parts (one fluorinated, one aromatic) should induce mesomorphism. Many molecular design efforts have been made in order to emphasize the molecular parameter which can govern the enantiotropic behavior. Among the compounds described in the literature (111-114), mainly monotropic, compounds **22, 23** (see Figure2) exhibit enantiotropic LC phase (115,116).

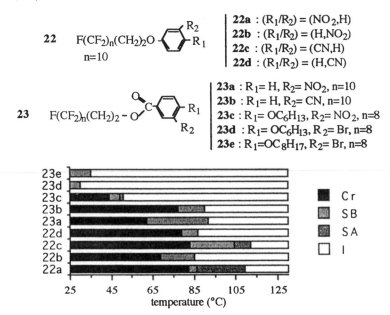

Figure 2. Phase transition temperatures for compounds 22 and 23 (115,116)

Dendrimeric Liquid Crystals

In the dendrimers **24**, the introduction of fluoroalkyl tails, **24b-d**, stabilize the enantiotropic mesophase while the hydrocarbon series, **24a**, exhibits the monotropic phase.

24

24a : $R_1=R_2=C_{10}H_{21}$
24b : $R_1=R_2=C_4F_9(CH_2)_6$
24c : $R_1=R_2=C_6F_{13}(CH_2)_4$
24d : $R_1=C_6F_{13}(CH_2)_4$; $R_2=C_{10}H_{21}$

The introduction of semifluorinated tails to tetrahedral pentaerythritol tetrabenzoate dramatically stabilizes the mesomorphic state above room temperature (117) (and where the crystallisation point is not observed) as compared to the first generation of carbosilane dendrimer derivatives where the smectic mesomophase is observed between -15 and -30°C (118).

The appearance of liquid crystal character is based on the fluorophobic effect which can produce the self assembled taper-shaped amphiphiles (119).

Conclusion

The description of molecular shapes in many recent studies has shown the great influence of a highly fluorinated tail on the lyotropic or thermotropic mesomorphism, as compared to hydrocarbon homologs. The multitude of liquid crystalline shapes : surfactant, discotic, calamitic (multiblock, twin, swallow, branched, polycyclic, monocyclic) and dendrimer, allows us to consider the appearance of the mesomorphism not exclusively due to the molecular shape but rather to the presence of incompatibility moieties (rigid - flexible, hydrophylic / lipophilic, hydrocarbon / fluorocarbon etc...). From its hydrophobic and lipophilic nature, but also from its intrinsic physical properties, the perfluorinated tail can be considered as one of these immiscible components which will allow achievement of LC behavior. This concept of incompatibility allows us to understand the liquid crystal properties of multi-blocks or of monocyclic derivatives and gives future prospects to further investigations. However, it is worth noting that the liquid crystalline behavior, more often smectogen in our case, is controlled, amplified or stabilized by the position of the fluoroalkyl tail

191

within the molecular arrangement but also by the nature of the spacer which bound this moiety to the main core. Lastly, the presence of the perfluorinated tail allows the synergetic use of their surface and anisotropic properties often required within liquid crystal display.

References

1. Tschierske, C. *J. Mater. Chem.* **1998**, *8*, 1485-1508.
2. Alig, I.; Braun, D.; Jarek, M.; Hellmann, G. P. *Macromol. Symp.* **1995**, *90*, 173.
3. Bracon, F.; Guittard, F.; Taffin de Givenchy, E.; Cambon, A. *J. Polym. Sci. Part A Polym. Chem.* **1999**, *37*, 4487-4496.
4. Percec, V.; Lee, V. *J.M.S.-Pure Appl. Chem.* **1992**, *A29*, 723-740.
5. Thornber, C. W. *Chem. Soc. Rev.* **1979**, *8*, 563-575.
6. Williams, D. E.; Houpt, D. J. *Acta Cryst.* **1986**, *B42*, 286-292.
7. Lenk, T. J.; Hallmark, V. M.; Hoffmann, C. L.; Rabolt, J. F.; Castner, D. G.; Erdelen, C.; Ringsdorf, H. *Langmuir* **1994**, *10*, 4610-6417.
8. Kim, H. I.; Koini, T.; Lee, T. R.; Perry, S. S. *Langmuir* **1997**, *13*, 7192-7196.
9. Schönherr, H.; Vancso, G. J. *Polym. Prep.* **1998**, *39*, 904-905.
10. Mach, P.; Huang, C. C.; Nguyen, H.T. *Langmuir* **1997**, *13*, 6357-6359.
11. Mach, P.; Huang, C. C.; Stoebe, T.; Wedell, E. D.; Nguyen, T.; Jeu, W. H. de; Guittard, F.; Naciri, J.; Shashidhar, R.; Clark, N.; Jiang, I. M.; Kao, F. J.; Liu, H.; Nohira, H. *Langmuir* **1998**, *14*, 4330-4341.
12. Thüenemann, A. F.; Lochhaas, K. H. *Langmuir* **1999**, *15*, 6724-6727.
13. Perutz, S.; Wang, J.; Kramer, E. J.; Ober, C. K.; Ellis, K. *Macromolecules* **1998**, *31*, 4272-4276.
14. Cambon, A.; Edwards, C. M.; Franke, R. P.; Lowe, K. C.; Reuter, P.; Rohlke, W.; Trabelsi, H.; Gambaretto, G. P.; Napoli, M.; Conte, L. WO Patent, 97/49715, 1997.
15. Trabelsi, H.; Geribaldi, S.; Rohlke, W.; Reuter, P.; Lowe, K. C.; Franke, R. P. FR Patent Appl.99/09707, 1999.
16. *Organofluorine Chemistry Principles and Commercial Applications*, Banks, R.E.; Smart, B.E.; Tatlow, J.C., Eds.; Plenum Press: New York NY, 1994.
17. Geelhaar, T. *Liq. Cryst.* **1998**, *24*, 91-98.
18. Hirschmann, H.; Reiffenrath, V. In *Handbook of Liquid Crystals*; Demus, D.; Goodby, J. W.; Gray, G.W.; Spiess, H. -W.; Vill, V., Eds.; Low Molecular Weight Liquid Crystals I, Wiley-VCH: Weinheim, 1998, 2A, pp 199.
19. Beresnev, L.; Haase, W. *Optical Mater.* **1998**, *9*, 201-211.
20. Bezborodov, V. S.; Petrov, V. F. *Liq. Cryst.* **1999**, *26*, 271-280.
21. Seomun, S. -S.; Gouda, T.; Takanishi, Y.; Ishikawa, K.; Takezoe, H. *Liq. Cryst.* **1999**, *26*, 151-161.
22. Petrov, V. F. *Liq. Cryst.* **1995**, *19*, 729-741.
23. S.T. Lagerwall, In *Handbook of Liquid Crystals*; Demus, D.; Goodby, J. W.; Gray, G.W.; Spiess, H. -W.; Vill, V., Eds.; Low Molecular Weight Liquid Crystals II, Wiley-VCH: Weinheim, 1998, 2B:, pp 654.
24. Guittard, F.; Taffin de Givenchy, E.; Geribaldi, S.; Cambon, A. *J. Fluorine Chem.* **1999**, *100*, 85-96.
25. Hird, M.; Toyne, K. J. *Mol. Cryst. Liq. Cryst.* **1998**, *323*, 1-67 and references therein.

192

26. Inoi,T. In *Organofluorine Chemistry: Principles and Commercial Applications*; Banks, R.E.; Smart, B.E.; Tatlow, J.C., Eds.; Plenum Press, New York NY, 1994; pp263-286.

27. *Fluorinated Surfactants. Synthesis, Properties, Applications*; E. Kissa, Ed.; Marcel Dekker: New York NY, 1994.

28. Monduzzi, M. *Current Opinion Colloid Interface Sci.* 1998, *3*, 467-477 and references therein.

29. Zur, C.; Miller, A.O.; Miethchen, R. *Liq. Cryst.* 1998, *24*, 695-699.

30. Ropers, M. -H.; Stebe, M. -J.; Schmitt, V. *J. Phys. Chem. B* 1999, *103*, 3468-3475.

31. Guittard, F.; Leaver, M. S.; Holmes, M. C. unpublished results.

32. Kratzat, K.; Guittard, F.; Taffin de Givenchy, E.; Cambon, A. *Langmuir* 1996, *12*, 6346-6350.

33. Tiddy, G. J. T.; Wheeler, B.A. *J. Colloid interface Sci.* 1974, *47*, 59-64.

34. Everiss, E.; Tiddy, G. J. T.; Wheeler, B.A. *J. Chem. Soc. Faraday Trans.* 1976, *72*, 1747-1752 and references therein.

35. Guo, W.; Brown, T. A.; Fung, B. M. *J. Phys. Chem.* 1991, *95*, 1829-1835.

36. Dahn, U.; Erdelen, C.; Ringsdorf, H.; Festag, R.; Wendorff, J. H.; Heiney, P. A.; Maliszewskyj, N. C. *Liq. Cryst.* 1995, *19*, 759-764 and reference therein.

37. Kelker, H.; Hatz, R. *Handbook of Liquid Crystals*, Verlag Chemie, 1979.

38. Vilalta, P.M.; Hammond, G.S.; Weiss, R.G. *Langmuir* 1993, *9*, 1910-1921.

39. Viney, C.; Twieg, R. J.; Russell, T. P. *Mol. Cryst. Liq. Cryst.* 1990, *182*, 291-294.

40. Viney, C.; Twieg, R. J.; Gordon, B. R.; Rabolt, J. F. *Mol. Cryst. Liq. Cryst.* 1991, *198*, 285-289.

41. Twieg, R.; Rabolt, J. F. *J. Polym. Sci. Polym. Lett. Ed.* 1983, *21*, 901-902.

42. Wang, J.; Ober, C. K. Liq. Cryst. 1999, *26*, 637-648.

43. El Abed, A.; Faure, M. C.; Hamdani, M.; Guittard, F.; Billard, J.; Peretti, P. *Mol. Cryst. Liq. Cryst.* 1999, *329*, 283-292.

44. Weissflog, W.; Richter, S.; Dietzmann, E.; Risse, J.; Diele, S.; Schiller, P.; Pelzl G. *Cryst. Res. Technol.* 1997, 32, 271-7.

45. Robinson, W. K.; Lehmann, P.; Coles, H. *J. Mol. Cryst. liq Cryst.* 1999, in press.

46. Guittard, F.; Sixou, P.; Cambon, A. *Liq. Cryst.* 1995, *19*, 667-673.

47. Dietzmann, E.; Weissflog, W.; Markscheffel, S.; Jakli, A.; Lose, D.; Diele, S. *Ferroelectrics* 1996, *180*, 341-354.

48. Lose, D.; Diele, S.; Pelzl, G.; Dietzmann, E.; Weissflog, W. *Liq. Cryst.* 1998, 24, 707-717.

49. Pelzl, G.; Diele, S.; Lose, D.; Ostrovski, B.I.; Weissflog, W. *Cryst. Res. Technol.* 1997, *32*, 99-109.

50. Small, A. C.; Hunt, D. K.; Pugh, C. *Liq. Cryst.* 1999, *26*, 849-857.

51. Dietzmann,E.; Weissflog, W. *Mol. Cryst. Liq. Cryst.* 1997, *299*, 419-426.

52. Arehart, S. V.; Pugh, C. *J. Am. Chem. Soc.* 1997, *119*, 3027-3037.

53. Pugh, C.; Bae, J. -Y.; Dharia, J.; Ge, J. J.; Cheng, S. Z. D. *Macromolecules* 1998, *31*, 5188-5200.

54. Misaki, S.; Takamatsu, S.; Suefuji, M.; Mitote, T.; Matsumura, M. *Mol. Cryst. Liq. Cryst.* 1981, *66*, 123-132.

55. Epstein, K. A.; Keyes, M. P.; Radcliffe, M. D.; Snustad, D. C. U.S. Patent 5,417,883, 1995.

56. Iannacchione, G. S.; Garland, C. W.; Johnson, P. M.; Huang, C.C. *Liq. Cryst.* 1999, *26*, 51-55.

57. Chen, B. -Q.; Yang, Y. -G.; Wen, J. -X. *Liq. cryst.* 1998, *24*, 539-542.

58. Kromm, P.; Cotrait, M.; Nguyen, H. T. *Liq. Cryst.* 1996, *21*, 95-102.

59. Kromm, P.; Cotrait, M.; Rouillon, J. -C.; Barois, P.; Nguyen, H. T. Liq. Cryst. **1996**, *21*, 121-131.
60. Yang, Y. -G.; Chen, B.-Q.; Wen, J.-X. *Liq. Cryst.* **1999**, *26*, 893-896.
61. Yang, X.; Abe, K.; Kato, R.; Yano, S.; Kato, T.; Miyazawa, K.; Takeuchi, H. *Liq. Cryst.* **1998**, *25*, 639-641.
62. Drzewinski, W.; Dabrowski, R.; Czuprynski, K.; Przedmojski, J.; Neubert, M. *Ferroelectrics* **1998**, *212*, 281-292.
63. Thiele, T.; Prescher, D.; Ruhmann, R., Wolff, D. *J. Fluorine Chem.* **1997**, *85*, 155-161.
64. Gray, G. W.; Harrison, K. J.; Nash, J. A. *Electron. Lett.* **1973**, *9*, 130-131.
65. Guittard, F.; Sixou, P.; Cambon, A. *Mol. Cryst. Liq. Cryst.* **1997**, *308*, 83-92.
66. Diele, S.; Lose, D.; Kruth, H.; Pelzl, G.; Guittard, F.; Cambon, A. *Liq. Cryst.* **1996**, *21*, 603-608.
67. Taffin de Givenchy, E.; Guittard, F.; Bracon, F.; Cambon, A. *Liq. Cryst.* **1999**, *26*, 1371-1377.
68. Pavlyuchenko, A. I.; Smirnova, N. I.; Kovshev, E. I.; Titov, V. V.; Purvanetskas, G.V. *Zh. Org. Khim.* **1976**, *12*, 1054-1057
69. Bailey, A. L.; Bates, G. S. *Mol. Cryst. Liq. Cryst.* **1991**, *198*, 417-421.
70. Pavlyuchenko, A. I.; Fialkov, Yu. A.; Shelyazhenko, S. V.; Yagupolískii, L. I.; Smirnova, N. I.; Petrov, V. F.; Grebenkin, M. F. *Zh. Fiz. Khim.* **1991**, *65*, 2249-2252.
71. Chiang, Y. H.; Ames, A. E.; Gaudiana, R. A.; Adams, T. G. *Mol. Cryst. liq. Cryst.* **1991**, *208*, 85-98.
72. Bartmann, E.; Dorsch, D.; Finkenzeller, U. *Mol. Cryst. Liq. Cryst.* **1991**, *204*, 77-89.
73. Doi, T.; Sakurai, Y.; Tamatani, A.; Takenaka, S.; Kusabayashi, S.; Nishihata, Y.; Terauchi, H. *J. Mater. Chem.* **1991**, *1*, 169-173.
74. Doi, T.; Takenaka, S.; Kusabayashi, S.; Nishihata, Y.; Terauchi, H. *Mol. Cryst. liq. Cryst.* **1991**, *204*, 9-14.
75. Moklyachuk, L. I.; Kornilov, M. Yu.; Fialkov, Yu. A.; Kremlev, M. M.; Yagupolískii, L. M. *Zh. Org. Khim.* **1990**, *26*, 1533-1539.
76. Koden, M.; Nakagawa, K.; Ishii, Y.; Funada, F.; Matsuura, M.; Awane, K. *Mol. Cryst. Liq. Cryst. Lett.* **1989**, *6*, 185-190.
77. Fialkov, Yu. A.; Zalesskaya, I. M.; Yagupolískii, L. M. *Zh. Org. Khim.* **1983**, *19*, 2055-2062.
78. Titov, V. V.; Zverkova, T. I.; Kovshev, E. I.; Fialkov, Yu. N.; Shelazhenko, S. V.; Yagupolískii, L. M. *Mol. Cryst. Liq. Cryst.* **1978**, *47*, 1-5.
79. Bartmann, E. *Ber. Bunsenges. Phys. Chem.* **1993**, *97*, 1349-1355.
80. Janulis, E. P.; Novack, J. C.; Papapolymerou, G. A.; Tristani-Kendra, M.; Huffman, W.A. *Ferroelectrics* **1988**, *85*, 375-384.
81. Ivashchenko, A. V.; Kovshev, E. I.; Lazareva, V. T.; Prudnikova, E. K.; Titov, V. V.; Zverkova, T. I.; Barnik, M. I.; Yagupolískii, L. M. *Mol. Cryst. Liq. Cryst.* **1981**, *67*, 235-240.
82. Fialkov, Yu. A.; Moklyachuk, L. I.; Kremlev, M. M.; Yagupolískii, L. M. *Zh. Org. Khim.* **1980**, *16*, 1476-1479.
83. Fialkov, Y. A.; Shelyazhenko, S. V.; Yagupolískii, L. M. *Zh. Org. Khim.* **1983**, *19*, 1048-1053.
84. Shelyazhenko, S. V.; Dronkina, M. I.; Fialkov, Y. A.; Yagupolískii, L. M. *Zh. Org. Khim.* **1988**, *24*, 619-625;
85. Janulis, E. P.; Osten, D. W.; Radcliffe, M. D.; Novack, J. C.; Tristani-Kendra, M.; Epstein, K. A.; Keyes, M.; Johnson, G. C.; Savu, P. M.; Spawn, T. D. *Liquid Crystal Materials, Devices, and Applications* **1992**, *1665*, 146-153.

194

86. Pavluchenko, A. I.; Smirnova, N. I.; Petrov, V. F.; Fialkov, Yu. A.; Shelyazhenko, S. V.; Yagupolsky, L. M. *Mol. Cryst. Liq. Cryst.* **1991**, *209*, 225-235.
87. Taffin de Givenchy, E.; Guittard, F.; Bracon, F.; Cambon, A. *Liq. Cryst.* **1999**, *26*, 1163-1170.
88. Taffin de Givenchy, E.; Guittard, F.; Geribaldi, S.; Cambon, A. *Mol. Cryst. Liq. Cryst.* **1999**, *332*, 1-7.
89. Liu, H.; Nohira, H. *Ferroelectrics* **1998**, *207*, 541-553.
90. Liu, H.; Nohira, H. *Liq. Cryst.* **1997**, *22*, 217-222.
91. Liu, H.; Nohira, H. *Liq. Cryst.* **1996**, *20*, 581-586.
92. Liu, H.; Nohira, H. *Liq. Cryst.* **1998**, *24*, 719-726.
93. Liu, H.; Nohira, H. *Mol. Cryst. Liq. Cryst.* **1997**, *302*, 247-252.
94. W.J. Cumming, R.A. Gaudiana, *Liq. Cryst.* **1996**, *20*, 283-286.
95. Taffin de Givenchy, E.; Guittard, F.; Geribaldi, S.; Cambon, A. *Mol. Cryst. Liq. Cryst.* **1999**, *332*, 9-16.
96. Gibbons, W.; Shannon, P. J.; Sun, S. -T. US Patent 5,929,201, 1999.
97. Manero, J.; Schmidt, W.; Hornung, B. DE Patent WO 9855458, 1998.
98. Hasegawa, M.; Keyes, M. P.; Radcliffe, M. D.; Savu, P. M.; Snustad, D. C.; Spawn, T. D. US Patent WO 9933814, 1999.
99. Tamura, N.; Takeuchi, H.; Matsui, S.; Mizazawa, K.; Hisatsune, Y.; Takeshita, F.; Nakagawa, E. JPN Patent WO 9912879, 1999.
100. Sarmento, S.; Simeao Carvalho, P.; Glogarova, M.; Chaves, M. R.; Nguyen, H. T.; Ribeiro, M. J. *Liq. Cryst.* **1998**, *25*, 375-385.
101. Pensec, S.; Tournilhac, F. -G.; Bassoul, P.; Durliat, C. *J. Phys. Chem. B* **1998**, *102*, 52-60.
102. Pensec, S.; Tournilhac, F. -G.; Bassoul, P. *J. Phys. II France* **1996**, *6*, 1597-1605.
103. Pensec, S.; Tournilhac, F. -G. *Chem. Commun.* **1997**, 441-2.
104. Tournilhac, F.; Blinov, L. M.; Simon, J.; Yablonsky, S. V. *Nature* **1992**, *359*, 621-623.
105. Tournilhac, F.; Bosio, L.; Nicoud, J.-F.; Simon, J. *Chem. Phys. Lett.* **1988**, *145*, 452-454.
106. Tournilhac, F.; Simon, J. *Ferroelectrics* **1991**, *114*, 283-287.
107. Shi, Y.; Tournilhac, F.G.; Kumar, S. *Phys. Rev. E* **1997**, *55*, 4382-4385.
108. Tournilhac, F.; Blinov, L. M.; Simon, J.; Yablonsky, S. V. *Nature* **1992**, *359*, 621-623.
109. Heppke, G.; Moro, D. *Science* **1998**, *279*, 1872-1873.
110. Bernhardt, H.; Weissflog, W.; Kresse, H. *Chem. Lett.* **1997**, 151-152.
111. Takenaka, S. *J. Chem. Soc. Chem. Commun.* **1992**, 1748-1749.
112. Okamoto, H.; Murai, H.; Takenaka, S. *Bull. Chem. Soc. Jpn.* **1997**, *70*, 3163-3166.
113. Johansson, G.; Percec, V.; Ungar, G.; Smith, K. *Chem. Mater.* **1997**, *9*, 164-175.
114. Guittard F. et al. (to be published).
115. Okamoto, H.; Yamada, N.; Takenaka, S. *J. Fluorine Chem.* **1998**, *91*, 125-132.
116. Duan, M.; Okamoto, H.; Petrov, V. F.; Takenaka, S. *Bull. Chem. Soc. Jpn.* **1999**, *72*, 1637-1642.
117. Pegenau, A.; Cheng, X. H.; Tshierske, C.; Göring, P.; Diele, S. *New J. Chem.* **1999**, *23*, 465-467.
118. Lorenz, K.; Frey, H.; Stühn, B.; Mülhaupt, R. *Macromolecules* **1997**, *30*, 6860-6868.
119. Percec, V.; Johansson, G.; Ungar, G.; Zhou, J. *J. Am. Chem. Soc.* **1996**, *118*, 9855-9866.

Chapter 14

First Nematic Calamitic Liquid Crystals with Negative Birefringence

Volker Reiffenrath and Matthias Bremer

Liquid Crystals Division, Merck KGaA, D-64271 Darmstadt, Germany

The combination of aliphatic core structures with lateral highly polarizable groups leads to nematic materials with very low or even negative birefringence.

Liquid crystalline behavior can be observed in certain, anisotropically shaped organic compounds.[1-3] A liquid crystal can flow like a liquid, but may posses other properties,[4,5] such as birefringence, which is normally characteristic of solid crystals, a prime example being rhombohedral calcite, $CaCO_3$. Without the birefringence of the liquid crystal, there would be no optical response to an applied voltage in the twisted nematic (TN)[6] mode of a liquid crystal display (LCD).

A uniaxial liquid crystal has two principal refractive indices. The ordinary ray n_o is defined as the light wave with the electric field perpendicular to the optical axis, whereas the extraordinary index n_e is observed for linearly polarized light with the electric field parallel to the optical axis. In nematic liquid crystals the optical axis is given by the director which ideally coincides with the long molecular axis. In nematic liquid crystals known so far the birefringence ($\Delta n = n_\| - n_\perp = n_e - n_o$) is always positive with a range of about +0.02 to +0.40.[1] We now report the first examples of nematic, calamitic liquid crystals with extremely small and even negative birefringence.[7,8]

The birefringence is related to the anisotropy of the molecular polarizability $\Delta \alpha = \alpha_\| - \alpha_\perp$ through the Vuks equation[9] (1a and 1b) where S is the Saupe orientational order parameter, ε_0 the static dielectric constant and N the number of molecules per unit volume.

$$\frac{n_e^2 - 1}{n^2 + 2} = \frac{N}{3\varepsilon_0} \left(\alpha + \frac{2\Delta \alpha S}{3} \right)$$

(1a)

$$\frac{n_o^2 - 1}{n^2 + 2} = \frac{N}{3\varepsilon_0}\left(\alpha - \frac{\Delta\alpha S}{3}\right)$$ (1b)

Since α can be calculated for isolated molecules with a quantum chemical method one can speculate what kind of molecular structure would lead to small or negative values of Δn. In order to derive *anisotropic* quantities from the calculation, the molecules are oriented so that the smallest molecular moment of inertia coincides with the x axis and the larger moments of inertia with the y and z axes, respectively.

The molecules are considered to be cylindrically symmetric, i.e. the long molecular axis ideally coincides with the x-axis and the perpendicular components of the polarizability tensor are averaged.[10,11]

Given the fact that nematic liquid crystals are rod-like molecules with a length to breadth ratio of typically 3 or larger, implying larger polarizability along the long axis, it is not surprising that Δn is greater than zero. One obvious way of lowering Δn is the introduction of small lateral, highly polarizable groups to a cylindrical core of low refractive power.

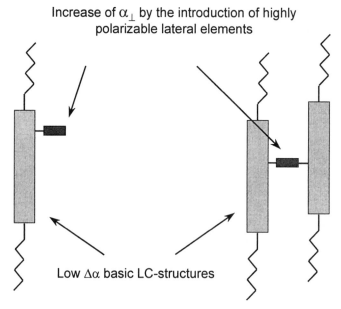

Increase of $\alpha_\perp$ by the introduction of highly polarizable lateral elements

Low $\Delta\alpha$ basic LC-structures

Figure 1. Strategy to achieve negative birefringence in calamitic materials

This led to the synthesis of bicyclohexanes with axial acetylenic substituents starting from axial cyanobicyclohexanes[12] or bicyclohexanones as outlined in Scheme 1; mesophases and extrapolated[13] ("virtual") clearing points and birefringence are given in Table I together with calculated (AM1)[14] optical anisotropies.

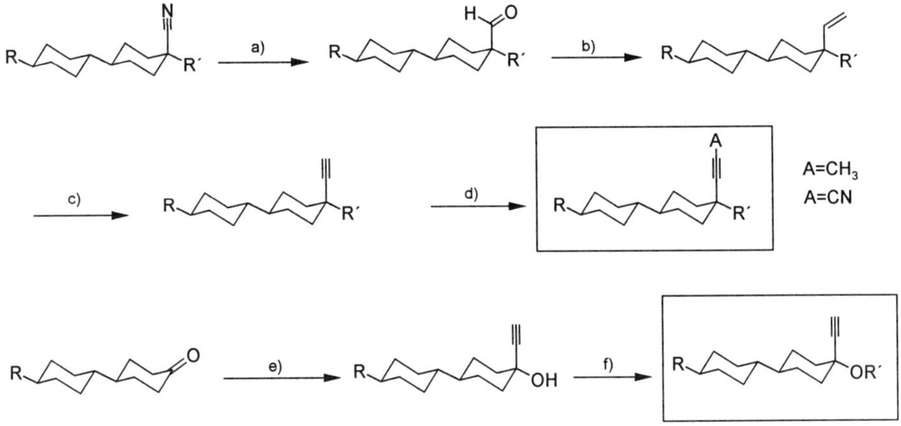

Scheme 1. Synthesis of axial alkynylbicyclohexanes. a) 1. DIBAL, Toluene, 25-50 °C; 2. H_3O^+ (90%). b) Methyltriphenylphosphoniumbromide, KOtBu, THF, 0 °C (90%). c) 1. Br_2, Et_2O, -10 to 0 °C; 2. Et_3N, RT (65%); 3. KOtBu, tBuOH, 60 °C (30%). d) A=CH_3: 1. BuLi, THF, -70 °C; 2. MeI, -70 °C to RT (70%). A=CN: 1. BuLi, THF, -70 °C; 2. TsCN, -70 °C to RT (70%). e) 1. Trimethylsilylacetylene, $Bu_4NF\cdot3H_2O$, THF, -30 °C; 2. KF, MeOH, RT (47%). f) 1. BuLi, THF, -30 °C to -5 °C; 2. R'I, DMSO, -5 °C to RT (73%).

Table I. Properties of axially substituted alkynylcyclohexanes

Entry	Structure	Phases	Virtual Clp.	Virtual Δn	Δn (calc)
1	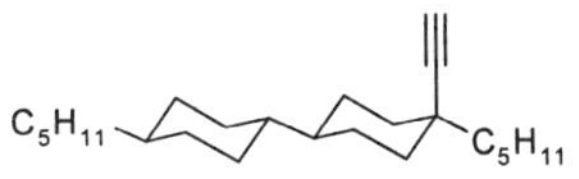	C 16 N 35 I	21	0.038	0.026
2	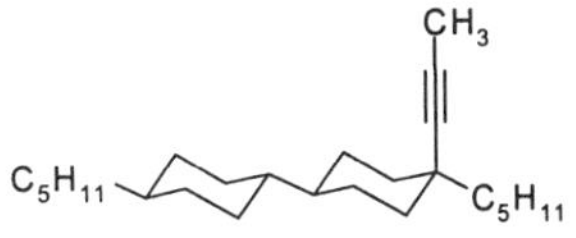	C 35 N 35 I	5	0.034	0.020
3		C 45 I	-122	0.033	0.028
4		C 55 I	-49	0.026	0.004

Note that the calculated values are all too small, especially so for the cyanoacetylene **4**. One possible explanation lies in the order parameter S. For the calculation a constant value of 0.7 is assumed, perhaps an oversimplification for compounds **1-4**, where S may vary strongly, depending on the lateral perturbation. Furthermore, the standard orientation used for the calculation might not be valid for molecules like **1-4** which significantly deviate from a cylindrical shape.

Figure 2 shows the standard orientation for the saturated bicyclohexane **5**, as a reference, and its alkynyl derivatives **6** and **7**. The calculated birefringence for the reference compound **5** (0.044) is in excellent agreement with the extrapolated experimental value (0.043). However, the values of Table I indicate that, for **1-4** and **6** and **7**, using the moments of inertia to define anisotropies is too crude an approximation to the behavior found in the condensed phase, where the long molecular axes must be tilted much stronger with respect to the macroscopic director.

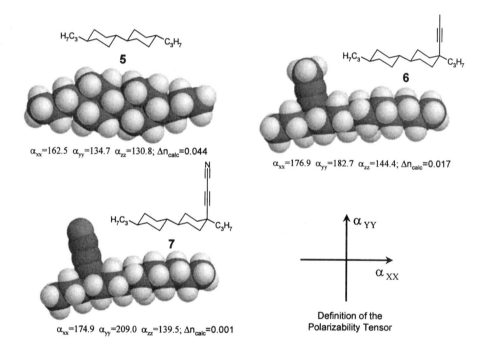

α_{xx}=162.5 α_{yy}=134.7 α_{zz}=130.8; Δn_{calc}=0.044

α_{xx}=176.9 α_{yy}=182.7 α_{zz}=144.4; Δn_{calc}=0.017

α_{xx}=174.9 α_{yy}=209.0 α_{zz}=139.5; Δn_{calc}=0.001

Definition of the
Polarizability Tensor

Figure 2. Orientation of LC molecules with the moments of inertia for the definition of anisotropies. The z-axis is not shown explicitly.

A relatively simple way to render the overall molecular shape more symmetrical and thus avoid a tilt of the molecules in the nematic phase is shown in Scheme 2 and Table II. One obtains liquid crystals **8** and **9** with high virtual clearing temperatures, although smectic phases are still present.

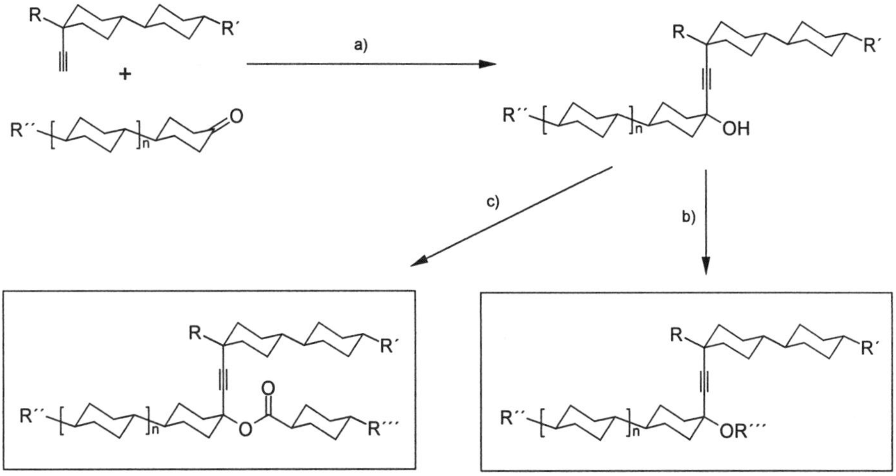

Scheme 2. Synthesis of bis(bicyclohexyl)acetylenes. a) 1. BuLi, THF, -70 °C; 2. Add 4-alkyl[bi]cyclohexanone, THF, -70 °C. b) 1. BuLi, THF, -30 °C to -5 °C; 2. R'''I, DMSO, -5 °C to RT (25%). c) 4-alkylcyclohexylcarbonylchloride, CH_2Cl_2, pyridine, 0 °C to RT (40%).

Table II. Properties of unsymmetrical bis(bicyclohexyl)acetylenes

Entry	Structure	Phases	Virtual Clp.	Virtual Δn
8	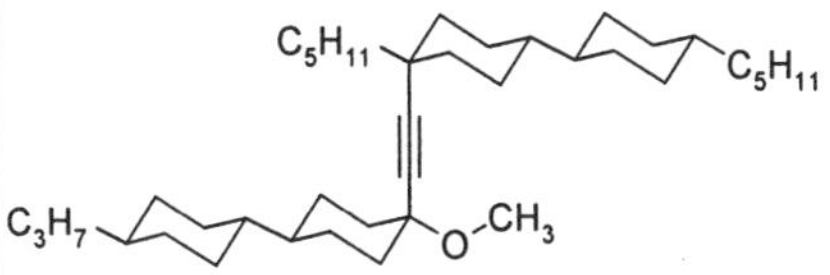	C 104 S_B (103) N 130 I	136	0.029
9	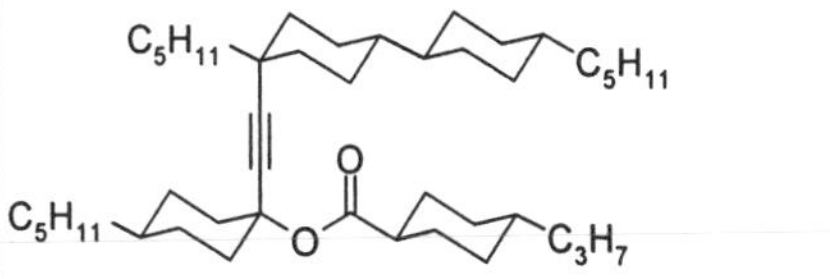	C 66 S_B 171 I	152	0.036

The acetylenes of types **1** and **3** can be coupled oxidatively[15] to yield symmetrical, and unsymmetrical dimers. These are readily separated by chromatography and the alkoxy-/alkylderivatives **10-16** of Table III now show pure (albeit monotropic) nematic phases. In these materials the extrapolated birefringence is around zero or slightly below.

Table III. Properties of symmetrical bis(bicyclohexyl)butadiynes

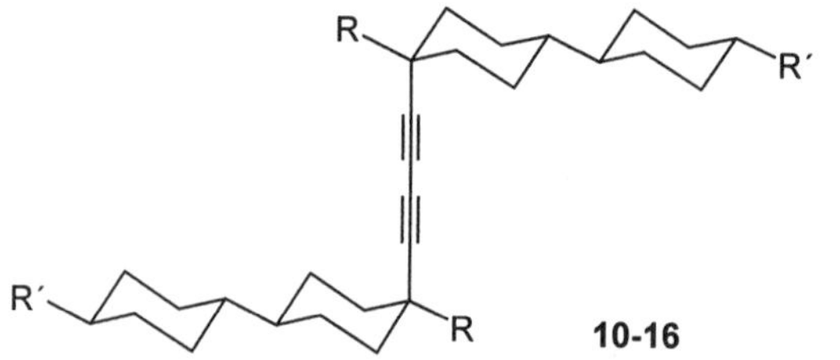

10-16

Entry	R	R'	Phases	Virtual Clp.	Virtual Δn
10	C_3H_7	C_3H_7	C 151 S_B 168 I	—	—
11	C_5H_{11}	C_5H_{11}	C 77 S_B 198 I	178	-0.004
12	C_7H_{15}	C_4H_9	C 61 S_B 208 I	154	-0.003
13	CH_3O	C_3H_7	C 121 N (80) I	—	—
14	CH_3O	C_5H_{11}	C 117 N (87) I	—	—
15	C_2H_5O	C_3H_7	C 143 N (117) I	103	0.001
16	C_2H_5O	C_5H_{11}	C 126 N (120) I	—	—

Table IV. Properties of unsymmetrical bis(bicyclohexyl)butadiynes

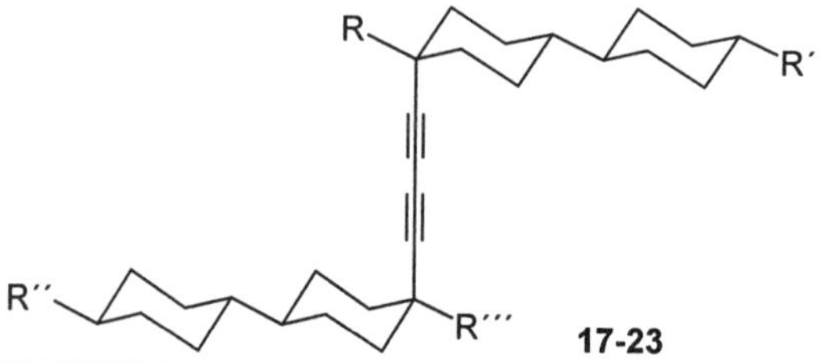

17-23

Entry	R	R'	R''	R'''	Phases	Virtual Clp.	Virtual Δn
17	CH_3O	C_3H_7	C_5H_{11}	C_5H_{11}	C 76 N 135 I	116	0.002
18	CH_3O	C_3H_7	C_4H_9	C_7H_{14}	C 100 S_B (63) N 128 I	105	0.001
19	CH_3O	C_5H_{11}	C_3H_7	C_3H_7	C 105 N 125 I	99	-0.005
20	CH_3O	C_5H_{11}	C_4H_9	C_7H_{14}	C 90 N 129 I	110	-0.004[a]
21	C_2H_5O	C_3H_7	C_4H_9	C_7H_{14}	C 78 S_B 134 N 145 I	126	-0.002
22	C_2H_5O	C_5H_{11}	C_3H_7	C_3H_7	C 99 S_B 117 N 142 I	123	-0.005
23	C_2H_5O	C_5H_{11}	C_4H_9	C_7H_{14}	C 66 S_B 135 N 144 I	130	0.003

a Δn (T/TN-I = 0.85): 0.0110

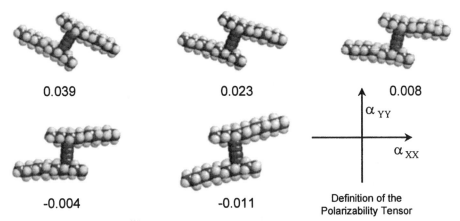

0.039   0.023   0.008

α_{YY}

α_{XX}

-0.004   -0.011

Definition of the
Polarizability Tensor

Figure 3. Calculated Δn values for **10** *at different orientations in Cartesian space*

Scheme 3. Bis(bicyclohexyl)hexatriynes from 1,4-dichlorobutyne. a) 1. 4 eqs. $NaNH_2$, liq. NH_3, -40 °C; 2. Add aldehyde, –40 °C, 2 hrs, then warm to RT (58%). b) 1. $SOCl_2$, CH_2Cl_2, RT; 2. 2 eqs. $NaNH_2$, liq. NH_3, -40 °C, 1hr, then warm to RT (28%).

Table V. Phase behavior and birefringence for bis(bicyclohexyl)acetylenes, -1,2-butadiynes and -1,2,3-hexatriynes.

Entry	n	Phases	Virtual Clp.	Virtual Δn	Δn ($T/T_{N-I} = 0.95$)
24	2	C 77 S_B 198 I	178	-0.004	
25	3	C 90 N (63) I	24	-0.012	-0.0104
26	4	C 163 N (29) I	-35	-0.046	

A further reduction of molecular symmetry leads to the materials **17-23** shown in Table IV, some of which now exhibit enantiotropic nematic phases, but even though the extrapolated birefringence of **20** is -0.004, a measurement of the pure material at a reduced temperature of 0.85 still gives a positive value of 0.011.

As indicated above, the orientation of the molecule determines the way anisotropies are defined, and thus the calculated birefringence. On the other hand, the orientation can easily be varied and the expected birefringence for any orientation be calculated. This is shown in Figure 3, and indeed, one obtains a value for Δn of around zero when the acetylenic unit is perpendicular to the x-axis (which corresponds to the director in the condensed phase).

A truly negative value for Δn in a monotropic nematic phase was finally obtained in the hexatriyne **25** shown in Table V. The introduction of a fourth triple bond in **26** decreases Δn even further but at the expense of lowering the clearing temperature. These compounds were synthesized (Scheme 3) following general literature procedures.[16]

The definitive proof for negative birefringence in **25** is shown in Figure 4 which depicts the diffusion zone of the nematic phase of **17**, which has positive birefringence, and **25** (left). The photograph shown in Figure 4 was taken at room temperature in sodium light. From both sides there is a gradual decrease in the absolute value of birefringence to the black zone at the center, where the mixture has zero birefringence. The reason for this lies in the continuous variation of Δn from a small negative value at the left to a small positive value at the right. Under white light, caused by the wavelength dependence of the birefringence, instead of the black center zone, there is a color change from white at the left side – via blue, indigo and red – to yellow at the right: A zero value for Δn is only found for a specific wavelength which therefore is missing from the rest of the visible spectrum. In addition, there is weak and wavelength dependent light scattering in materials with a very low absolute value of the birefringence, resulting e.g. in a faint blue color of the nematic phase of **29**.

Figure 4. Diffusion zone of 17 (right) and 25 under sodium light

The central triple bond can be replaced by an aromatic ring without losing nematogenic character. By palladium catalyzed coupling[17] of a (substituted) 1-iodo-4-bromobenzene with acetylenes of types **1** and **3**, the asymmetric derivatives **27-29** shown in Table VI were synthesized. The liquid crystal **29** is the second example for a (monotropic) nematic material with negative birefringence measured for the pure material close to its clearing point.

Table VI. Properties of materials with 1,4-dialkynylbenzene bridges

Entry	X	Y	Phases	Virtual Clp.	Virtual Δn	Δn ($T/T_{N-I} = 0.95$)
27	H	H	C 123 N (30) I	-13	0.005	
28	H	F	C 121 N (33) I	0	0.006	
29	F	H	C 117 N (36) I	0	0.006	-0.002

We have shown that the popular view of the birefringence in nematic materials always being positive is wrong. It is relatively straightforward to synthesize nematic

204

liquid crystals with very small and negative birefringence. Many possible applications can be envisaged for such materials, e.g. the use as a birefringence lowering dopant to standard TN mixtures. Furthermore, the interesting question of biaxiality in the materials presented here has not been addressed and will be the subject of further work.

Acknowledgment. We thank T. Mergner and R. Tavakoli for experimental assistance, and Dr. J. Krause, J. Haas and H. Heldmann for the physical evaluation of the new substances.

References

1. *Handbook of Liquid Crystals*; Demus, D., Goodby, J., Gray, G. W., Spiess, H.-W., Vill, V., Eds.; Wiley-VCH: Weinheim, **1998**.
2. Demus, D.; Demus, H.; Zaschke, H. *Flüssige Kristalle in Tabellen*; VEB Deutscher Verlag für Grundstoffindustrie: Leipzip, **1974**.
3. Vill, V. *LiqCryst 3.1- Databases of Liquid Crystalline Compounds*; LCI Publisher GmbH: Hamburg, **1998**.
4. de Gennes, P. G.; Prost, J. *The Physics of Liquid Crystals*; Oxford University Press: London: **1993**.
5. de Jeu, W. *Physical Properties of Liquid Crystalline Materials*; Gordon and Breach: New York **1980**.
6. Schadt, M.; Helfrich, W. *Appl. Phys. Lett.* **1971**, *18*, 127.
7. For negative birefringence in smectic thallium soaps see: Pelzl, G.; Sackmann, H. *Mol. Cryst. Liq. Cryst.* **1971**, *15*, 75.
8. *Smectic* liquid crystals with negative birefringence have been reported previously: (a) Thurmes, W. N.; Wand, M. D.; Vohra, R. T.; Crandall, C. M.; Xue, J.; Walba, D. M. *Liq. Cryst.* **1998**, *25*, 149. (b) Walba, D. M.; Dyer, D. J.; Sierra, T.; Cobben, P. L.; Shao, R.; Clark, N. A. *J. Am. Chem. Soc.* **1996**, *118*, 1211. We have prepared a compound structurally similar to those reported by Thurmes et al. but do not observe negative birefringence when measuring in our nematic host, in fact, the material exhibits a rather strong positive value(+0.104) in agreement with a calculated value of +0.096.
9. Vuks, M. F. *Opt. & Spectroscopy* **1966**, *20*, 361.
10. Bremer, M.; Tarumi, K. *Adv. Mater.* **1993**, *5*, 842.
11. Klasen, M.; Bremer, M.; Götz, A.; Manabe, A.; Naemura, S.; Tarumi, K. *Jpn. J. Appl. Phys.* **1998**, *37*, L945.
12. Eidenschink, R.; Haas, G.; Römer, M.; Scheuble, B. S. *Angew.Chem.* **1984**, *96*, 151.
13. Clearing temperatures [°C] and birefringence were determined by linear extrapolation from a 10% w/w solution in the commercially available Merck mixture ZLI-4792 (T_{NI} = 92.8°C, $\Delta\varepsilon$ = 5.3, Δn = 0.0964). For the pure substances the mesophases were identified by optical microscopy, and the phase transition temperatures [°C] by differential scanning calorimetry (DSC). The DSC traces did

not indicate thermal decomposition for any of the reported acetylenes; this is probably due to the steric bulk of the cyclohexane substituents. Optical anisotropies were determined by using a conventional Abbé refractometer (20 °C, λ=589.3 nm).

14. Dewar, M. J. S.; Zoebisch, E. G.; Stewart, J. J. P. *J. Am. Chem. Soc.* **1985**, *107*, 3902.
15. Eglington, G.; Galbraith, A. R. *Chem. Ind.* **1956**, 737.
16. (a) Bohlmann, F. *Chem. Ber.* **1951**, *84*, 785. (b) Bohlmann, F.; Viehe, H. G. *Chem. Ber.* **1954**, *87*, 712.
17. (a) Godt, A. *J. Org. Chem.* **1997**, *62*, 7471. (b) Alami, M.; Ferri, F.; Linstrumelle, G. *Tetrahedron Lett.* **1993**, 6403.

Chapter 15

Liquid Crystalline Trioxadecalins: The Mesogenic Chirality as Sensor for Molecular Conformation and Orientation

Volkmar Vill[1], Bruno Bertini[2], and Denis Sinou[2]

[1]Institute of Organic Chemistry, University of Hamburg, Martin-Luther-King-Platz 6, 20146 Hamburg, Germany
[2]Laboratoire de Synthèse Asymetrique, associé au CNRS, CPE Lyon, Université Claude Bernard Lyon 1, 43, boulevard du 11 novembre 1918, 69622 Villeurbanne cédex, France

Dedicated to Professor Günther Wulff on the occasion of his 65th birthday

Traditionally, chirality is introduced to mesogens by a sterically disturbing substituent. Oxygen heterocycles offer a different concept: the exchange between isosteric $-CH_2-$ and $-O-$ groups causes chirality without steric hindrance. The macroscopic chiral properties of trioxadecalin-based liquid crystals are therefore extremely sensitive to small changes in the chemical structure and the chemical environment. The helical inversion phenomenon can be explained by small changes to the main axis of molecules.

Chirality is an important factor for molecular properties. The liquid crystalline state is a mediator between microscopic molecular structures and macroscopic appearance of matter. Chiral additives to liquid crystalline phases can cause the following types of 'mesogenic chirality': helical ordering of the molecules (helical pitch in N^*, S_C^*, TGB_A, BP), polar properties (ferroelectricity) or phase transitions to phases which require chirality ($N \rightarrow BP$, $S_A \rightarrow TGB_A$). Therefore liquid crystals are well established as sensors for chirality [1,2,3]. The next step is to use this 'mesogenic chirality' as a monitor or sensor for small conformational and orientational changes of molecules.

Today, more than 80 000 mesogenic compounds are known and 16 000 of them are chiral [4]. Most of these compounds have a chiral center induced by the exchange of a $-CH_2-$ group by a $-CHMe-$ group in the flexible wing (Figure 1). Thus, chirality is induced by a steric hindrance which disturbs the mesogenic order. For example, compound **B** has a clearing temperature 47K lower than compound **C**.

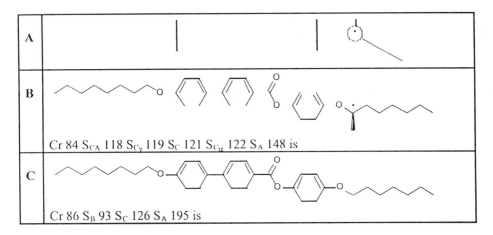

A		
B		
	Cr 84 S$_{CA}$ 118 S$_{Cγ}$ 119 S$_C$ 121 S$_{Cα}$ 122 S$_A$ 148 is	
C		
	Cr 86 S$_B$ 93 S$_C$ 126 S$_A$ 195 is	

Figure 1. Classical conception of chiral liquid crystals.
A: the general model, B: chiral, methyl-branched compound 'MHPOBC' with low clearing temperatures, C: non-chiral compound with high clearing temperature. All data are from [4].

Chirality Induced by Oxygen Heterocycles

Our aim is to separate chiral effects from mesogenic effects by the isosteric interchange of -CH$_2$- and -O- groups in conformationally rigid units. Figure 2 demonstrates both methods of this replacement. The non-chiral 2,5-*trans*-substituted cyclohexane changes to the chiral tetrahydropyran unit by the replacement of -CH$_2$- with -O-, whereas the *meso* form of the tetraoxadecalins is transformed to the chiral trioxadecalins by the replacement of -O- with -CH$_2$-. Basic phase types and transition temperatures of chiral and non-chiral compounds are very similar, but the 'mesogenic chirality' of the former is huge: short-pitch cholesteric phases, high spontaneous polarization, blue phases and TGB$_A$ phases [5,6]. Of special interest are the trioxadecalins, [7] which show cholesteric phases with temperature and concentration

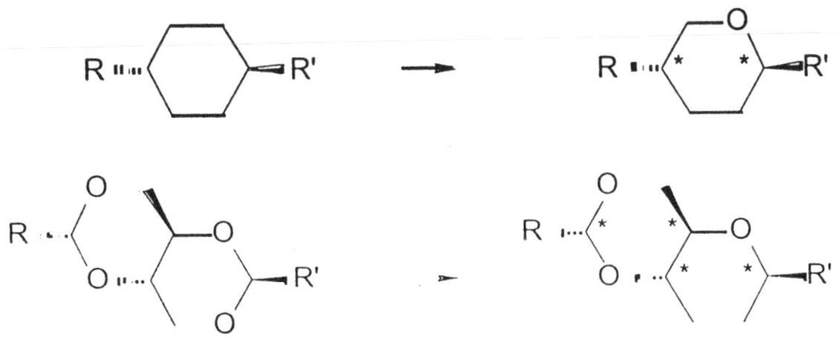

Figure 2. Chirality induced by -CH$_2$- to -O- exchange.

208

dependent helical inversions. The previously used synthetic approach [6] was limited in the substitution pattern and a new pathway will be used here.

Figure 3 shows the synthesis of chiral diols 5 from triacetyl glucal 1. These diols can be condensed with aldehydes to trioxadecalins 6-8, with boronic acids to boratrioxadecalins 9-11, or they can be esterified to cone-shaped mesogens with a tetrahydropyran core 12-13 as shown in Figure 4. Details of the preparation will be discussed elsewhere [8].

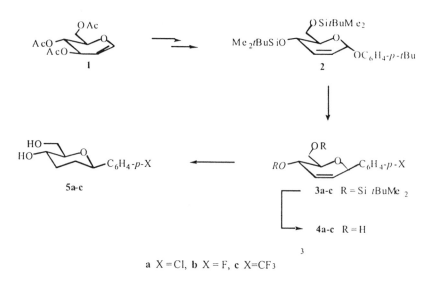

a X = Cl, b X = F, c X=CF₃

Figure 3: Synthesis of chiral diols from glucal.

The boron compounds 9-11 exhibit a clear behavior pattern: short-pitch cholesteric phases, blue phases and TGB_A phases. The cone-shaped compounds 12-13 behave similarly, while the trioxadecalins 6-8 are extremely variable in their macroscopic chirality.

Table 1 lists the new compounds together with previously published compounds. Compounds a and b show only a cholesteric phase with a short pitch. Compounds c, l, p exhibit a cholesteric phase with a temperature-dependent inversion. However, compounds f and q show no inversion at all in the pure form, but only in mixtures with non-chiral nematics. That is, they induce a helical pitch of opposite sign in the mixture to that which they exhibit themselves in the pure state.

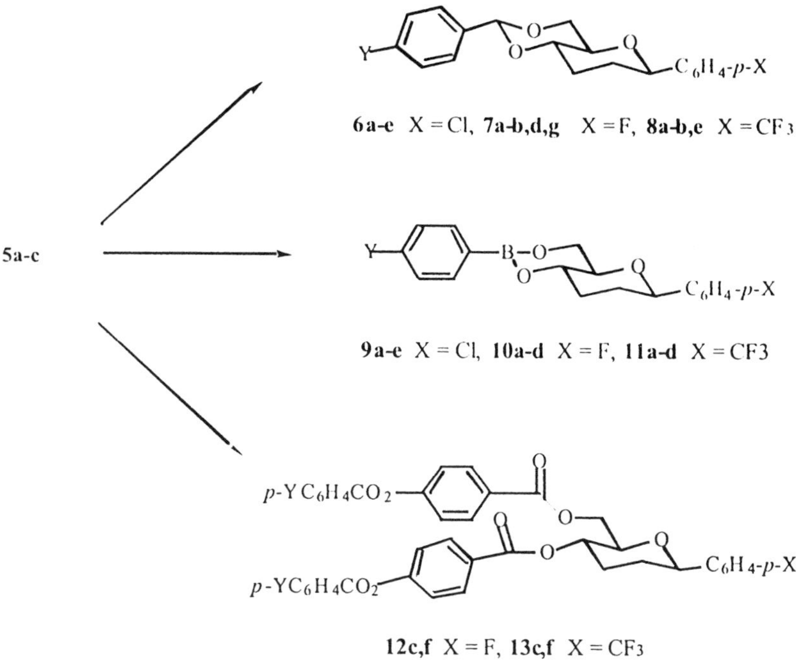
5a-c
Y
O
O
O
C6H4-p-X
6a-e X = Cl, 7a-b,d,g X = F, 8a-b,e X = CF3
Y
B
O
O
O
C6H4-p-X
9a-e X = Cl, 10a-d X = F, 11a-d X = CF3
p-YC6H4CO2
p-YC6H4CO2
O
O
O
O
O
C6H4-p-X
12c,f X = F, 13c,f X = CF3

a Y = O CH_3 ; **b** Y = O C_4H_9 ; **c** Y = O C_6H_{13} ; **d** Y = O C_8H_{17} ;

e Y = O$C_{10}H_{21}$; **f** Y = O$C_{14}H_{29}$; **g** Y = C≡CH

Figure 4: Synthesis of trioxadecalins and cone-shaped tetrahydropyrans.

All compounds have the same rigid chiral core, but modifying the aromatic substituent, which is far away from the stereo centers, influences the mesogenic chiral properties by magnitudes. Compounds **b** and **c** differ only by the exchange of CH_3O- for NC- terminal group. Hence it is not to be expected that both compounds have a different conformational equilibrium in the decalin system. Nevertheless only **c** has a temperature-dependent helical inversion.

Mesogenic Chirality as Sensor for Molecular Orientation

The explanation of all these unusual chirality effects is quite simple. The helical pitch is caused by the asymmetry of the molecular surface relative to the main axis of the molecule. This pitch is perpendicular to the main axis for the cholesteric phase as shown in Fig. 5. The twist from molecule to molecule is small. For example, a helical pitch of 500 nm and a molecular diameter of 0.5 nm would yield a twist angle

Table 1. Transition Temperatures of Trioxadecaline Compounds

$$Y - \text{(structure: phenyl ring — trioxadecaline with two O and central O — chain with double bonds)} - X$$

	X	Y	Transition temperatures [°C]						Comment
a	CH_3O	OCH_3	Cr 195.0				Ch 206	is	
b	$C_8H_{17}O$	OCH_3	Cr 134.7				Ch 151.7	is	
c	$C_8H_{17}O$	CN	Cr 123.4				Ch 182.2	is	112[a]
d	Cl	OCH_3	Cr 190.3				Ch 182.4	dec.	
e	Cl	OC_4H_9	Cr 155.6				"N" 171.1	is	[b]
f	Cl	OC_6H_{13}	Cr 142.0	S_A 115.4			Ch 163.3	is	[c]
g	Cl	OC_8H_{17}	Cr 128.6	S_A 132.9			Ch 158.8	is	
h	Cl	$OC_{10}H_{21}$	Cr 122.1	S_A 134.3	TGB_A 134.5		Ch 141.2	is	
i	F	OCH_3	Cr 181.2				Ch 153	is	
j	F	OC_4H_9	Cr 146.5				Ch 155	is	
k	F	OC_8H_{17}	Cr 116.4				Ch 102	is	
l	F	$C{\equiv}CH$	Cr 188.0				Ch 190	is	144[a]
m	CF_3	OCH_3	Cr 201.8					is	
n	CF_3	OC_4H_9	Cr 184.7					is	
o	CF_3	OC_8H_{17}	Cr 156.7					is	
p	$C_8H_{17}O$	Br	Cr 132.5				Ch 140.1	is	132[a]
q	$C_8H_{17}O$	I	Cr 134.5				Ch 132.2	is	[c]

a) inversion temperature of helix, b) no visible helical ordering, c) helical inversion in contact with non-chiral nematics

between two molecules of about 360°/1000 = 0.36°. Thus, a classical drawing of a cholesteric phase as shown in Figure 5 will always overemphasise the twisting on the molecular level. On the other hand the macroscopic effects of this 'small' twisting are significant and easy to study, e.g. finger print textures and selective reflection of light.

A change of the helical pitch can be induced by change of the molecular surface (→ conformation), by aggregation, or by change of the main axis (→ orientation). The main axis is defined by order parameters of the rigid core and the flexible wing groups. Orientation of the latter is more affected by changes in temperature. Figure 6 demonstrates that in some cases the main axis will shift, if the average order of the wing groups changes and if the mesogenic group has a zigzag form. Part A displays the chemical structure of the compound. Part B symbolizes the conformation of the compound at low temperature, including the stiff cyano group, the zigzag-shaped core and the flexible chain on the right side. The paraffin chain is extended and the avarage conformation is represend by a cone parallel to the Ar-O bond. The main axis (the black line) is bonded at the cyano group and the center of the paraffin cone. Part C demonstrates the conformation at higher temperatures. The cyano group and the core are stiff and are nearly unchanged, whereas the paraffin chains are more disordered and the average conformation is seen as a broader and shorter cone. This will shift only one of the base points of the main axis yielding a reorientation of the latter.

The total rotation of the main axis might be smaller than 0.1°, but even this can explain the change of the pitch. Thus, the helical pitch, which is a simply accessible macroscopic parameter, is much more sensitive to the molecular shape and orientation than spectroscopic methods (polarized UV) or molecular modeling. Changes to the helical pitch can be induced by changing temperature, pressure, dielectric environment or by impurities.

Unusual dependence of the helical pitch on temperature and concentration has been observed before, e.g. for cholesteryl halides.[9] Compared to these, the trioxadecalins are more sensitive and much easier to analyse, because of the very simple chemical structure. Important for such sensitive behavior is

- that the chiral centers are located inside of the mesogenic group and not in the wing group,
- that the chiral centers are not introduced by groups disturbing mesogenic phases.

Figure 2 shows tetrahydropyrans and trioxadecalins. The former are linear and flexible, showing little sensitivity, whereas the latter are rigid and zigzag shaped; both parameters seem to enhance sensitivity.

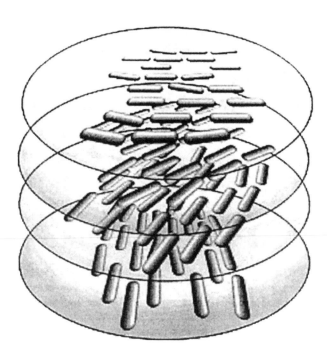

Figure 5: Molecular alignment in the cholesteric phase.

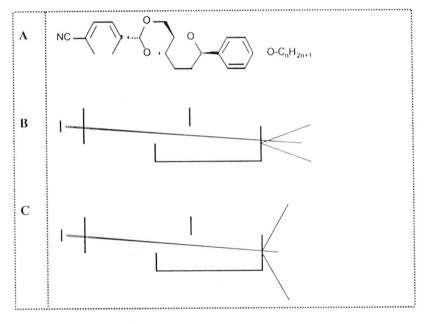

Figure 6. Shift of the main axis of trioxadecalin compounds upon temperature change. A: chemical sketch, B: low temperature conformation, C: higher temperature conformation

A better understanding of this behavior might give us more insight in general molecular structures. The use of these effects in lyotropic environments will perhaps give way to the development of methods to study cell-cell recognition processes through the 'mesogenic chirality' of ordered micelles.

Acknowledgement

Financial support by DAAD/Procope is gratefully acknowledged. We thank Matthias Wulf for drawing the cholesteric phase.

References

1. Solladié, G., Zimmermann, R.G. *Angew. Chem.* **1984**, *96*, 335-349.
2. Stegemeyer, H. *Nachr. Chem., Tech. Lab.* **1988**, *36*, 360-364.
3. Goodby, J.W. *J. Mater. Chem.* **1991**, *1*, 307-318.
4. Vill, V. *LiqCryst 3.4 - Database of Liquid Crystals*, LCI Publisher, Hamburg, **2000**; Fujitsu FQS, Fukuoka, **2000**; http://liqcryst.chemie.uni-hamburg.de.
5. Vill, V.; Tunger, H.-W.; Borwitzky, A. *Ferroelectrics* **1996**, *180*, 227-231.
6. Vill, V.; Tunger, H.-W. *Liebigs Ann.* **1995**, 1055-1059.
7. Vill, V.; von Minden, H.M.; Bruce, D. W. *J. Mater. Chem.* **1997**, *7*, 893–899.
8. Bertini, B.; Moineau, Ch.; Sinou, D.; Gesekus, G.; Vill, V. *submitted*; Bertini, B., Ph.D. Thesis, Lyon, **2000**.
9. Leder, L.B. *J. Chem. Phys.* **1971**, *55*, 2649-2657.

Chapter 16

Spontaneous Polarization in Ferroelectric Liquid Crystals: Role of the Molecular Architecture and Optical Purity

D. Guillon[1], M. Z. Cherkaoui[1], P. Sebastião[2], S. Méry[1], J. F. Nicoud[1], and Y. Galerne[1]

[1]Institut de Physique et Chimie des Matériaux de Strasbourg, Groupe des Matériaux Organiques, 23 rue du Loess, 67037 Strasbourg Cedex, France
[2]Centro de Fisica da Matéria Condensada, Av. Prof. Gama Pinto 2, 1649-003 Lisboa, Portugal

The role of several molecular factors is analyzed with respect to the value of the spontaneous polarization measured in ferroelectric liquid crystal compounds containing a sulfinate group as unique chiral center. First, it is shown that different molecular contributions are involved in the distribution of the transverse dipole moments (orientational fluctuations, dilution effect with increasing of the aliphatic chain length of the molecules, tilt angle). Second, it is observed that the presence of a siloxane group in the molecular architecture strongly contributes to the stabilization of the ferroelectric phases with an increase of the spontaneous polarization. Finally, it is shown that the thermotropic behaviour, the polar order and, consequently, all the related electro-optical properties of ferroelectric liquid crystals may depend significantly on the optical purity of the materials.

Introduction

With the development of optoelectronics, the transposition of ferroelectricity, found in Rochelle'salts (potassium and sodium tartrate) (*1*) at the beginning of the 20th century, to anisotropic liquids and, in particular, to thermotropic liquid crystals has been a matter of great interest for the last two decades (*2*). The molecular design of these compounds was directly deduced from the original work of R.B. Meyer who predicted, using a symmetry argument, that tilted smectic phases obtained with chiral molecules having a transverse dipole moment, should exhibit an electric spontaneous polarization (3). Among the different tilted smectic phases encountered with calamitic liquid crystals, the so-called smectic C* phase, which results from the stacking of liquid layers of chiral molecules, revealed to be the most appropriate to build up fast electro-optical devices (4). In this context, a large number of compounds exhibiting the smectic C* phase has been synthesized with various polar groups and diverse chiral centers (5-9). More recently, other kinds of chirality have been explored in the design of smectic C* ferroelectric materials, such as axial chirality (10) or planar chirality (11).

One of the main objectives which has driven the molecular design and the synthesis of smectic C* materials was the preparation of high spontaneous polarization materials exhibiting in the same time a good thermal stability (among other properties). Indeed, the absolute value of the spontaneous polarization is one of the key parameters in the electro-optical properties. However, the synthesis of a large number of compounds has shown that the measured values of the spontaneous polarization were fairly low with respect to the expected values from the magnitude of the molecular dipole moment (12). In other words, the yield factor - which is an indicator of how efficient are the molecular parameters in the chemical structure – most usually appears rather low. The purpose of the present paper is to clarify how this yield factor can be improved by investigating in detail the role of different molecular parameters in the establishment of the spontaneous polarization, and by analyzing how the distribution of the transverse dipole moments could be the most anisotropic.

Molecular design for smectic C* phases

It is now well known that the strength of the transverse dipole moment has direct consequences on the value of the spontaneous polarization. In addition, it is also clear that in order to maximize the polar order, this transverse dipole moment and the chiral center need to be strongly coupled, i.e. they should be situated as close as possible to each other near the rigid core of the molecule; the ideal case would be that both be localized at the same position in the molecular architecture (13, 14). In the following, will only be considered ferroelectric compounds containing a sulfinyl group as unique chiral center (15). The sulfinyl group directly attached to the rigid core combines both properties of being chiral and strongly polar at the same time.

The first part of this section concerns the homologous series represented above and designated as *m-BTS-On* series, *m* and *n* being the number of carbon atoms in the achiral and chiral aliphatic chains respectively. The second part is dealing with the same type of molecules in which a siloxane group has been introduced.

Role of the terminal aliphatic chains

When varying the aliphatic chain lengths of the molecules, the macroscopic properties are typically modified. For example, the rotational viscosity as well as the tilt angle of the S_C^* molecules increase with increasing chain length. The case of the variation of the spontaneous polarization, *Ps*, as a function of chain length is less obvious; indeed, increasing the length of the non-polar moiety contributes to dilute the transverse dipole moment in the material and thus to decrease the value of *Ps*, whereas in the same time, the increase of the tilt angle should have a tendency to increase *Ps*.

In order to avoid any effect of the enantiomeric excess, *ee* (see next section), only the terms with the same *n* corresponding to the same *ee* (about 80%) have been considered, namely the *m-BTS-O10* and *m-BTS-O12* series for which *m* was varied between 10 and 18. The results are similar for both series (*16*). The values of *Ps* as a function of *m* for the *m-BTS-O10* series are shown in figure 1.

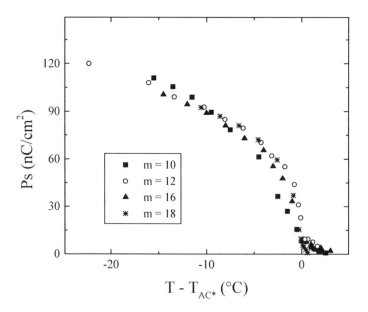

Figure 1. Spontaneous polarization, Ps, in the smectic C* phase as a function of temperature, T-T$_{AC}$*, for different terms of the m-BTS-O10 series (Reproduced with permission from reference 16).

It is interesting to notice that the value of *Ps* does not seem to depend significantly upon the length of the aliphatic chain *m*, even far from the S_C^*-S_A transition. For the same compounds, the variation of the corresponding tilt angle, θ, of the molecules within the smectic C* layers is shown in figure 2.

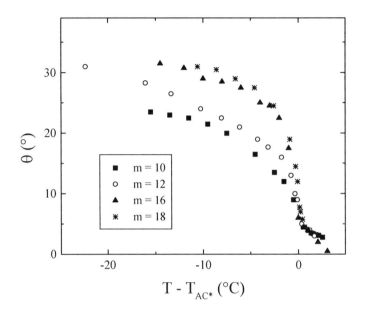

Figure 2. Tilt angle, θ, in the smectic C phase as a function of temperature, T-T_{AC}*, for different terms of the m-BTS-O10 series (Reproduced with permission from reference 16).*

From this figure, it is clear that the values of θ depend significantly upon the length of the aliphatic chain. Such a change of θ as a function of *m* should have induced an increase of *Ps*, since *Ps* and θ are generally believed to be proportional in a first approximation. However, as pointed above, the variation of *Ps* with *m* is not significant and therefore, needs to be investigated in more detail.

Let us analyze *Ps* according to the expression given by de Gennes and Prost (*17*): $Ps = f \, n_b \, \mu_T$, where μ_T is the transverse dipole moment, n_b the number of molecules per unit volume, and f a yield factor representing in fact the anisotropic distribution of the molecular dipole moments, usually of the order of a few percents. From the variation of the yield factor, f, represented in figure 3 as a function of the tilt angle of the molecules within the smectic layers, two main features can be identified. First, f increases regularly with θ, almost linearly in the domain that excludes the regions very close to the transition and very far from the transition where θ has a tendency to saturate. Second, for the same tilt angle θ, it is quite striking to note that f decreases strongly with the increasing chain lengths (up to 50% from n = 10 to n = 18). This important effect is attributed to larger orientational fluctuations of the chiral group for

218

longer aliphatic chain lengths, which results from the larger volume given to it, then inducing a smaller anisotropy of the dipole moments distribution.

The above results show that different molecular contributions are involved in the distribution of the transverse dipole moments. On one hand, dilution effect and increased orientational fluctuations, occuring when the aliphatic chain length is increased, contribute to decrease the values of the spontaneous polarization. On the other hand, the increase of tilt angle with increasing m contributes to increase Ps. The latter behaviour is probably due to a better stabilization of the dipolar vectors, induced by a better defined building of the different smectic sublayers, with a pronounced microsegregation in space of aliphatic and aromatic moities of the molecules, when the aliphatic sublayer is thickened.

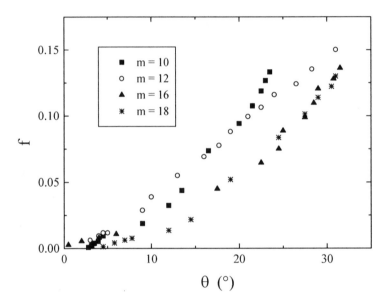

Figure 3. Yield factor, f, as a function of tilt angle, θ, for different terms of the m-BTS-O10 series (Reproduced with permission from reference 16).

In the case of the homologous series studied here, all these antagonistic effects act simultaneously in order to keep the spontaneous polarization practically independent of the aliphatic chain length. However, it remains that it is possible to modify more significantly the molecular architecture in order to favor one of the contributions acting on the spontaneous polarization behavior. For example, the introduction of a bulky siloxane group at one end of the molecule should contribute to a significant increase of the tilt angle and, at the same time, should contribute to a well defined stacking of the different smectic sublayers (due to the amphiphilic interactions), leading to high values of Ps. Let us discuss this possibility in the next section.

Influence of siloxane groups

The introduction of siloxane groups into the molecular architecture of calamitic liquid crystal molecules was initially performed in order to reach a compromise between the good mechanical properties of polymers and the fast switching times of low molar mass liquid crystals (*18*, *19*). The special feature of these compounds is that the molecules contain three distinct parts incompatible with each other. As a result, the siloxane moieties localize themselves in separate sublayers, resembling side-chain polymers for which the backbones are mostly inserted between the layers. The elemental smectic layer is thus constituted by the superposition of three separate sublayers formed by the three molecular moieties (*20*), as represented in figure 4.

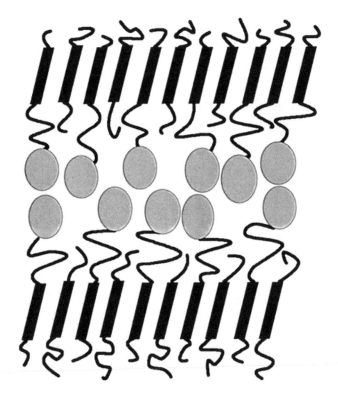

Figure 4. Schematic representation of the smectic C layers of organosiloxane compounds.

In figure 4, the central part of the smectic layers has been arbitrarily chosen to be formed by the siloxane sublayers, with the siloxane groups (ellipses) arranged in a partially bilayered structure. This central part is fringed with the disordered paraffin chains (wavy lines) arranged in single layers. The aromatic cores (rectangles) are confined in a distinct mono-arrangement sublayer, and tilted with respect to the normal layers.

In order to make a meaningful comparison, there will be presented now the electro-optical characterization of two ferroelectric organosiloxane liquid crystals for which a short siloxane part is connected to the sulfinate mesogenic moiety discussed in the previous section. These compounds, abbreviated in the following as Si_nM, have the following chemical structure and polymorphic behavior:

$$n = 1, \quad Si_2M : K \xrightarrow{36.1°C} S_{C*} \xleftarrow{68.2°C} I$$

$$n = 2, \quad Si_3M : K \xrightarrow{29.1°C} S_{C*} \xleftarrow{53.6°C} I$$

Their properties will be compared to the ones exhibited by the previuosly described *10-BTS-O10* compound.

$$K \xrightarrow{81.4°} S_{C*} \xleftarrow{93.5°C} S_A \xleftarrow{111.9°C} I$$

First, as it can be seen directly on their thermotropic behavior, the presence of the siloxane group strongly contributes to the increase of the ferroelectric smectic C* phase temperature range by more than a factor of 2. Moreover, since the S_C* phase can be supercooled, both siloxane compounds present ferroelectric properties at room temperature. There is, therefore, a stabilization effect on the S_C* phase.

In figure 5 and 6 are reported respectively the variation of the spontaneous polarization and the optical tilt angle as a function of temperature for the two organosiloxane compounds Si_2M and Si_3M, together with that of the *10-BTS-O10* derivative. Far from the transition, the organosiloxane derivatives exhibit a spontaneous polarization much higher (by a factor larger than 2) than for the similar compound without siloxane group. It can also be observed that *Ps* varies with the size of the siloxane moities. As regard to the variation of the tilt angle, both materials, Si_2M and Si_3M, have similar tilt angle values, whatever the bulkiness of the siloxane moiety. Moreover, the optical tilt angle is almost temperature independent with values between 38° and 44°. These values are much higher than the ones (between 20 and 25°) noted for the non-siloxane *10-BTS-O10* compound. This can be easily explained by the fact that the siloxane units have a molecular area larger than that of the aliphatic chains and of the mesogenic cores, thus forcing the tilting of the mesogenic cores with respect to the layer normal.

In summary, the values of *Ps* and *θ* are much higher for *Si₂M* and *Si₃M* than for *10-BTS-O10*, confirming the strong relationship between the two parameters. Nevertheless, the switching times for the siloxane derivatives are of the same order of magnitude (a few tens of μs) as those for the *10-BTS-O10*, indicating that they are still competitive for fast display devices in spite of their higher molecular weight (*21*). Finally, it has also to be mentioned that the yield factor, *f*, defined in the previous section , is about 35% for *Si₂M* and 25% for *Si₃M* whereas it is only 12% for *10-BTS-O10* far from the transition (*21*). This represents a considerable improvement and clearly indicates that the effective dipole moment is much more efficient in the siloxane derivatives than in the mesogen deprived of siloxane moiety.

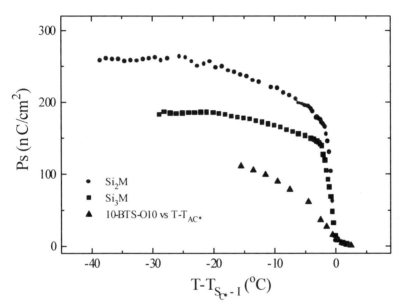

Figure 5. Spontaneous polarization as a function of temperature for the Si2M, Si3M, and 10-BTS-O10.

Although there is a large dilution effect due to the addition of a bulky siloxane group to the *m-BTS-On* mesogenic moiety, there is a considerable increase of transverse dipolar order. This is due, of course, to the increase of the tilt angle, but undoubtedly also to the decrease of orientational fluctuations in relation with a well defined layered structure.

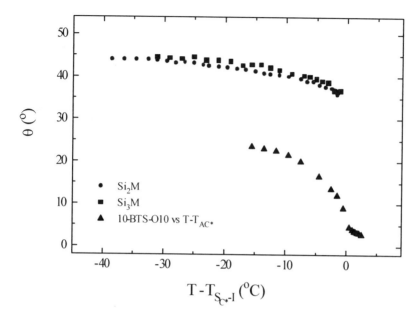

Figure 6. Optical tilt angle as a function of temperature for the Si2M, Si3M, and 10-BTS-O10.

Role of the enantiomeric excess

For a very long time, the interest in ferroelectric liquid crystals was focused on the synthesis of stable, broad-range room temperature smectic C* materials with efficient electro-optical properties. However, it has to be reminded that the main prerequisite for ferroelectricity is the introduction of chirality with optical activity into Sc compounds. The ferroelectric parameters should then be quantitatively related to the optical purity of the materials. Therefore, in addition to their chemical stability, great care should be taken about the optical stability and the enantiomeric excess of the mesogens. In other words, for a given smectic C* compound, it is important to know for which corresponding enantiomeric excess (*ee*) the ferroelectric properties have been measured (*22*). When changing the enantiomeric excess, the initial polar order which is driven by different kinds of interactions between neighboring molecules, may be affected. Consequently, the anisotropic distribution of the effective transverse dipole moments, responsible for the spontaneous polarization, can also be affected. Hence, the extrapolation of the spontaneous polarization value from a given enantiomeric concentration to the optically pure form should not only take into account the linear dipole moment compensation effect (between the two pure R and S enantiomers), but also the behavior related to the variation of the polar order.

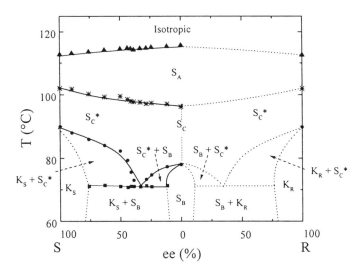

Figure 7. Binary phase diagram between the two pure enantiomers (R) and (S)-12-BTS-O8 (Reproduced with permission from reference 23).

In order to obtain further insight in the effects of the enantiomeric excess on both mesomorphic and electro-optical properties of ferroelectric liquid crystals, let us consider now a pair of pure enantiomers *12-BTS-O8*, still belonging to the same homologous series as described in the first section. The phase diagram between both enantiomers is represented in figure 7 (*23*). It is worth noting that the stability of the smectic A phase is the highest for the racemate, and also that a smectic B phase is present at lower temperature around the racemic composition. This is an indication of the stabilisation of non-tilted smectic phases for *ee* ~ 0. But most interesting is to remark the large effect on the stability of the smectic C* phase induced by the enantiomeric excess. Thus, at the eutectic point (*ee* = 33.4%) the S_C^* mesophase is stable over a large temperature domain up to 27 K. This stability decreases considerably with increasing *ee* (13 K for the pure enantiomer). On the contrary, for compositions between the racemate and the eutectic point, the stability of the S_C^* mesophase decreases with decreasing the enantiomeric excess. From these results, it is clear that, at least in the case of the series under consideration, the thermodynamic behavior is strongly dependent on the enantiomeric excess, when looking for example at the variation of the transition temperatures.

The dependence of the spontaneous polarization on the enantiomeric excess in binary mixtures of *(R)* and *(S)-12-BTS-O8* is shown for several temperatures in figure 8. For mixtures leading to *ee* below 30%, the measured values of *Ps* are very small. In fact for three compositions investigated (*ee* = 6, 11 and 27%), the values of Ps do not exceed 3 to 4 nC x cm^{-2}. For the pure enantiomer, high spontaneous polarization of about 270 nC x cm^{-2} was measured near the transition to the crystalline phase (T_{AC^*} -

T = 12.5°C). Here it is important to stress out that a decrease of the enantiomeric excess of only 10% induces a drastic decrease of the spontaneous polarization of about 50%. Such a non-linear dependence of *Ps* as a function of *ee* shown in figure 8 is explained by a difference of interactions between the pairs of chiral molecules with equal and opposite handedness, respectively. This chiral discrimination has been recently proposed to be mainly due to electrostatic forces between effective atomic charges in the vicinity of the chiral center (*24*).

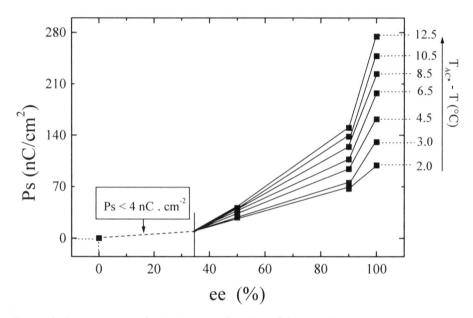

Figure 8. Spontaneous polarization as a function of the enantiomeric excess at different temperatures for mixtures of (R) and (S)-12-BTS-O8 (Reproduced with permission from reference 23).

The results presented in this section show that the variation of the enantiomeric excess has large effects on the thermodynamic properties of the investigated ferroelectric chiral sulfinyl compounds. These effects manifest themselves by distinct thermotropic sequences in all the enantiomeric concentration range. The transition temperatures are in fact greatly affected, and even an additional orthogonal and ordered smectic B phase is observed for the racemate and its neighboring compositions. The polar order is consequently drastically affected and, in turn, leads to a non-linear dependence of the electro-optic quantities on enantiomeric excess. Then, the most important conclusion of this section is that the thermotropic behavior, the polar order, and consequently all the related electro-optical properties of ferroelectric liquid crystals may depend significantly on the optical purity of the materials.

Conclusion

The different possibilities described in this paper to tailor chemically materials in order to tune the electro-optical properties of ferroelectric liquid crystals, show unambiguously how rich can be the molecular engineering for producing pure liquid crystals compounds with the expected properties. However, it should not be forgotten the role of mixtures in tuning the physical properties. For example, in the case of the organosiloxanes derivatives, the tilt angle (and consequently the spontaneous polarization) within the smectic layers can easily be controlled by mixing them with an appropriate amount of the corresponding mesogenic moities. Indeed, an increased density of the polarizable and chiral moieties would contribute to reduce the mismatch between the bulkiness of the mesogenic parts and that of the siloxane groups, and consequently to reduce the tilt angle. Let us stress out also that this paper shows how sensitive can be the values of the spontaneous polarization according to the optical purity of chiral ferroelectric compounds. All these factors, namely the location of the chiral and polar group in the chemical structure, the strength of the transverse molecular dipole, the length of the aliphatic chains, the nature and the bulkiness of the different parts of the molecules, and the optical purity represent a variety of tools that the chemical engineer can use to produce liquid crystals materials with given electro-optical properties.

Acknowledgments:

P. Sebastião wishes to thank Praxis XXI (project 3/3-1/MMA/1769/95) for financial support and D. Guillon for a fellowship under the same Praxis XXI project.

References

1. Valasek, J. *Phys. Rev.* **1921**, *17*, 475.
2. *Ferroelectric Liquid Crystals: Principles, Properties and Applications*; Goodby, J.W., Ed.; Gordon and Breach Science Publishers: New York, 1991.
3. Meyer, R.B.; Liébert, L.; Strzelecki, L.; Keller, P. *J. Phys. (Paris) Lett.* **1975**, *36*, 69.
4. Clark, N.A.; Lagerwall, S.T. *Appl. Phys. Lett.* **1980**, *36*, 899.
5. Walba, D.M.; Vohra, R.T.; Clark, N.A.; Hanndschy, M.A.; Xue, J.; Parmar, D.S., Lagerwall, S.T.; Skarp, K. *J. Am. Chem. Soc.* **1986**, *108*, 17424.
6. Scherowsky, G.; Gay, J. *Liq. Cryst.* **1989**, *5*, 1253.
7. Kasumuto, T.; Hannamoto, T.; Hiyama, T.; Takehara, S.; Shoji, T.; Osawa, M.; Kuriyama, T.; Nakamura, K.; Fujisawa, T. *Chem. Lett.* **1991**, 311.
8. Sierra, T.; Ros, M.B.; Omenat, A.; Serrano, J.L. *Chem. Mater.* **1993**, *5*, 938.
9. Sakashita, K.; Shindo, M.; Nakauchi, J.; Uematsu, M.; Kageyama, Y.; Hayashi, S.; Ikemoto, T.; Mori, K. *Mol. Cryst. Liq. Cryst.* **1991**, *199*, 119.

10. Lunkwitz, R.; Tschierske, C; Langhoff, A.; Giesselmann, F.; Zugenmaier, P. *J. Mater. Chem.* **1997**, *7*, 1713.

11. Jacq, P.; Malthête, J. *Liq. Cryst.* **1996**, *21*, 291.

12. Skarp, K.; Handschy, M. *Mol. Cryst. Liq. Cryst.* **1988**, *165*, 439.

13. Walba, D.M.; Slater, S.C.; Thurmes, W.N.; Clark, N.A.; Hanndschy, M.A. and Supon, F. *J. Am. Chem. Soc.* **1986**, *108*, 5210.

14. Goodby, J.W.; Patel, J.S. and Chin, E. *J. Phys. Chem.* **1987**, *91*, 5151.

15. Cherkaoui, M.Z.; Nicoud, J.F. and Guillon, D. *Chem. Mater.* **1994**, *6*, 2026.

16. Cherkaoui, M.Z.; Nicoud, J.F.; Galerne, Y. and Guillon, D. *J. Chem. Phys.* **1997**, *106*, 7816.

17. de Gennes, P.G. and Prost, J.; *The Physics of Liquid Crystals*, 2nd edition: Clarendon, Oxford, 1993.

18. Newton, J; Coles, H.J.; Hodge, P. and Hammington, J. *J. Mater. Chem.* **1994**, *4*, 869.

19. Naciri, J.; Ruth, J.; Crawford, G.; Shashidhar, R. and Ratna, B.R. *Chem. Mater.* **1995**, *7*, 1397.

20. Ibn-Elhaj, M.; Skoulios, A.; Guillon, D.; Newton, J.; Hodge, P. and Coles, H.J. *J. Phys. France* **1996**, *6*, 1807.

21. Sebastião, P.; Méry, S.; Sieffert, M.; Nicoud, J.F.; Galerne, Y. and Guillon, D. *Ferroelectrics* **1998**, *212*, 133.

22. Goodby, J.W. In *Ferroelectric Liquid Crystals*; Gordon & Breach Science Publishers: New York, 1991; p. 158.

23. Cherkaoui, M.Z.; Nicoud, J.F.; Galerne, Y. and Guillon, D. *Liq. Cryst.* **1999**, *26*, 1315.

24. Osipov, M.A. and Guillon, D. *Phys. Rev. E* **1999**, *60*, 6855.

Chapter 17

Optical Switching of a Ferroelectric Liquid Crystal Spatial Light Modulator Using Chiral Thioindigo Dopants

Robert P. Lemieux, Liviu Dinescu, and Kenneth E. Maly

Department of Chemistry, Queen's University, Kingston, Ontario K7L 3N6, Canada

The spontaneous polarization of a surface-stabilized ferroelectric liquid crystal (SSFLC) can be photomodulated using visible light, with no appreciable destabilization of the SmC* phase, via the *trans-cis* photoisomerization of a chiral thioindigo dopant. When the latter is used in combination with a photoinert chiral dopant of opposite polarization, *trans-cis* photoisomerization can trigger the switching of a SSFLC spatial light modulator intrinsically via a polarization sign inversion.

Introduction

Over the past twenty years, the search for new chiral materials exhibiting or inducing a chiral smectic C (SmC*) liquid crystal phase has been motivated by the potential application of such materials in surface-stabilized ferroelectric liquid crystal (SSFLC) displays (*1-4*). In 1980, Clark and Lagerwall discovered that the helical structure of a SmC* liquid crystal phase spontaneously unwinds between polyimide-coated glass slides with a layer thickness on the order of 5 μm (*5*). As predicted by Meyer, the dissymmetry of this *surface-stabilized* SmC* phase gives rise to a spontaneous polarization (P_S) oriented perpendicular to the glass slide, i.e., transverse to the molecular long axis (*6*). By virtue of its spontaneous polarization, a SSFLC can be switched between two degenerate states corresponding to opposite tilt orientations using an applied electric field (Goldstone-mode switching, see Figure 1), thus producing a spatial light modulator (SLM) when the SSFLC film is placed between crossed polarizers (*5*). SSFLC-SLM devices are characterized by a high viewing contrast, wide viewing angle, bistability and a switching time about a thousand times faster than that typically achieved in commercial twisted nematic LCD devices (*1-4*).

Considerable efforts have been made over the past decade to develop photonic analogues to the SSFLC-SLM because of their potential use in optical computing, dynamic holography, telecommunications and optical data storage (*7*). Most of the optical switching mechanisms developed for optically addressed SSFLC spatial light modulators (SSFLC-OASLM) have been extrinsic in nature, using a photoconductive or photodiode layer to trigger the Goldstone-mode switching of an underlying SSFLC film upon illumination. Other studies have focused on an alternative *intrinsic* approach to optical switching of SSFLC-OASLM devices which is based on the photomodulation of P_S above and below a switching threshold using a photochromic dopant (*8-19*).

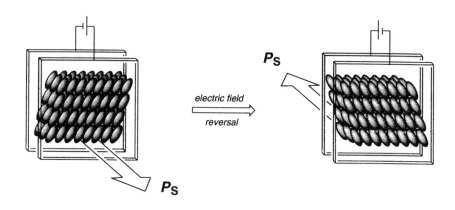

Figure 1. Goldstone-mode switching of a surface-stabilized SmC* liquid crystal.

Several groups have shown that the spontaneous polarization of a SSFLC can be photomodulated in the near-UV range via the reversible *trans-cis* photoisomerization of chiral azobenzene dopants and azobenzene-containing side-chain SmC* copolymers (*8-15*). Ikeda and co-workers have shown that this effect can trigger the Goldstone-mode switching of a SSFLC-SLM by lowering the coercive force (switching threshold voltage) of the SSFLC below a d.c. voltage applied in a direction opposite to that of P_S (*8,12,13,15*). The origin of P_S modulation is thought to be a disruption in the order of the SmC* phase caused by the photoinduced change in shape of the azobenzene dopant from rod-like (*trans*) to bent (*cis*). This so-called photomechanical effect causes a decrease in P_S and a downward shift of the SmC*-SmA* or SmC*-N* phase transition temperature, producing, in some cases, an isothermal phase transition to a non-ferroelectric mesophase (*10-12,14*).

Recently, we demonstrated a new approach to photomodulate P_S in the visible range, without concomitant destabilization of the SmC* phase, which is based on the *trans-cis* photoisomerization of chiral thioindigo dopants (see Figure 2) (*17,18*). According to this approach, a photoactive SmC* phase is obtained by doping an achiral SmC liquid crystal host with a chiral thioindigo dopant. The resulting spontaneous polarization is a function of the thioindigo dopant and depends, in large part, on the degree of steric coupling between the chiral centers and the neighboring functional groups contributing to a dipole moment transverse to the molecular long axis (stereopolar coupling) (*20*). In the *trans*-form, the thioindigo chromophore is non-polar and P_S arises from polar ordering of the two alkoxy groups; in the *cis*-form, the thioindigo chromophore possesses a transverse dipole moment which can also contribute to P_S

provided that some degree of steric coupling exists between the chromophore and the chiral side-chains.

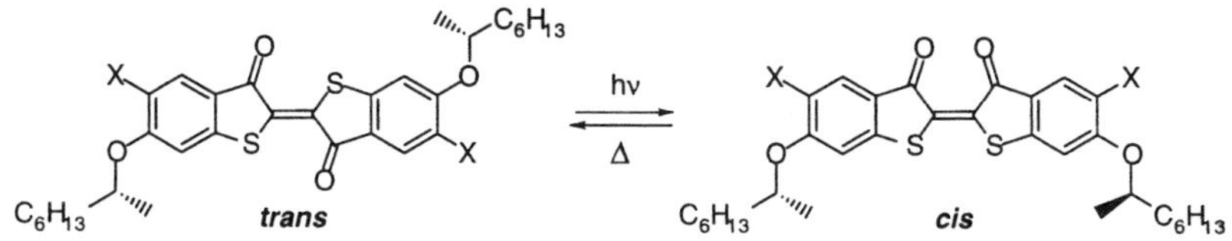

Figure 2. Trans-cis photoisomerization of chiral thioindigo dopants.

Hence, *trans-cis* photoisomerization of the thioindigo dopant **1** in the achiral SmC liquid crystal host MX6120 (Displaytech Inc., Boulder, CO) was shown to have a negligible effect on the spontaneous polarization of the induced SmC* phase due to a lack of steric coupling between the chiral 2-octyloxy side-chains and the thioindigo core (*16*). In order to harness the photoinduced change in transverse dipole moment of the thioindigo core to produce a measurable change in P_S, the dopants **2** and **3** were designed. In these molecules, the chiral side-chains are strongly coupled to the thio-indigo core via steric and dipole-dipole interactions with the adjacent nitro and chloro substituents.

Photomodulation of P_S Using Chiral Thioindigo Dopants

Synthesis

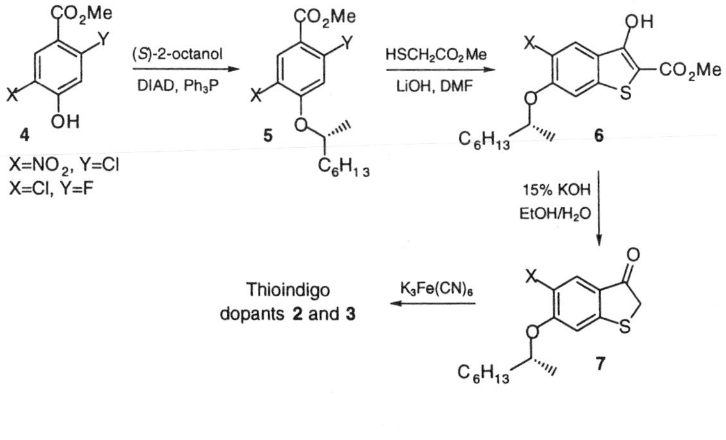

Scheme 1. Synthesis of the chiral thioindigo dopants 2 and 3.

The thioindigo dopants **2** and **3** were synthesized following the approach outlined in Scheme 1 (*18*). The key step in this synthesis is the conversion of the functionalized methyl benzoate **5** to the corresponding enol ester **6** via a one-pot nucleophilic aromatic substitution/Dieckmann condensation sequence using methyl thioglycolate under basic conditions. Ester hydrolysis and decarboxylation to give the corresponding benzothiophenone is followed by oxidation with potassium ferricyanide to give the chiral thioindigo dopants in optically pure form.

Ferroelectric Properties

Doping the SmC liquid crystal host **PhB** with either compound **2** or **3** induces a ferroelectric SmC* phase with a positive spontaneous polarization. In the case of the dinitro dopant **2**, a plot of the reduced polarization P_o (spontaneous polarization normalized for tilt angle variations, $P_o = P_S/\sin\theta$) vs. dopant mole fraction x_d shows a leveling of P_o beyond 3 mol%, which indicates a saturation of the SmC* phase due to the low solubility of the dopant (see Figure 3). By contrast, a similar plot for the dichloro dopant **3** shows a positive deviation from linearity which is consistent with that observed for some dopants with rigid chiral cores (Type II dopants) (*21*). Such behavior, which is not observed with dopant **1** (*16*), is consistent with a strong coupling of the chiral side-chains with the thioindigo core. Unlike **2**, the dichloro dopant **3** dissolves readily in **PhB** up to 10 mol% without any appreciable shift in the SmA-SmC phase transition temperature.

PhB; phase sequence: Cr 35 S_C 70.5 S_A 72 N 75 I

At a given mole fraction, the spontaneous polarization induced by dopant **2** or **3** is significantly less than that induced by dopant **1**. These results are consistent with steric coupling of the chiral 2-octyloxy side-chains with the thioindigo core in **2** and **3**, which forces the alkoxy groups in an anti-coplanar conformation with respect to the *ortho*-substituents (*22*), and orients the corresponding transverse dipoles in opposite directions with respect to the polar axis when the thioindigo core is in the *trans*-configuration. The spontaneous polarization induced by dopant **1** is roughly twice that induced by dopants bearing a single 2-octyloxy side-chain, which suggests that the chiral side-chains in **1** are fully decoupled from the thioindigo core (*23*).

Photomodulation of P_S

Irradiation of dopants **2** and **3** dissolved in benzene (10^{-4} M) at wavelengths corresponding to the visible absorption bands of the *trans*-isomers—514 nm (Ar laser) for **2** and 532 nm (Nd:YAG laser) for **3**—gives photostationary states enriched with the

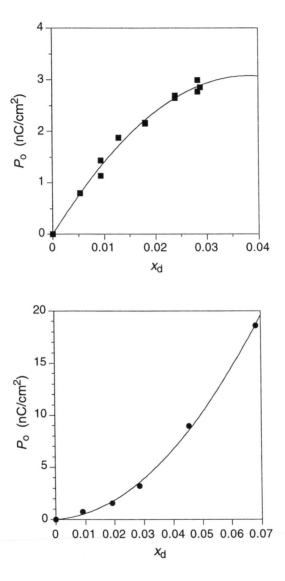

Figure 3. Reduced polarization P_o measured as a function of dopant mole fraction x_d at 10 K below the SmC-SmA* phase transition temperature $(T–T_C = –10$ K) for the SmC* mixtures **2/PhB** (top) and **3/PhB** (bottom). (Reproduced with permission from reference 18. Copyright 1999 The Royal Society of Chemistry.)*

corresponding *cis*-isomers, λ_{max} = 455 and 465 nm, respectively. Upon standing in the dark for 24 h, the two dopants thermally isomerize back to the more stable *trans*-isomers. Irradiation of 4 μm films of **2/PhB** and **3/PhB** mixtures in the SmC* phase at λ = 514 and 532 nm, respectively, causes an increase in spontaneous polarization that is consistent with an increase in transverse dipole moment of the chiral dopants via *trans-cis* photoisomerization (see Figure 4). Irradiation of the **2/PhB** SmC* phase at 514 nm causes an increase in P_S by a relatively constant factor of 1.7-2.0, with the P_S values leveling off as x_d approaches the dopant solubility limit of 3 mol%. Irradiation of the **3/PhB** SmC* phase at 532 nm causes an increase in P_S by a factor which decreases with increasing dopant mole fraction—from 4.0 at x_d = 0.019 to 1.9 at x_d = 0.10. This phenomenon is thought to be related to either the non-linear dependence of the induced polarization on x_d (vide supra), or to antiparallel coupling of the transverse dipoles via dopant aggregation. In both mixtures, the initial P_S value is restored within a period of one minute after shutting off the laser, which indicates that thermal relaxation to the *trans*-isomer is much faster in the SmC* phase than in isotropic liquid solution. This irradiation/thermal relaxation cycle can be repeated a number of times without any decrease in spontaneous polarization and/or P_S photomodulation.

In order to determine the effect, if any, of *trans-cis* photoisomerization on the thermal stability of the SmC* phase in **2/PhB** and **3/PhB** mixtures, P_S was measured as a function of temperature with the 4 μm films kept in the dark, and under constant irradiation at 514 and 532 nm, respectively (see Figure 5). Extrapolation of the P_S vs. temperature plot to P_S = 0 for a 2.6 mol% **2/PhB** mixture shows that *trans-cis* photoisomerization does not cause an appreciable shift in the SmC*-SmA* phase transition temperature; at the same dopant mole fraction, a **3/PhB** mixture gives the same result. However, at higher dopant mole fractions, e.g., x_d = 0.07, *trans-cis* photoisomerization causes a relatively small depression (by 1 K) of the SmC*-SmA* phase transition temperature. At a comparable dopant mole fraction, the *trans-cis* photoisomerization of an azobenzene-containing side-chain SmC* copolymer causes a 15 K drop in the SmC*-SmA* phase transition temperature (*11*).

Optical Switching of a SSFLC-SLM

Principle of Photoinduced Polarization Inversion

Optical switching of a thioindigo-containing SSFLC-SLM can be achieved in the presence of a d.c. electric field by photoinduced polarization inversion using a SmC* liquid crystal mixture consisting of the SmC host **PhB**, the chiral thioindigo dopant **3** and a photoinert chiral dopant such as (*S,S*)-4,4'-bis[(2-chloro-3-methylbutanoyl)oxy] biphenyl (**8**), which induces a negative spontaneous polarization (*24*). The dopant mole fractions are adjusted to produce a net negative polarization when the SSFLC is in the dark (i.e., P_S induced by **8** is greater than P_S induced by **3** in the *trans*-form), and a net positive polarization when the SSFLC is under irradiation at 532 nm (i.e., P_S induced by **3** in the *cis*-enriched photostationary state is greater than P_S induced by **8**). In the presence of an applied d.c. electric, P_S photoinversion results in Goldstone-mode switching of the SSFLC-SLM (see Figure 6).

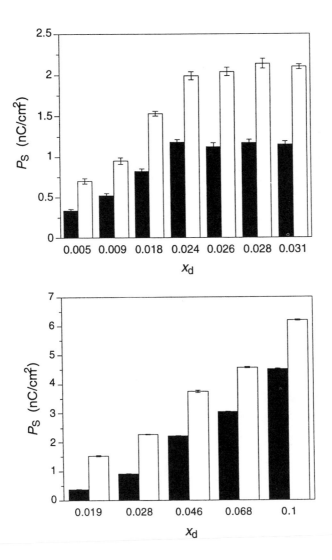

Figure 4. Spontaneous polarization P_S measured as a function of dopant mole fraction x_d at $T–T_C = –10$ K for the SmC mixtures 2/**PhB** (top) and 3/**PhB** (bottom) under constant irradiation at $\lambda = 514$ nm and 532 nm (white columns) and in the dark (black columns). (Reproduced with permission from reference 18. Copyright 1999 The Royal Society of Chemistry.)*

234

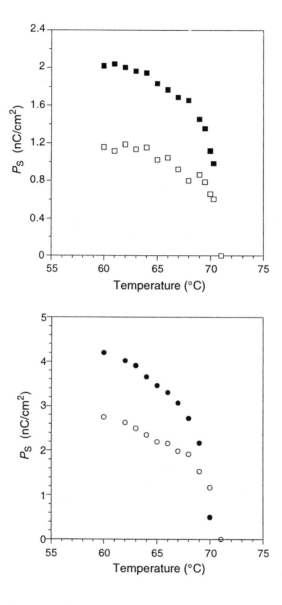

Figure 5. Spontaneous polarization P_S as a function of temperature for the SmC* mixtures 2/**PhB** (top) and 3/**PhB** (bottom) under constant irradiation at $\lambda = 514$ and 532 nm (filled symbols), and in the dark (open symbols). The dopant mole fractions are 0.026 and 0.07, respectively. (Reproduced with permission from reference 18. Copyright 1999 The Royal Society of Chemistry.)

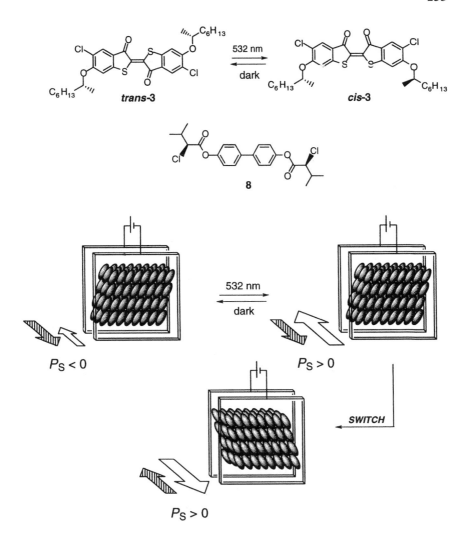

Figure 6. Optical switching of a SSFLC spatial light modulator by photoinduced polarization inversion. The unshaded and shaded arrows represent P_S induced by dopants 3 and 8, respectively. (Reproduced with permission from reference 19. Copyright 1999 Wiley-VCH Verlag GmbH.)

Demonstration of Optical Switching

The P_S photoinversion switching mechanism was demonstrated using the experimental setup described in Figure 7 (*19*). The SmC* mixture containing 3 (3.0 mol%) and 8 (1.3 mol%) was introduced in a 4 μm ITO glass cell and aligned by slow cooling to $T-T_C = -10$ K in a hot stage positioned vertically between two crossed polarizers. The intensity of a 665 nm *reading* beam passing through the polarizers/SSFLC

236

assembly was measured as a function of irradiation conditions using a photodiode detector. A 532 nm *writing* beam was generated from a 50 mW Nd:YAG laser and positioned to irradiate the entire addressed area of the SSFLC cell; a 17 V d.c. voltage was maintained across the cell throughout the experiment.

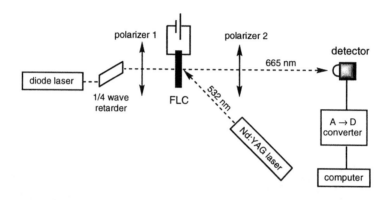

Figure 7. Experimental setup for the time-resolved detection of optical switching in the thioindigo-doped SSFLC-SLM. (Reproduced with permission from reference 19. Copyright 1999 Wiley-VCH Verlag GmbH.)

Irradiation of the SSFLC cell for 2.5 sec caused an increase in transmittance corresponding to Goldstone-mode switching of the SSFLC-SLM (see Figure 8, top). The optical switching time (0-90% transmittance change) was 0.70 sec at T–T$_C$ = –10 K and decreased with increasing temperature to 0.31 sec (T–T$_C$ = –5 K) and 0.12 sec (T–T$_C$ = –1 K)—a behavior that is consistent with the corresponding decrease in viscosity of the SmC* phase. When the temperature reached the SmC*-SmA* phase transition point (71 °C), the photoinduced Goldstone-mode switching stopped. After the writing beam was turned OFF, the SSFLC-SLM eventually switched back to its original state as a result of the thermal isomerization of *cis*-3 back to *trans*-3.

In order to show that the observed change in transmittance (measured as photodetector voltage change, ΔV) corresponds to photoinduced Goldstone-mode switching, the experiment was repeated using a square wave a.c. voltage (17 V, 0.01 Hz) instead of a constant d.c. voltage across the SSFLC cell. Under these conditions, Goldstone-mode switching is caused by the periodic inversion of the electric field. With the SSFLC-SLM kept in the dark, electric field inversion resulted in a change in transmittance that is ca. 20% less than that obtained in the photoinduced switching experiment (see Figure 8, bottom). Upon constant irradiation of the SSFLC-SLM at 532 nm, electric field inversion resulted in the same change in transmittance, but with an upward baseline shift that approximately makes up the difference between ΔV measured in this and the previous switching experiment. These results are consistent with a photoinduced switching of the SSFLC-SLM. Furthermore, it was shown that the photoinduced switching is achieved without any depression of the SmC*-SmA* phase transition temperature, although the baseline shift in Figure 8 suggests that the photoisomerization of 3 does disrupt the SmC* layer structure to some extent.

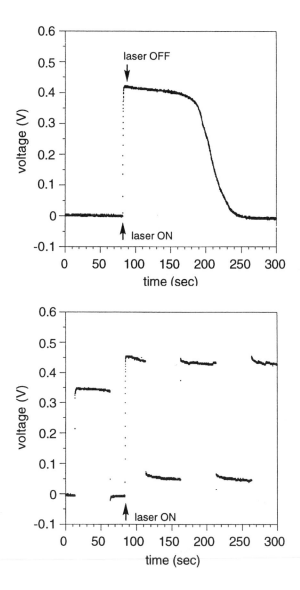

Figure 8. Change in transmittance (photodetector voltage) of the thioindigo-containing SSFLC-SLM positioned between crossed polarizers: (i) upon irradiation at 532 nm with a 17 V d.c. voltage applied across the cell (top), and (ii) upon irradiation at 532 nm with a 17 V square-wave a.c. voltage (0.01 Hz) applied across the cell (bottom). (Reproduced with permission from reference 19. Copyright 1999 Wiley-VCH Verlag GmbH.)

238

Acknowledgments

We are grateful to the Natural Sciences and Engineering Research Council of Canada and to the donors of the Petroleum Research Fund, administered by the American Chemical Society, for financial support of this work.

References

1. Lagerwall, S. T. In *Handbook of Liquid Crystals*; D. Demus, J. W. Goodby, G. W. Gray, H. W. Spiess and V. Vill, Ed.; Wiley-VCH: Weinheim, 1998; Vol. 2B.
2. Walba, D. M. *Science* **1995**, *270*, 250.
3. Goodby, J. W.; Blinc, R.; Clark, N. A.; Lagerwall, S. T.; Osipov, M. A.; Pikin, S. A.; Sakurai, T.; Yoshino, K.; Zeks, B. *Ferroelectric Liquid Crystals: Principles, Properties and Applications*; Gordon & Breach: Philadelphia, 1991.
4. Dijon, J. In *Liquid Crystals: Applications and Uses*; B. Bahadur, Ed.; World Scientific: Singapore, 1990; Vol. 1; Chapter 13.
5. Clark, N. A.; Lagerwall, S. T. *Appl. Phys. Lett.* **1980**, *36*, 899.
6. Meyer, R. B.; Liebert, L.; Strzelecki, L.; Keller, P. *J. Phys. (Paris) Lett.* **1975**, *36*, L69.
7. Moddel, G. In *Spatial Light Modulator Technology: Materials, Devices and Applications*; U. Efron, Ed.; Marcel Dekker: New York, 1995; pp 287-359.
8. Ikeda, T.; Tsutsumi, O.; Sasaki, T. *Synth. Met.* **1996**, *81*, 289.
9. Blinov, L. M.; Kozlovsky, M. V.; Nakayama, K.; Ozaki, M.; Yoshino, K. *Jpn. J. Appl. Phys.* **1996**, *35*, 5405.
10. Negishi, M.; Tsutsumi, O.; Ikeda, T.; Hiyama, T.; Kawamura, J.; Aizawa, M.; Takehara, S. *Chem. Lett.* **1996**, 319.
11. Oge, T.; Zentel, R. *Macromol. Chem. Phys.* **1996**, *197*, 1805.
12. Sasaki, T.; Ikeda, T. *J. Phys. Chem.* **1995**, *99*, 13013.
13. Sasaki, T.; Ikeda, T.; Ichimura, K. *J. Am. Chem. Soc.* **1994**, *116*, 625.
14. Walton, H. G.; Coles, H. J.; Guillon, D.; Poeti, G. *Liq. Cryst.* **1994**, *17*, 333.
15. Ikeda, T.; Sasaki, T.; Ichimura, K. *Nature* **1993**, *361*, 428.
16. Dinescu, L.; Lemieux, R. P. *Liq. Cryst.* **1996**, *20*, 741.
17. Dinescu, L.; Lemieux, R. P. *J. Am. Chem. Soc.* **1997**, *119*, 8111.
18. Dinescu, L.; Maly, K. E.; Lemieux, R. P. *J. Mater. Chem.* **1999**, *9*, 1679.
19. Dinescu, L.; Lemieux, R. P. *Adv. Mater.* **1999**, *11*, 42.
20. Walba, D. M. In *Advances in the Synthesis and Reactivity of Solids*; T. E. Mallouck, Ed.; JAI Press, Ltd.: Greenwich, CT, 1991; Vol. 1; pp 173-235.
21. Stegemeyer, H.; Meister, R.; Hoffmann, U.; Sprick, A.; Becker, A. *J. Mater. Chem.* **1995**, *5*, 2183.
22. Walba, D. M.; Ros, M. B.; Clark, N. A.; Shao, R.; Robinson, M. G.; Liu, J.-Y.; Johnson, K. M.; Doroski, D. *J. Am. Chem. Soc.* **1991**, *113*, 5471.
23. Wand, M. D.; Vohra, R.; Walba, D. M.; Clark, N. A.; Shao, R. *Mol. Cryst. Liq. Cryst.* **1991**, *202*, 183.
24. Bomelburg, J.; Hansel, C.; Heppke, G.; Hollidi, J.; Lotzsch, D.; Scherf, K.-D.; Wuthe, K.; Zaschke, H. *Mol. Cryst. Liq. Cryst.* **1990**, *192*, 335.

Chapter 18

Order–Disorder and Order–Order Transitions in Smectic C* Liquid Crystalline Diblock Copolymers

Mitchell Anthamatten and Paula T. Hammond

Department of Chemical Engineering, Massachusetts Institute of Technology, Cambridge, MA 02139

Block copolymers containing side-chain liquid crystalline (LC) moieties enable materials to have a combination of polymeric and liquid crystalline properties. If the two blocks are incompatible, microphase segregation can occur resulting in isolated LC and amorphous domains. We are interested in how LC ordering overlaps with block copolymer microstructure. A series of diblocks with an amorphous, polystyrene block and a methacrylate based side-chain LC block were prepared. Room temperature and morphological phase diagrams were evaluated using a combination of SAXS and TEM, and microstructure transitions were identified using SAXS at elevated temperatures. These results are compared to LC transitions studied with microscopy and calorimetry. The LC phase effectively changes the Flory-Huggins interaction parameter. Effort is also being made to model the observed transitions and to optimize electro-optic responses.

Thermotropic liquid crystals are rod-like organic molecules that form an ordered fluid with anisotropic optical and electro-magnetic properties. Of particular interest are chiral smectic C* liquid crystals which arrange into layers and have a net spontaneous polarization.[1] These materials are ferroelectric, and an electric field coupled to the spontaneous polarization can be used to switch the material between two stable states. Following their discovery, ferroelectric liquid crystals received widespread attention because of their microsecond response times and bistable switching capabilities. Many applications can be envisioned including light valves, optical switches, and memory storage devices. However, to realize any application, the helical twist associated with the smectic C* phase must be unwound; this creates engineering challenges such as maintaining a constant cell thickness and controlling mesogen orientation across the electo-optic cell.

Polymer liquid crystals (PLC's) are formed by incorporating liquid crystals into the polymer backbone or as a side-chain moiety.[2] Their thermotropic phase behavior depends not only on mesogen characteristics such as shape and polarity, but also the flexibility, length, and tacticity of the polymer backbone. Mesogenic ordering is

240

affected strongly by interactions with the polymer backbone.[3] Generally, side-chain PLC's have broader smectic phases than their small molecule analogs.

Diblock copolymers are formed if two polymer chains A and B are covalently joined at their ends. If A and B are dissimilar enough, and if the total molecular weight is large enough, microphase segregation occurs resulting in a microstructure on the scale of the radius of gyration of the polymer. The extent of phase segregation can be quantified by the product χN where χ is the Flory-Huggin's interaction parameter, describing the enthalpic penalty of mixing, and N is the copolymer degree of polymerization ($N = N_A + N_B$), reflecting the entropic drive of an N dimensional chain.[4] Upon microphase segregation, periodic domains rich in A and B form a self-assembled morphology, and, as Figure 1 suggests, the type of microstructure depends largely on the composition of the diblock.

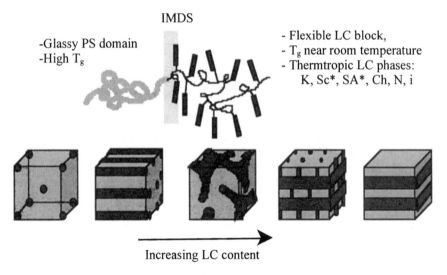

Figure 1. *Cartoons illustrating the molecular architecture of LC diblock copolymers (top) and possible phase-segregated microstructures (bottom)*

This paper relates to diblock copolymers with a high T_g amorphous block and a more flexible, side-chain ferroelectric liquid crystalline (LC) block. This architecture allows a self-supportable LC material, e.g. a glassy solid with electro-optical properties. The idea is to combine the mechanical integrity associated with the PS block with the electro-optic properties arising from the LC block. Additionally, it has been shown that the block copolymer intermaterial dividing surface (IMDS) is capable of stabilizing liquid crystalline side-chain moieties much as a buffed electro-optic cell stabilizes small molecule LC's.[5,6] With sufficient control over the type and orientation of microdomain morphology it should be possible to achieve macroscopic spontaneous polarization over large domains. A number of research groups, including ours, have already taken significant steps toward developing this technology.[7-18]

Previously we reported the synthesis and characterization of a series of well-defined smectic C* diblock copolymers.[14] Room temperature morphologies and mesomorphic superstructures in solvent cast, roll-cast, and fiber drawn samples were also reported.[18] The objective of this paper is to probe how thermotropic LC order transitions affect diblock copolymer morphology. The approach is to systematically measure and compare LC transition temperatures to morphological order-order transition (OOT) and order-disorder transition (ODT) temperatures.

Materials

The molecular structure of the diblock copolymers considered here is shown in Figure 2. Synthesis followed the literature procedure that involves direct anionic polymerization of styrene followed by addition of a mesogenic methacrylate monomer.[14] This technique enables control of both the total molecular weight and the LC volume fraction ϕ_{LC} which determine the phase segregated microstructure as shown in Figure 3. For ϕ_{LC} less than ~ 0.25, dispersed morphologies (DS) are observed that consist of unaligned LC spheres in a continuous PS domain. At intermediate volume fractions the diblocks form layered mophologies that can be subdivided into completely lamellar (L) and predominately lamellar (PL) morphologies. The PL morphology consists of lamellae that coexist with PS cylinders which are arranged either hexagonally (PL/HCP) or in a modified layer (PL/ML) fashion. Cylinders in the ML form retain periodic spacing between planes of cylinders but not necessarily between cylinders themselves. Finally, at high ϕ_{LC}, only hexagonally-close-packed PS cylinders (HCP$_{PS}$) self-assemble in a continuous LC phase. Note the phase diagram is asymmetric in that microstructures with continuous LC phases occupy a much larger part of phase space than do structures with a continuous PS phase.

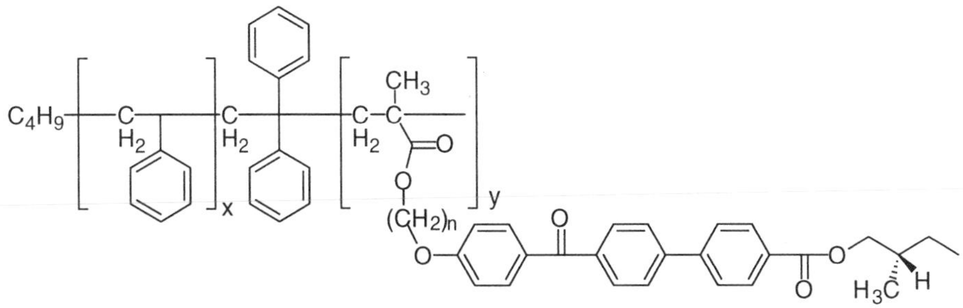

Figure 2. Molecular structure of PS-HBPB diblock copolymers; n = 6 for all materials in this study.

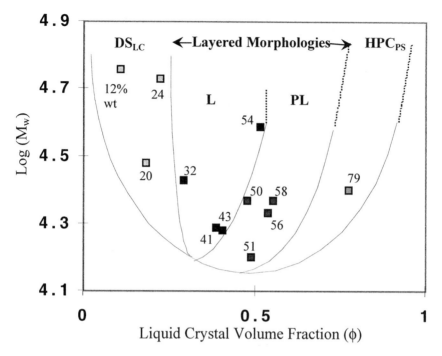

Figure 3. *Experimentally determined morphological phase diagram for PS-PHBPB diblock copolymers. Morphologies observed include: dispersed LC spheres (DS_{LC}), lamellae (L), predominately lamellae (PL), and hexagonally-packed PS cylinders (HPC_{PS}).*

Experimental

Liquid crystalline phase transition temperatures were assigned using differential scanning calorimetry (DSC) and polarized optical microscopy (POM).[6] A Perkin-Elmer DSC-7 with a cooling accessory was used at scanning rates of 20 °C / min. Microscopy observations were made using a Leitz optical microscope equipped with a CCD camera and a Metler FP-82 heating stage and controller (heating/ cooling rate of 10 °C / min). LC phases were assigned by examining textures after samples were annealed for hours just beneath their phase transition temperature.

Small angle X-ray scattering (SAXS) was used to confirm phase assignments and to measure block copolymer order-disorder transition and order-order transition (ODT & OOT) temperatures. Before X-ray analysis, in order to encourage phase segregation, all samples were annealed in a vacuum oven for 48 hours at 110 °C, just above the T_g of the PS block. The X-ray sample chamber was evacuated, and samples were suspended vertically, using polyimide adhesive tape, in an Instec temperature controlled hotstage (model: HS250). X-rays were generated from a rotating copper anode producing Cu Kα radiation (λ = 1.54 Å) and operating at 40 kV x 30 mA. X-rays were scattered on to a detector, 54.1 cm from the sample, that consists of a pressurized xenon chamber with a wire grid assembly (512 x 512). Data were acquired as scattered X-ray intensity, I, taken as a function of the scattering vector $\mathbf{q}$ = $4\pi/(\lambda \sin \theta)$ and temperature. To avoid sample degradation, shorter scattering times were used for high-temperature scans. Preceding data analysis, all intensities were normalized to one hour to allow direct comparison between temperatures. However, due to differences in sample thickness, direct comparison of intensity between samples of different composition is not appropriate. ODT and OOT transitions were determined by plotting both inverse maximum intensity $1/I_{max}$ and the wavelength of concentration fluctuations D vs. inverse temperature T^{-1}. In the disordered state, $1/I_{max}$ is predicted to decrease linearly with T^{-1} and D should be nearly independent of T^{-1}; the theory behind this analysis is addressed in reference 19. Other observations were also considered when assigning ODT's, such as decreases in film viscosity and the disappearance of higher-order SAXS peaks.

Transmission electron micrographs were obtained using a JOEL 200CX electron microscope operating at 160kV. Preceding microscopy, samples were cut into 40 nm sections using a Reichert-Jung FC4E microtome and were then stained with RuO_4 vapor for 20 minutes.

Results & Discussion

Six LC diblock copolymers representing different regions of the phase diagram were studied at elevated temperatures. Table I is a summary of the material characteristics and LC phase transitions—systematic synthesis & characterization of LC phases is described in elsewhere.[6] Notice that for samples with intermediate to high LC volume fractions, PS-HBPB50, 56, & 58, that the smectic mesophases appear destabilized. These samples have isotropization temperatures (T_{iso}) as low as 159 °C and only exhibit the smectic C* mesophase. For samples with lower LC fractions typically have smectic phases that are stable up to 180 - 200 °C.[6] The destabilization of smectic phases in PS-HBPB50, 56, & 58 results from an entropic frustration arising from conformationally asymmetric diblocks situated at a lamellar interface. This idea will be discussed in more detail later; however, note that these three sample are the only ones that exhibit the PL morphology. Conversely, PS-HBPB79, with a higher LC content (ϕ_{LC} = 0.77), has a very stable smectic phase and forms a HPC microstructure.

The temperature dependent SAXS scattering patterns for PS-HBPB32 are shown in Figure 4. The smaller angle reflections at q $\approx$ 0.34 nm^{-1} correspond to block copolymer microdomain ordering of about 186 Å. Higher-order peaks are difficult to discern since only two-hour scans were acquired at each temperature. The room temperature morphology of these samples was confirmed in an earlier study using much longer scans.[18] Weaker signals, not shown in the figure, are present at wider angles (q $\approx$ 2.2 nm^{-1}) and arise from smectic ordering. These peaks appeared broader than expected for polymer smectics[20,21], however, upon prolonged annealing, they became narrower and more defined. Additionally, some broadening in the wider-angle regime arises from equipment limitations. PS-HBPB43 had nearly identical thermal behavior and thus is not shown.

Table I. Material Characteristics and LC phase transitions

Diblock	Block M_n (kg/mol)		Total M_n (kg/mol)	M_w/M_n	RT Morph.	LC Phase Transitions (°C)
	PS	LC				
PS-HBPB32	17.7	7.3	24.6	1.07	LAM	H: S_C*(163)S_A(**181**)I C: I(171)S_A(156)S_C*
PS-HBPB43	10.3	7.4	17.7	1.08	LAM	H: S_C*(136)S_A(158)Ch(**177**)I C: I(166)S_C*
PS-HBPB50*	8.4	11.0	19.4	1.14	PL/ML	H: S_C*(**161**)I C: I(145)S_C*
PS-HBPB56*	10.5	11.0	21.5	1.07	PL/HPC	H: S_C*(**159**)I C: I(153)S_C*
PS-HBPB58*	8.8	12.0	20.8	1.11	PL/ML	H: S_C*(**177**)I C: I(171)S_C*
PS-HBPB79*	4.8	16.3	21.1	1.08	HPC	H: S_C*(176)S_A(**212**)I C: I(203)S_A(173)S_C*

Both PS-HBPB32 & 43 have ODT temperatures that are slightly higher than their isotropization temperature T_{iso}. This correlation indicates that the isotropization of the LC domain triggers the ODT transition. This phenomenon can be explained by slightly modifying the criterion for phase segregation to account for LC order as well as segmental interactions. For amorphous-amorphous diblock copolymers, phase segregation occurs if the product χN exceeds a critical value. For the LC diblocks considered here, there is an additional increment of free energy $\chi_{LC} N$ involved with mixing ordered mesogens and disordered PS chain segments. Thus, as suggested in our earlier papers[6,22], phase segregation occurs when $(\chi_{LC} + \chi)N$ exceeds a critical value. The quantity $(\chi_{LC} + \chi)$ represents the thermodynamic incompatibility between the two blocks. For PS-HBPB32 & 43, the χ_{LC} term disappears as the sample is heated through the LC isotropization point, thus reducing $\chi_{eff}N$ below its critical value. In a study of symmetrical PS-LC diblocks ($\phi \sim 0.5$) Yamada et. al. observed a similar phenomena where for lower molecular weight samples (< 10 kg/mol), T_{ODT} was coincident with T_{iso}.[23]

PS-HBPB50, 56, and 58 all exhibit PL microphase segregated morphologies at room temperature.[18] PS-HBPB58 was unique since it showed easily discernible scattering peaks for both lamellae and the cylindrical defects in the shorter (2 hr) scans used at elevated temperatures. The SAXS profiles for PS-HBPB58 change significantly as temperature is increased as shown in Figure 5. The morphology at lower temperatures is predominately lamellae with cylindrical defects arranged in modified layer fashion (PL/ML). At lower temperatures the first-order Bragg peak at $q \approx 0.39$ nm^{-1} corresponds to a lamellar spacing of 156 Å, and the less intense peak $q \approx 0.50$ nm^{-1} (123 Å) signifies the layers of ML cylinders. Both peaks have higher-order signals that correspond to a scattering ratio of 1:2:3. Smectic layer ordering is apparent from Bragg scattering at wider angles ($q \sim 2.1$ nm^{-1}). Between 160 and 180 °C this reflection disappears which is consistent with S_C* to isotropic transition at 177

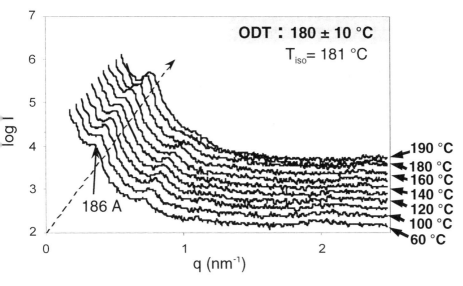

Figure 4. *SAXS 1D diffraction patterns at elevated temperatures for PS-HBPB32. Data are plotted as the logarithm of relative scattered intensity log I(q) vs. the scattering factor* ***q****.*

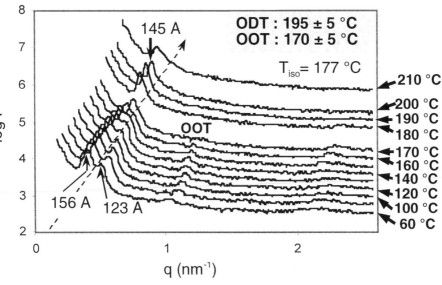

Figure 5. *SAXS 1D diffraction patterns at elevated temperatures for PS-HBPB58. Data are plotted as the logarithm of relative scattered intensity log I(q) vs. the scattering factor* ***q****.*

°C observed using POM. Between 120 °C and 160 °C, the first-order peaks gradually move closer together until they are completely merged at 180 °C, where, based on analysis of SAXS intensities, the morphology is strongly phase-segregated. At temperatures exceeding 190 °C, the intensity of the remaining peak decreases, indicative of an ODT. Upon cooling, two separate sets of peaks reappeared, indicating that this transition is reversible; this has been explained in detail in a recent publication.[22]

To confirm PS-HBPB58's morphology, a sample was annealed at 110 °C for 72 hours. As discussed in reference 22, resulting TEM micrographs combined with SAXS diffraction patterns confirm that alternating lamellae are observed coexisting with PS cylinders. To examine the morphology at the higher-temperature state, a SAXS specimen was quickly cooled from 210 °C to room temperature, and another TEM was acquired; this time a completely lamellar morphology was observed. Based on these and other experiments[22], the sample undergoes a thermoreversible OOT, between 160-170 °C, from a predominately lamellar morphology to a completely lamellar one. This type of transition was an unexpected result; based on the theory of amorphous-amorphous block copolymers, one would expect the opposite: an order transition upon heating from a lamellar morphology to a cylindrical morphology.[19,24-26]

Figure 6 is a cartoon that explains how liquid-crystalline order triggers the OOT in PS-HBPB58 (PL/ML → L). When the LC phase is ordered, the polymer molecule's natural shape is 'wedge-shaped', and it is conformationally asymmetric in the x-y plane. Consequently, it natural for this shape to self-assemble into a structure with a curved IMDS[27,28] with the LC domain on the convex side. However, the diblock's volume fraction ($\phi \sim 0.56$) is more compatible with a lamellar microstructure. Upon heating, as the LC order disappears, the diblock is less asymmetric and becomes more suited for a lamellar interface. In summary, if conformationally asymmetric diblock are situated at a planar interface the LC chains become entropically frustrated. This explains 1) the existence of a near-equilibrium microstructure with coexisting lamellae and PS cylinders 2) the destabilization of the smectic phases observed in PS-HBPB50,56, & 58, and 3) the OOT that occurs in PS-HBPB58.

Free Energy Description of Side-Chain LC Diblock Copolymers

Theories for diblock copolymer phase segregation are now advanced and agree well with experiment. For a recent review of this topic the reader is referred to Hamley.[4] Theories accurately predict microphase structures as a function of diblock copolymer composition ϕ and the product χN. A classic example is Semenov's mean-field theory for highly incompatible blocks[25]; it predicts microdomain structure by balancing interfacial free energies between the two blocks with elastic free energies associated with stretching polymer chains away from the IMDS. The Semenov theory enables the ODT phase boundary to be estimated however it predicts the phase boundaries between morphologies to be independent of χN.

Figure 7a shows a simplified phase diagram for amorphous-amorphous block copolymers. If one block contains, as part of its monomer repeat unit, a liquid crystalline moiety, then one would expect the morphological phase diagram to change

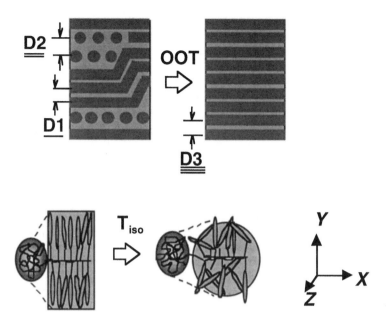

Figure 6. *Top: order-order transition observed in PS-HBPB58 from a predominately lamellar morphology to a completely lamellar one. Three different Bragg periods: D1 (156 Å), D2 (123 Å), & D3 (145 Å) were observed in SAXS (see Figure 5). Bottom: cartoon showing how LC smectic order can lead to conformationally asymmetric diblocks in the x-y plane—upon isotropization, conformational asymmetry is reduced.*

significantly. Figures 7b and 7c are projections of phase diagrams for main-chain and side-chain liquid crystalline diblock copolymers. The upper regions of Figures 7b and 7c represent lower, anisotropic states and are asymmetric about $\phi_{LC} = 0.5$. Experimentally, this is indeed the case; we recently confirmed that morphological phase boundaries as well as ODT temperatures are strongly affected by LC ordering.[6] The lamellar samples PS-HBPB32 & 43 are cases in which ODT's are initiated by isotropization of the LC phase. Other researchers have observed similar phenomena.[23] However, it is less frequent for transitions in LC order to trigger an OOT.[29] In PS-HBPB58 the sample's morphology transforms, upon heating, from a PL/ML morphology to a completely lamellar one; this transition is unusual because curvature is *lost* upon heating; as Figure 7a indicates, this transition is counter to theoretical predictions for amorphous-amorphous block copolymers.

Theoretical studies have focused on diblocks with mesogens incorporated tandem within the LC main chain. Consequently, the extension of the LC backbone perpendicular to the IMDS is facilitated below the isotropization temperature by LC ordering. This results in a bias of morphological phases toward microstructures with the LC phase residing in micelle cores[30,31] as shown in Figure 7b. To our knowledge, there have been no theoretical phase diagram models formulated for LC diblocks that have an amorphous block and a side-chain LC block. An extensive analysis of order in nematic side-chain homopolymers was carried out by Wang and Warner.[3] This model involved self-consistent calculation of the mesogenic and LC backbone order parameters, which in turn were used to predict backbone chain configurations, both parallel and perpendicular to mesogenic ordering.

We are seeking a free energy expression for side-chain LC diblock copolymers that captures key features of the morphological phase diagram such as LC induced ODT's and OOT's. Four free energy components are considered: $F_{TOT} = F_{LC} + F_{AM} + F_{IMDS} + F_{AM,LC}$ where F_{TOT} is the total free energy per diblock copolymer, F_{LC} is the total free energy associated with the LC block (including stretching), F_{AM} is due to entropic stretching of the amorphous block, F_{IMDS} is a surface tension term, and $F_{AM,LC}$ represents Flory-Huggins type interactions between the LC and amorphous repeat units. F_{LC} is described using Wang and Warner's model[3] and includes elastic stretching of the LC backbone. Since this model was intended for LC homopolymers, F_{LC} is independent of diblock copolymer morphology. Calculation of the amorphous stretching energy is similar to that of Gido[28] where the amorphous block is considered a polymer brush grafted to a surface with mean and Gaussian curvatures H and K. A 'wedge-like' geometry is assumed subjecting both domains to the volume constraint:

$$\frac{\frac{1}{3} K h_i^3 + H h_i^2 + h_i}{V_i} = \sigma \qquad (1)$$

Where h_i is the brush height and σ is the graft density at the IMDS. The brush energy per chain is calculated by progressively adding chains to a surface. By differentiating eqn 1 $d\sigma/dh$ can be calculated and the stretching free energy obtained using:

$$F_{AM} = \frac{1}{\sigma} \int_0^{\alpha_{AM}} d\sigma' \, s(\sigma) = \frac{1}{\sigma} \int_0^{h_{AM}} dh' \frac{d\sigma}{dh} s(h') \qquad (2)$$

This equation is exact for lamellae and for grafting to a concave side of the IMDS; it is a good approximation for convex surfaces with lower curvatures. However, equation 2 requires an expression for h_{AM}. This is obtained by applying the wedge-like volume constraint (eqn. 1) to each phase, equating the two resulting graft densities, and solving for h_{AM} as a function of H, K, and h_{LC}. The third term, the interfacial surface energy F_{int}, is the interfacial tension γ divided by the graft density σ. For the case of a homogeneous melt F_{int} is assumed zero. Lastly, $F_{AM,LC}$ describes the Flory-Huggins interactions between amorphous and LC monomers: $F_{AM,LC} = \chi N \langle \phi_{LC} \phi_{AM} \rangle$.

In summary, a free energy description has been outlined to reconcile our experimental observation of LC triggered ODT and OOT transitions. By minimizing the F_T at a given temperature, the equilibrium morphology for a pre-specified architecture can be determined. By iteration over a temperature and composition range, phase diagrams resembling those in Figures 7b & 7c should be predicted. Ongoing work includes refinement and validation of this model.

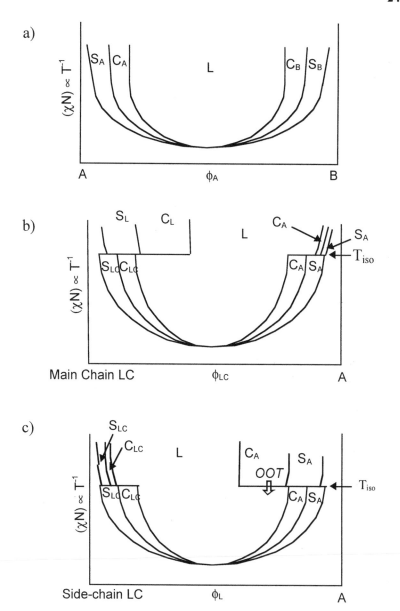

Figure 7. *Morphological phase space predictions for a) amorphous-amorphous, b) (main-chain LC)-amorphous, and c) (side-chain LC) - amorphous diblock copolymers. Regions include LC spheres (S_{LC}) and cylinders (C_{LC}), lamellae (L), amorphous cylinders (C_A) and spheres (S_A).*

Conclusions

A series of LC side-chain diblock copolymers were investigated The data presented indicate that LC isotropization can induce ODT's for the lower molecular weight samples. PS-HBPB58 exhibits a thermoreversible OOT that progresses, upon heating, from a predominately lamellar morphology to a completely lamellar one. Conformational asymmetry explains 1) the destabilization of smectic phases in samples with $\phi_{LC} \sim 0.50 - 0.58$, 2) the presence of curvature at unusually low LC volume fractions, and 3) the loss of curvature in PS-HBPB upon heating through its OOT. Efforts are currently underway to describe the interplay between mesogenic and morphological order by using a free energy modeling approach.

Acknowledgements

The research was supported by the National Sciece Foundation Polymer Program for funding under Grant Number DMR-9526394. We are also thankful to Professor Ned Thomas for use of his microtome equipment.

Literature Cited

1) Clark, N. A.; Lagerwall, S. T. *Appl. Phys. Lett.* **1980**, *36*, 899-901.
2) Blackwood, K. M. *Sciencs* **1996**, *273*, 909-912.
3) Wang, X. J.; Warner, M. *J. Phys. A: Math Gen.* **1987**, *20*, 713-731.
4) Hamley, I. W. *The Physics of Block Copolymers*; Oxford University Press: New York, 1998.
5) Mao, G.; Wang, J.; Clingman, S. R.; Ober, K.; Chen, J. T.; Thomas, E. L. *Macromolecules* **1997**, *30*, 2556-2567.
6) Zheng, W. Y.; Hammond, P. T. *Macromolecules* **1998**, *31*, 711-721.
7) Ruokolainen, J.; Saariaho, M.; Ikkala, O.; ten Brinke, G.; Thomas, E. L.; Serimaa, R.; Torkkeli, M. *Macromolecules* **1999**, *32*, 1152-1158.
8) Chiellini, E.; Gallo, G.; Angeloni, A. S.; Laus, M.; Bignozzi, C. *Macromol. Symp.* **1994**, *77*, 349-358.
9) Adams, J.; Gronski, W. *Makromol. Chem., Rapid Commun.* **1989**, *10*, 553-557.
10) Zaschke, B.; Frank, W.; Fischer, H.; Schmutzler, K.; Arnold, M. *Polymer Bulletin* **1991**, *27*, 1-8.
11) Bohnert, R.; Finkelmann, H. *Macromol. Chem. Phys.* **1994**, *195*, 689-700.
12) Yamada, M.; Iguchi, T.; Hirao, A.; Nakahama, S.; Watanabe, J. *Macromolecules* **1995**, *28*, 50-58.
13) Fischer, H.; Poser, S. *Acta Polymer.* **1996**, *47*, 413-428.
14) Zheng, W. Y.; Hammond, P. T. *Macromol. Rapid Commun.* **1996**, *17*, 813-824.
15) Omenat, A.; Hikmet, R. A. M.; Lub, J.; van der Sluis, P. *Macromolecules* **1996**, *29*, 6730-6736.

16) Mao, G.; Wang, J.; Ober, C. K.; Brehmer, M.; O'Rourke, M. J.; Thomas, E. L. *Chem. Mater.* **1998**, *10*, 1538-1545.
17) Brehmer, M.; Mao, G.; Ober, C. K. *Macromol. Symp.* **1997**, *117*, 175-179.
18) Anthamatten, M.; Zheng, W. Y.; Hammond, P. T. *Macromolecules* **1999**, *32*, 4838-4848.
19) Leibler, L. *Macromolecules* **1980**, *13*, 1602-1617.
20) Lipatov, Y. S.; Tsukruk, V. V.; Shilov, V. V. *Rev. Macromol. Chem. Phys.* **1984**, *C24*, 173-238.
21) Shilov, V. V.; Tsukruk, V. V.; Lipatov, Y. S. *Journal of Polymer Science* **1984**, *22*, 41-47.
22) Anthamatten, M.; Hammond, P. T. *Macromolecules* **2000**, *32*, 8066-8076.
23) Yamada, M.; Iguchi, T.; Hirao, A.; Nakahama, S.; Watanabe, J. *Polymer J.* **1998**, *30*, 23-30.
24) Fredrickson, G. H.; Helfand, E. *J. Chem. Phys.* **1987**, *87*, 697-705.
25) Semenov, A. N. *Sov. Phys. JETP* **1985**, *61*, 733-742.
26) Lescanec, R. L.; Muthukumar, M. *Macromolecules* **1993**, *26*, 3908-3916.
27) Milner, S. T. *Macromolecules* **1994**, *27*, 2333-2335.
28) Gido, S. P.; Z.G., W. *Macromolecules* **1997**, *30*, 1997.
29) Saenger, J.; Gronski, W.; Maas, S.; Stuehn, B.; Heck, B. *Macromolecules* **1997**, *30*, 6783-6787.
30) Sones, R. A.; Petschek, R. G. *Physical Review E* **1994**, *50*, 2906-2912.
31) Williams, D. R. M.; Halperin, A. *Phys. Rev. Letters* **1993**, *71*, 1557-1560.

Chapter 19

Ferroelectric Liquid Crystalline Polymers

R. Shashidhar and J. Naciri

Center for Bio/Molecular Science and Engineering, Code 6900, Naval Research
Laboratory, Washington, DC 20375

We present here results of our studies on the structure-property
relationship of side chain liquid crystalline polymers exhibiting
ferroelectric behavior. In particular the polarization properties of
linear as well as cyclic siloxane-based ferroelectric polymers and
the nature of the self-assembly of linear siloxane-based systems in
monolayers and multilayers are also discussed.

Since the early work on chiral smectic-C side chain liquid crystalline polymer
exhibiting ferroelectric behavior[1-7], there has been considerable activity aimed at
synthesising new polymeric and oligomeric materials with ferroelectric properties.
An excellent recent review[8] of the behavior of ferroelectric liquid crystalline
polymers (FLCPs) summarizes the different types of chemical structures, phase
behavior and the associated ferroelectric properties. The Liquid Crystal Group at the
Naval Research Laboratory (NRL) has been working on the structure-property
relationship of different types of FLCPs like homopolymers, copolymers, terpolymers
and also on cyclic siloxane oligomers. The NRL group has also focussed on studying
the ability of FLCPs to form monolayers at the air-water interface and Langmuir
Blodgett (LB) multilayers on surfaces. This chapter is essentially an overview of the
several important results of the NRL group. The chapter has been organized into two
main parts. The first part is devoted to a discussion of the ferroelectric properties of
bulk FLCPs while the second part deals with the structure and functionality of LB
multilayers.

Ferroelectric Properties of Bulk FLCPs.

The characteristic feature of the bulk FLCP is that it possesses a layered
structure and the constituent chiral molecules are tilted within each layer[8]. The most
important characteristic of FLCP is the existence of spontaneous polarization (P_s) due

to the non-vanishing component of the polarization vector. However a finite value of P_s can be obtained only on unwinding the helix resulting from the chirality of the molecules and its propagation along the layer normal. This is usually achieved by surface stabilization geometry[9] wherein the alignment layer of the surface interacts with the liquid crystal molecules to suppress the azimuthal degeneracy of the tilt. The molecular features that dictate polarization of FLCPs are the same as those that determine P_s in low molar mass ferroelectric liquid crystals (FLCs). However, in FLCPs there are some additional factors - dilution of the side groups and nature of spacer that attaches the mesogenic unit to the backbone.

When an internal alternating electric field is applied to an FLCP, the mesogenic units switch from one state to another state of equal and opposite tilt. This switching is known to be very fast ($\sim$ 100 μs or less) for low molar mass FLCs where as FLCPs can exhibit switching times a few hundred microseconds to a few hundred milliseconds depending on the coupling of the mesogenic unit to the backbone and also the population of the mesogenic units in the polymer which dictates the packing of the mesogenic groups. In this section we discuss the polarization as well as the switching characteristics of different types of FLCPs. We shall also present results concerning collective mode as well as magnetic field induced order in FLCPs.

Polarization and Tilt Angle Studies

The effect of varying the number of mesogenic groups (attached to a partially methylated siloxane backbone) on the polarization has been studied. The structures of the homopolymer (where in mesogenic units are attached to all available substituting units of the backbone), the copolymer (where in the mesogenic units are attached to only some sites, the other contain methyl groups) are shown in Figure 1 along with the structure of the mesogenic unit[10]. Data on the temperature dependence of P_s for the homopolymer and copolymer[11] (which bear the same mesogenic unit) are shown in Figure 2. It is seen that any given relative temperature (T-T_{AC^*}), where T_{AC^*} is the smectic A (Sm-A) – smectic C* (Sm-C*) transition temperature, P_s values are similar for both homopolymer and copolymer. Hence the dilution of the mesogenic units in the backbone does not appear to affect P_s.

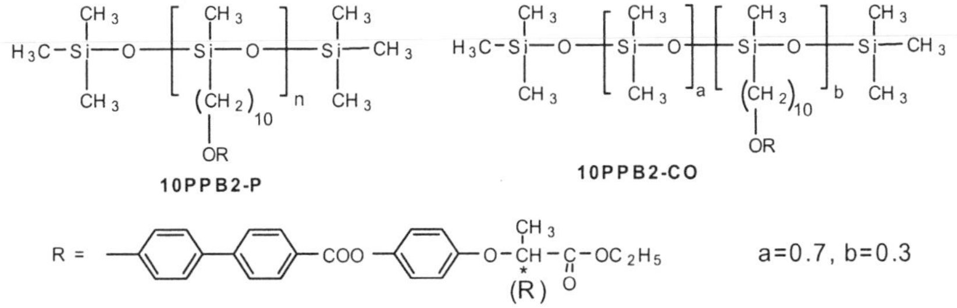

*Figure 1. Chemical structures of FLCPs **10PPB2-P** (homopolymer) and **10PPB2-CO** (copolymer). R represents the mesogenic unit attached to the polysiloxane backbone. (R*) defines the configuration of the chiral center. From ref. [10].*

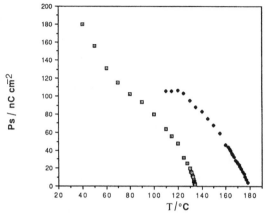

*Figure 2.   The temperature dependence of P_s for the homopolymer **10PPB2-P** (diamond) and   the copolymer **10PPB2-CO** (square). T_{AC*} values are: 178 °C for **10PPB2-P** and 136 °C for **10PPB2-CO**. From ref [11].*

$DP_n = 30$ $b/(a+b) = 0.3$

Series I R = -C_2H_5, -C_3H_7, -C_4H_9 **10PPBm-CO** where m is the number of carbon atoms in R
Series II R = -CH_2CF_3, -$CH_2C_2F_5$,-$CH_2C_3F_7$ **10PPBFm-CO** where m is the number of carbon atoms in R

Figure 3. Molecular structures of the ferroelectric liquid crystalline copolymers. From ref. [13]

On the other hand, the unattached mesogen (which is a low molar mass FLC) has a much higher P_s than the copolymer. Mixtures of the copolymer with its side group antecedent mesogen essentially scale with the concentration of the copolymer[12]. We have also investigated the effect of changing the chiral end chain on P_s. The materials studied, whose chemical structures[13] are given in Figure 3, essentially consist of two classes of materials – Series I have different lengths of hydrocarbon chains close to the chiral center while Series II have instead fluorocarbon chains. Figure 4 compares P_s data for two corresponding members (with the same number of carbon atoms in the chain) of Series I and II. Clearly, P_s of the fluorocarbon copolymer is significantly higher than the corresponding hydrocarbon analog. This result is likely due to the increase in crystallinity and hindrance to rotation of the fluorinated chains. However, for each of the series, there is hardly any dependence of P_s for successive homologs within the series[13]. We also studied P_s of terpolymers[14] where in two different types of mesogenic units are attached to the siloxane backbone in addition to the methyl group (See Figure 5). This is a particularly useful approach if the two mesogens are structurally incompatible and hence a direct mixing of the two polymers in the bulk state is difficult. Interestingly, this leads to unusual polarization dependence on temperature[14] (Figure 6). P_s varies more or less linearly right up to the transition temperature to the smectic A phase.

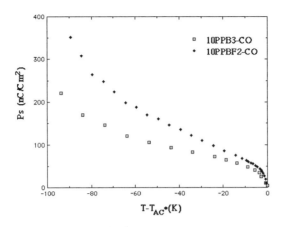

*Figure 4. Comparison of the polarization values for the copolymers **10PPB3-CO** and **10PPBF2-CO**. From ref. [13].*

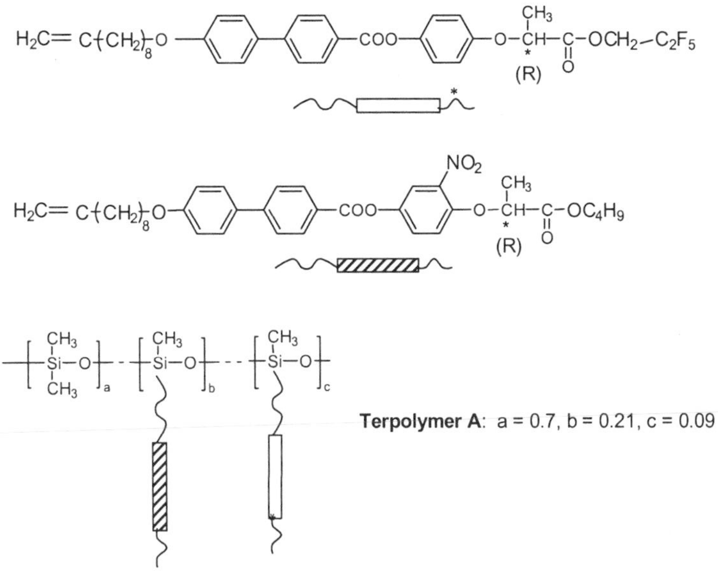

H₂C= C(CH₂)₈O biphenyl-COO-phenyl-O-CH(CH₃)-C(=O)-OCH₂-C₂F₅ (R)

H₂C= C(CH₂)₈O biphenyl-COO-phenyl(NO₂)-O-CH(CH₃)-C(=O)-OC₄H₉ (R)

Si(CH₃)₂-O structure with a, b, c units

Terpolymer A: a = 0.7, b = 0.21, c = 0.09

Figure 5. Structure of Terpolymer A.

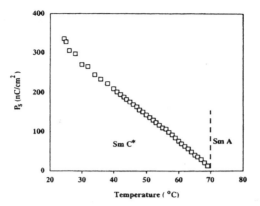

Figure 6. Temperature variation of the spontaneous polarization for terpolymer A. The vertical dashed line signifies T_{AC*} transition. From ref. [14].

Electro-optic Switching

The response time (τ) associated with switching of molecules of FLCP from one tilt to the other upon reversing the field can be determined by measuring the optical transmission through an FLCP sample in a surface-stabilized geometry. To a first

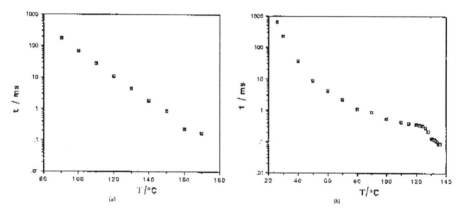

Figure 7. The temperature dependence of (a) the switching times (τ) in the Sm-C phase of the polymer **10PPB2-P** (10 μm cell, E = ±34V) and (b) the Sm-A and Sm-C* phases of the copolymer **10PPB2-CO** (15μm cell, E = ±100V). From ref [11].*

approximation, τ is given by γ/P_sE, where γ is the rotational viscosity and E the applied field. Hence with increasing P_s, τ should decrease. However, this is not necessarily the case since there is always an increase in viscosity associated with an increase of P_s. Hence, generally with increasing P_s, usually a slower switching (higher τ) is observed. This problem can be overcome to some extent by the copolymer approach. The switching times of the copolymer **10PPB-2CO** are much lower than those of the corresponding homopolymer[11] **10PPB-2P** over a wide range of temperatures in the Sm-C* phase (see Figures 7 a and b). We see that the copolymer exhibits a switching time of less than 1 ms over 40° range of temperatures. Generally, FLCPs also exhibit electroclinic switching in the higher temperature Sm-A phase. Without an applied field the molecules are normal within the layer in the Sm-A phase but are tilted within the layer upon application of the electric field. The induced molecular tilt is a linear function of the applied field. Electroclinic switching times of 100 µm or less[11] are seen for **10PPB-2CO** (see Figure 7b). It should also be pointed out that unlike low molar mass FLCs, rather a high voltage (~100V) needs to be applied to FLCP to achieve switching. From the display application point of view, FLCPs that can switch fast (~100 µm) at ambient temperatures are required. None of the FLCP materials known so far can claim to satisfy this requirement.

Ferroelectric Properties of Cyclic Siloxane Oligomers

The existence of a nematic phase in cyclic siloxane-based oligomers has been known[15-21]. In this case the mesogenic unit is attached to a cyclic (instead of linear) siloxane backbone. The first detailed study of cyclic siloxane oligomers exhibiting ferroelectric properties has been carried out only recently[22-24]. Choosing the same mesogenic unit (**10PPB2**) as that discussed earlier, we have compared the properties

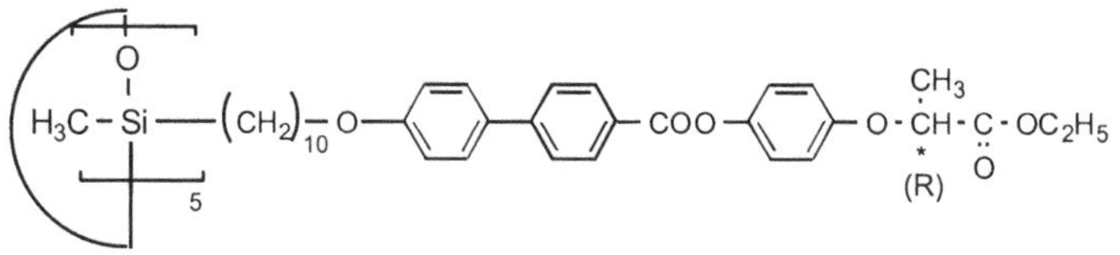

Figure 8. Structure of cyclic siloxane material NJ1. From ref. [23].

of the unattached monomer to its linear and cyclic siloxane counterparts. The structure[23] of the cyclic siloxane material is given in Figure 8. The polarization data[24] are compared in Figure 9 while the switching times[24] are given in Figure 10. It is seen that the P_s values of the cyclic siloxane pentamer is very similar to the linear siloxane polymer, and is significantly lower than the P_s values of the unattached monomer. On the other hand, the response times of the cyclic-siloxane pentamer are very similar to those of the monomer and significantly faster than the response times of the linear siloxane polymer (Figure 10). We speculate that this result is likely due to the fact that the cyclic-backbone (consisting of 5 siloxane units) does not affect the cooperative motion of the mesogens that are practically decoupled from the backbone due to the long flexible spacer. It is also important to make some comments on the layer structure in the ferroelectric phase of the cyclic-siloxane pentamer. X-ray studies on highly oriented monodomains have shown that the layer structure is represented by a monolayer of periodicity of 3.44 nm which is very close to the

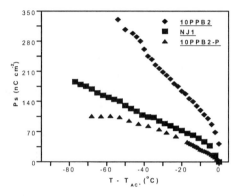

*Figure 9. Temperature dependence of the spontaneous polarization of the cyclic pentamer **NJ1**, the homopolymer **(10PPB2-P)** and the unattached monomer **(10PPB2)**. From ref. [24].*

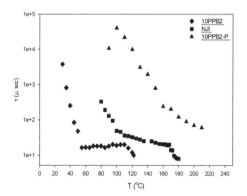

Figure 10. Temperature dependence of the switching time in the Sm-A and the Sm-C phase of **NJ1, 10PPB2-P** and **10PPB2**. From ref. [24].*

length of the molecule calculated from a simple molecular model[24]. This implies that the mesogenic groups of adjacent cyclic units must be highly interdigitated as shown in Figure 11. Interestingly, such an efficient packing does not affect the switching process as seen by low τ values exhibited by this material. It should also be mentioned that there have been reports that indicate that cyclic siloxanes exhibiting a nematic phase can be drawn into fibers and can also be cast into films. This would make ferroelectric cyclic-siloxanes extremely interesting for applications. Such efforts have not yet begun.

Figure 11. Interdigitated side groups in NJI. From ref [23].

Dynamics of Linear Siloxane FLCP

We have carried out two types of studies on the dynamics of FLCP – (i) a comparative study of the soft-mode dynamics of FLCP in comparison with that of the unattached monomeric mesogen and (ii) study of the decay time associated with the dipolar order of FLCP subjected to a strong magnetic field. These results are presented below.

Dielectric Studies

The dynamics associated with the switching of the monomeric mesogens exhibiting ferroelectric switching is well known[8]. We carried out a comparative study of the collective mode dynamics of the ferroelectric copolymer **10PPBF2-CO** in relation to the dynamics of the unattached mesogenic unit (**10PPF2**).

Apart from the high-frequency relaxation modes, the ferroelectric phase shows two collective modes associated with the two component tilt order parameter - the Goldstone mode which is associated with phase (azimuthal) fluctuations and the soft mode (that is normally observed only close to the transition to the Sm-A phase) corresponding to molecular tilt fluctuations. The soft mode persists throughout the Sm-A phase while the Goldstone mode is seen in the Sm-A phase only close to the transition to the Sm-C* phase. Figure 12 shows temperature dependence of the soft mode relaxation frequency for the monomer **10PPBF2** as well as the copolymer **10PPBF2-CO**[25]. Clearly, the relaxation frequency is two order of magnitude smaller for the copolymer. This is to be expected in view of the higher viscosity of the copolymer. The dielectric strength, as determined from the frequency dependence of the static dielectric constants at different temperatures, has been plotted for both the monomer and the polymer as a function of reduced temperature in the Sm-C* phase (Figure 13)[25]. It is remarkable that the data for both the monomer and the copolymer fall on the same curve. Thus the interesting result emerges that while the process of polymerization affects the dynamic properties, the static properties are relatively unaffected.

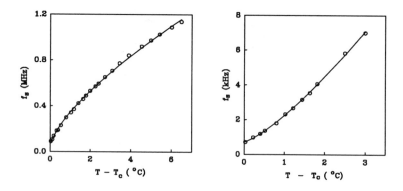

*Figure 12. Temperature dependence of the soft mode frequency for left) **10PPBF2** monomer and, right) **10PPBF2-CO** copolymer. The circles are data and the line is a fit to equations derived from Landau theory. From ref. [25].*

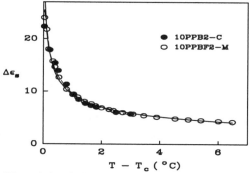

*Figure 13. Plot of the dielectric strength of the soft mode for **10PPBF2** monomer and **10PPBF2-CO** copolymer. The solid and dashed lines represent the fits to Lanadau theory for the two materials respectively. From ref. [25].*

Stability of Magnetic Field-induced Dipolar Order

In order for FLCP to be useful in non-linear optic applications, it is essential that the dipolar order of the FLCP is stable after the orienting electric or magnetic field is removed. We have investigated the stability of in-plane dipolar order by probing the optical birefringence (Δn) produced by aligning the mesogenic unit of FLCP under a high magnetic field.

Data on Δn versus time (with the magnetic field being removed) are shown in Figure 14 at a few temperatures above and below the glass transition temperature (T_g)[26]. The data have been analyzed using a phenomenological model for the dynamic decay of Δn which is given by the stretched exponential $\Delta n\,(t) = \Delta n_{max} \exp\,[-(t/\tau)^{\beta}]$, where Δn_{max} is the saturated birefringence in the presence of the field, $\Delta n\,(t)$ is the

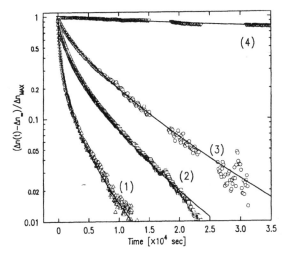

Figure 14. Normalized Δn(t) vs. t for various temperatures t after removal of a 17.3 T field. The field is zero at t =30s. The solid lines are fits of the zero-field data to an equation describing the dynamic decay. The values of T shown are (1) 51.7, (2) 27.0, (3) 11.6 and (4) 2.5 °C. $T_g = 8$ °C. For convenience of display, the data for (1) and (2) have been scaled in t by factors of 10 and 2.5, respectively. From ref. [26].

birefringence after a time t upon removal of the field, τ is the average orientational decay time and β is the exponent. When the relaxation is sharp (a single relaxation process), β = 1. The experimental data have been fit to the above expression. Data for T>T_g can be fitted to a single exponential β = 1, while for T<T_g, the data cannot be fit to a single exponential. Hence the decay of the dipolar order has very different dynamics above and below T_g. We observe that for T < T_g (plot marked (4) in Figure 14), the value of (Δn (t) - Δn)/ $Δn_{max}$) remains high (~0.8) even 15 hours after removal of the magnetic field. Hence the dipolar order can be significantly stabilized after removal of the field as long as the temperature is below T_g. However, the long-term stability needed for practical applications is yet to be established.

Self-assembling Monolayers and Multilayers of FLCPs

Multilayer polymer films are of considerable practical interest since they can possess a wide range of unique optical, mechanical and electronic properties. Depending on applications, very thin films (~100nm films for IR sensing applications) to very thick films (1-2μm films for NLO applications) need to be fabricated. Such films should be defect free and should retain the functionality of the bulk material. One of the ways of fabricating such films is to self-assemble a well formed monolayer at the air-water interface (Langmuir film) and then fabricate a multilayer film (Langmuir-Blodgett film) on a surface by successive transfers of the monolayers onto the surface. Hence the structure and stability of the Langmuir films, the order and functionality of the LB films are all important aspects in the approach to the development of functional multilayer materials. We have investigated the formation of Langmuir films and their stability. We have also fabricated LB films and studied their structure as well as functionality. These results are summarized in the following paragraphs.

The Langmuir film properties of a LC homopolymer, its corresponding mesogenic side group moiety as well as two copolymers have been studied[27-29]. It was observed that while the monolayer of the monomer is significantly less stable than that of the copolymer, the molecular area at dense packing was lower for the copolymer. Using alternate dipping technique and a computer-controlled film balance, multilayer films were formed on a substrate treated with octadecyl trichlorsilane (OTS). These multilayers, formed at typical surface pressures of 0.010 – 0.027 Nm^{-1} showed remarkable layer order along a direction perpendicular to the layer normal. In fact, x-ray reflectivity studies[30-31] showed several orders of Bragg reflection maxima showing the high layer order (Figure 15). The smaller fringes, seen between the Bragg peaks, are referred to as "Keissig fringes" and result from constructive interference of the layer surfaces. The existence of the fringes indicates smooth interfaces between layers. The reason for the high degree order can be ascribed to the molecular level phase separation – the siloxane backbone prefers to pack such that the backbones of the adjacent layers are bunched together, while the mesogens attached to the backbone interdigitate and pack as shown in the inset of Figure 15. Such an efficient layer order has also been seen in LB films formed by spin-coating the copolymer film from a solvent showing thereby the strong self-assembling nature of the copolymer material. It should also be mentioned that spin-coating of the homopolymer solution does not lead to films of higher layer order. This is because the copolymer has enough space (due to the fact that the mesogen group is attached only to a fraction of the substituting sites of the backbone) for the mesogenic groups to adjust their orientation during deposition and form a tightly packed, well-ordered layer structure. This result also implies therefore that the deposition of each layer should be a dynamic process – the side groups have the freedom to interdigitate and pack with respect to the next layer. This hypothesis is also supported by the fact that the layer spacing is independent of the number of depositions as well as of the nature of the deposition (x or y type).

The question arises as to when the multilayer structure attains the structure of the bulk state. In other words, when does 2-d monolayer structure become bulk-like 3-d film?. To answer this question, we have carried out a quantitative study of the structure and correlation of films consisting of 2, 3 and 6 monolayers of a copolymer. Specular reflectivity of these films is shown in Figure 16 along with the data for the OTS film (which is the underlying layer on the substrate) onto which the monolayers have been transferred by successive layer-by-layer deposition. An important feature is that the reflectivity data for n=6 shows the development of Bragg-like features resembling that the bulk-copolymer. This is confirmed by election density profiles (Figure 16) calculated on the basis of a model expressed as a Fourier series involving four terms in the summation and also including Fourier phases since the unit cell is asymmetric[32].

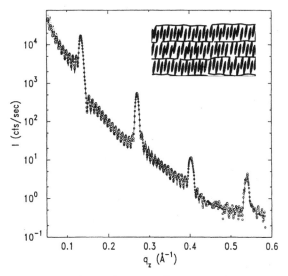

Figure 15. X-ray specular reflectivity from a 16-layer LB multilayer film of 10PPB2-CO. The four Bragg peaks confirm the presence of a well ordered layer structure within the film. The subsidiary maxima (Kiessig fringes) extend into the fourth Brillouin zone. The solid line is a best fit to the electron density model described in the text. The inset represents the backbone and side group packing. From ref. [30].

Two important points emerge from Figure 16. First, the unit cell density profile for the polymer layers consists of a large and a small maximum indicative of phenyl rings of the interdigitated side chains and the confinement of the siloxane backbone respectively. Secondly, the smaller maximum becomes clear for each unit cell for n=6 showing thereby that the bulk-like feature of the backbone confinement becomes distinct for this layer. It is also important to demonstrate that a thin FLCP film has indeed the same functionality of the bulk material. This has been done[33] by fabricating a 30-layer film on a specially designed substrate consisting of 500 gold interdigitated electrodes (Figure 17). The material, a copolymer, was formed on such an electrode structure by successive transfer of monolayers. The deposition of the side group mesogens and the siloxane backbone within the electrodes are represented in Figure 14. The OTS at the bottom of the surface ensures that the mesogens are vertical with respect to the surface while the self- assembly and the microphase separation ensures that the siloxane backbone is horizontal. This in-plane geometry enables us to apply an electric field along the layer plane of the LB layers and monitor the development of the polarization of the film. The polarization- current traces of the 30-layer film and of a bulk sample of the same material (in a regular surface-stabilized geometry) are shown in Figure 18. Clearly, the multilayer film exhibits ferroelectric polarization. Comparing the current traces at different frequencies shows an interesting feature. At 28 Hz, the switching process for both the LB film and the bulk material is complete before the reversal of the applied

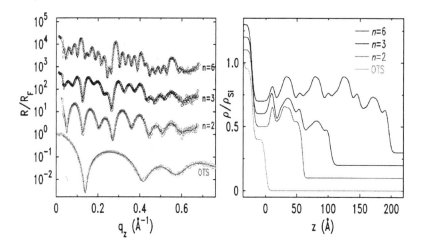

Figure 16.   Left) Specular reflectivity for LB films and the bare substrate.   n denotes the number of transferred monolayers.   Solid lines are best fits to the model described in the text.   Right) Electron density profiles determined from the experimental data of left).   These profiles correspond to (from the bottom) OTS, 2,3 and 6 layers respectively. From ref. [32].

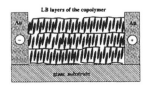

Figure 17.   Cross-sectional view of the LB layer deposited between a pair of goldelectrodes. From ref. [33].

voltage. On the other hand, at 120Hz, the signal from the LB film regains the base line while that from the bulk still lags the field showing thereby that the switching process for the LB film is faster. This is likely due to the different geometries adopted in the two cases. We have also calculated the total polarization for the LB film by integrating the current from the 250 pairs of interdigitated electrodes. The polarization of the film is found to be very similar to that to the bulk and exhibits a similar temperature dependence. Thus we have established that a 120 nm thick 30-layer film exhibits functionality similar to that of the bulk material.

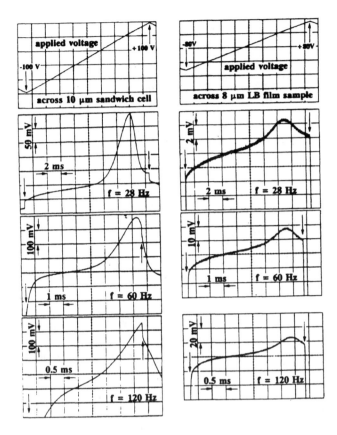

Figure 18. Oscilloscope traces of the polarization current in a 30-layer LB film (right) and in the bulk material in a sandwich cell (left) on applying a triangular wave electric field of different frequencies. All traces are obtained at 70 °C with a field of 10 V/μm. The polarization current for the LB film follows the field to higher frequencies than the bulk. The vertical arrows indicate minimum and maximum of the applied voltage. From ref. [33].

Potential Applications of FLCPs

We saw that the response times of FLCP are much slower than their low molar mass counterparts. Hence the prospects of FLCP for displays with video rates are unlikely. However, the properties should be adequate for applications like electronic billboards where slower response times can be tolerated[34,35]. For non-linear optic applications[36-38], the efficiency obtained so far for frequency doubling applications is too low to be of practical use. Perhaps the most likely areas in which FLCPs can make an impact are pyroelectric IR detectors and optical memory devices. However much work needs to be done for realizing this potential.

Acknowledgements

The work reviewed here was done in collaboration with large number of coworkers. In particular, we are indebted to Drs. R. Geer, S. Mery, S. Sprunt, and R. Duran for their work. We are also grateful to the Office of Naval Research for financial support throughout this work.

References

1. Shibaev, V. P.; Beresnev, L. A.; Blinov, L. M.; Plate, N. A. *Polym. Bull.* **1984**, 12, 299-301.
2. Decobert, G.; Soyer, F.; Dubois, J. C. *Polym. Bull.* **1985**, 14, 179.
3. Decobert, G.; Debois, J. C.; Esselin, S.; Noel, C. *Liq. Cryst.* **1986**, 1, 307.
4. Esselin, S.; Bosio, L.; Noel, C.; Decobert, G.; Dubois, J. C. *Liq. Cryst.* **1987**, 2, 505.
5. Esselin, S.; Noel, C.; Decobert, G.; Debois, J. C. *Mol. Cryst. Liq. Cryst.* **1988**, 155, 371.
6. Shivaev, V.; Kozlovski, M.; Plate, N.; Beresnev, L.; Blinov, L. *Vissokolol Soedin* **1987**, 29 (7), 1470.
7. Zentel, R. *Liq. Cryst.* **1988**, 3, 531.
8. Dubois, J. C.; Le Barny, P.; Mauzac, M.; and Noel, C.; in *Handbook of Liquid Crystals*; Eds: Demus, D.; Goodby, J.; Gray, G. W.; Spiess, H.-W.; Vill, V.,.; Wiley-VCH: New York, **1998**, 3, Chapt. 4, p. 207.
9. Clark, N.; Lagerwall, S. *Appl. Phys. Lett.* **1980**, 36, 899.
10. Naciri, J.; Mery, S.; Pfeiffer, S.; Shashidhar, R. *Journal of the SID* **1994**, 2/4, 175.
11. Naciri, J.; Pfeiffer, S.; Shashidhar, R. *Liq. Cryst.* **1991**, 10, 585.
12. Ruth, J.; Naciri, J.; Shashidhar, R. *Liq. Cryst.* **1994**, 16, 883. See also Pfeiffer, S.; Shashidhar,R.; Naciri, J.; Mery, S. *Proc. SPIE* **1992**, 1665, 166.
13. Naciri, J.; Crawford, G. P.; Ratna, B. R.; Shashidhar, R. *Ferroelectrics* **1993**, 148, 297.
14. Naciri, J.; Ratna, B. R.; Baral-Tosh, S.; Keller, P.; Shashidhar, R. *Macromolecules* **1995**, 28, 5274.
15. Patnaik, S. S.; Pachter, R.; Bunning, T. J.; Crane, R. L.; Adams, W. W. *Liq. Cryst.* **1994**, 16, 911.
16. Kreuzer, F. H.; Gawhary, M.; Winkler, R.; Finckelmann, H. *E.P. 603335* **1981.**
17. Pinsl, J.; Brauchle, C.; Kreuzer, F. H. *J. Mol. Electronics* **1987**, 3, 9.
18. Tsai, M.; Chen, S.; Jacobs, S. *Appl. Phys. Lett.* **1989**, 54, 2395.
19. Hahn, B.; Percec, V. *Mol. Cryst. Liq. Cryst.* **1988**, 157,125.
20. Kreuzer, F. H.; Andrejewsky, D.; Haas, W.; Haberle, N.; Riepl, G.; Spes, P. *Mol. Cryst. Liq. Cryst.* **1991**, 199, 345.

21. Bunning, T. J.; Klie, H. E.; Samulski, E. T.; Adams, W. W.; Crane, R. L. *Mol. Cryst. Liq. Cryst.* **1993**, 231, 163.

22. Walba, D.M.; Zummach, D.A.; Wand, M.D.; Thurmes, W.N.; More, K.M.; Arnett, K.E. *Proc. SPIE* **1993**, 1911, 21.

23. Gruneberg, K.; Naciri, J.; Wolff, D.; Shashidhar, R. *Proc. SPIE,* **1996**, 2651, 186.

24. Gruneberg, K.; Naciri, J.; Shashidhar, R. *Chem. of Materials* **1996**, 8, 2486.

25. Prasad, K.; Shankar Rao, D. S.; Khened, S. M.; Chandrasekhar, S.; Naciri, J.; Shashidhar, R. *Physica A* **1996**, 224, 24.

26. Sprunt, S.; Nounesis, G.; Naciri, J.; Ratna, B. R.; Shashidhar, R. *Appl. Phys. Lett.* **1994**, 65, 2681.

27. Rettig, W.; Naciri, J.; Shashidhar, R.; Duran, R.S. *Macromolecules* **1991**, 24, 6539.

28. Rettig, W.; Naciri, J.; Shashidhar, R.; Duran, R.S. *Thin Solid Films* **1992**, 210/211, 114.

29. Adams, J.; Rettig, W.; Duran, R. S.; Naciri, J.; Shashidhar, R. *Macromolecules* **1993**, 26, 2871.

30. Geer, R.; Qadri, S.; Shashidhar, R.; Thibodeaux, A. F.; Duran, R.S. *Liquid Crystals* **1994**, 16(5), 869.

31. Geer, R.; Qadri, S.; Shashidhar R. *Ferroelectrics* **1993**, 149, 147.

32. Geer, R.; Thibodeaux, A. F.; Duran, R. S.; Shashidhar, R. *Europhys. Lett.* **1995**, 32(5), 419.

33. Pfeiffer, S.; Shashidhar, R.; Fare, T. L.; Naciri, J.; Adams, J.; and Duran, R. S. *Appl. Phys. Lett.* **1993**, 63(9), 1285.

34. Hachiya, S.; Tomoike, K.; Yuasa, K.; Togawa, S.; Sekiya, T.; Takahashi, K.; **Kawaski, K,** *J. SID* **1993**, 1/3, 295.

35. Yuasa, K.; Uchida, S.; Sekiya, T.; Hashimoto, K.; Kawasaki, K. *Proc. SPIE* **1992**, 1665, 154.

36. Ozaki, M.; Utsumi, M.; Yoshino, K.; Skarp, K.; *Jpn. J. Appl. Phys.* **1993**, 32, L852.

37. Wischerhoff, E.; Zentel, R.; Redmond, M.; Mondain-Monval, O.; Coles, H. *Macromol. Chem. Phys.* **1994**, 214, 125.

38. Walba, D. M.; Zummach, D. A.; Wand, M. D.; Thurmes, W. N.; Moray, K. M.; Arnett, K. E. *Proc. SPIE* **1993**, 1911, 21.

Chapter 20

Design of Smectic Liquid Crystal Phases Using Layer Interface Clinicity

David M. Walba[1], Eva Körblova[1], Renfan Shao[2], Joseph E. Maclennan[2], Darren R. Link[2], Matthew A. Glaser[2], and Noel A. Clark[2]

Departments of [1]Chemistry and Biochemistry and [2]Physics, Ferroelectric Liquid Crystal Materials Research Center, University of Colorado, Boulder, CO 80309

The design and synthesis of a ferroelectric bow-phase (banana-phase) composed of racemic molecules is described. To date many hundreds of bent-core molecules have been screened for liquid crystallinity, motivated by the recent discovery of very unusual and interesting chiral, antiferroelectric smectic phases from achiral molecules. Based upon a model for bow-phase antiferroelectric structure driven by layer interface clinicity, incorporation of one racemic methylheptyloxycarbonyl tail into the prototype bow-shaped mesogen structure was accomplished. This led to a material showing the unmistakable B7 texture by polarized light microscopy. This B7 phase was shown to be ferroelectric by a combination of electrooptic properties and current response in 4μm transparent capacitor liquid crystal cells. To our knowledge, this is the first ferroelectric smectic LC composed of racemic molecules.

Since Meyer's revolutionary work of 1974, the paradigm for ferroelectric smectic liquid crystals has required enantiomerically enriched molecules[1, 2]. The recent observation of antiferroelectric switching in smectic liquid crystals (LCs) composed of achiral molecules has changed the polar smectic LC landscape, however. First, Soto-Bustamante and Blinov et al., working in the Haase labs in Darmstadt, reported antiferroelectric behavior in an achiral polymer/monomer blend.[3] Shortly thereafter, Takezoe and Watanabe et al. reported a smectic liquid crystal phase composed of achiral bent-core, or bow-shaped (also known as banana-shaped) molecules exhibiting high-susceptibility electrooptic (EO) behavior characteristic of ferroelectric or antiferroelectric liquid crystals (FLCs or AFLCs)[4]. The observed EO behavior was ascribed to a ferroelectric supermolecular structure with C_{2V} symmetry.

This phase, known as B2 or SmCP (see discussion below for a detailed description of the SmCP nomenclature), indeed possesses a polar layer structure, wherein the molecular bows are oriented with their imaginary "arrows" along a polar axis within each layer, as suggested by Takezoe and Watanabe. The phase, however, is *antiferroelectric*, with antiparallel orientation of the polar axis in adjacent layers[5, 6, 7]. Adding to the novelty of the B2 system, a tilt of the molecular bows about the polar axis leading to a chiral layer structure, was discovered by Clark et al.[7]. The global free energy minimum, the SmC_SP_A structure, is a macroscopic racemate, where adjacent layers are heterochiral. Most interestingly, however, an apparently metastable, though easily observed, second antiferroelectric phase, the SmC_AP_A, composed of chiral macroscopic domains, was also described in the "B2 phase" temperature range of the classic materials[7]. This system represents the first known liquid conglomerate[8].

Since the initial report, many hundreds of mesogens in the bent-core family have been characterized, and seven "banana phases[9]," B1-B7, have been identified based mainly upon observed textures in the polarized light microscope and by X-ray diffraction[5]. In the literature the SmC_SP_A and SmC_AP_A phases are typically, though incorrectly, lumped together under the name "B2 phase". Both B2 and B7 phases show electrooptic switching. T date these materials have been shown to exhibit behavior characteristic of *antiferroelectric* supermolecular structures. That is, a ferroelectric state is accessible by application of an electric field, but this state rapidly reverts to an antiferroelectric structure upon removal of the field. Herein we report the directed design, synthesis, and characterization of a ferroelectric bow-phase liquid crystal system. The new phase is composed of racemic molecules, and, being macroscopically chiral, forms a ferroelectric liquid crystal conglomerate. This ferroelectric smectic LC, composed of racemic molecules, breaks the Meyer paradigm for formation of FLCs. No new paradigm is involved, however. The mesogen showing the target phase was *designed* using phenomena well known in the FLC literature.

The First Liquid Conglomerate

The "classic" bow-phase mesogenic compound, which we term **NOBOW** (NonylOxy BOW-phase mesogen) has the bis-Schiff-base diester structure shown in Figure 1. The C_2-symmetrical series of alkoxy homologues containing this material was first prepared and characterize by Matsunaga[10]. Based upon differential scanning calorimetry and X-ray studies (experimental measures of the layer spacings), in combination with molecular lengths of the series of homologues measured from molecular models, Matsunaga proposed that materials of this type show mesophases of the "smectic C type." His analysis and conclusions were elegant and substantially correct.

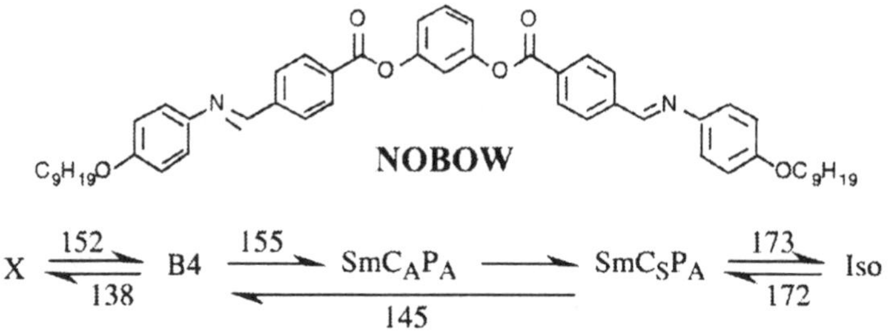

NOBOW

$$X \underset{138}{\overset{152}{\rightleftharpoons}} B4 \xrightarrow{155} SmC_AP_A \longrightarrow SmC_SP_A \underset{172}{\overset{173}{\rightleftharpoons}} Iso$$

145

Figure 1.   Structure and phase sequence of NOBOW, a prototypical bow-phase mesogen.

As shown in Figure 1, **NOBOW** exhibits the "B4" phase at temperatures below the smectic LC temperature range. This phase, also termed "blue crystal," is actually a crystal modification showing a strong circular dichroism, and clear indications of helical layer deformations by atomic force microscopy[11]. This crystalline phase is composed of macroscopic chiral domains of "random" handedness. The opposite sign of the CD exhibited by enantiomeric domains is easily observed.

While the blue crystal bow-phase is certainly extremely interesting, there is nothing new in finding chiral crystals growing from achiral or racemic molecules. Pasteur first documented this type of spontaneous breaking of achiral symmetry in his classic 1848 study of the "racemic acid" sodium ammonium salt conglomerate[12]. Since then a very large number of examples of spontaneous chiral symmetry breaking in the formation of <u>crystals</u> have been reported, including good evidence for the formation of 2-dimensional crystal conglomerates on graphite[13], and at the air-water interface[14].

The discovery of the **NOBOW** *liquid* conglomerate[7, 8], however, represented a paradigm shift. Until then, a central dogma in LC design held that chiral LC phases only occurred with enantiomerically enriched molecules. For example, when Feringa observed a chiral nematic phase after irradiation with circularly polarized light (CPL) of an achiral nematic doped with specially designed photoactive molecules, he confidently concluded that the dopant molecules <u>must</u> be partially resolved by the CPL[15]. Meyer's prediction of the ferroelectric nature of the chiral SmC* phase used the same argument by relying upon compounds composed of enantiomerically enriched molecules to break the macroscopic achiral (and nonpolar) symmetry of the SmC phase.

The Metastable SmC_AP_A Phase of NOBOW

Clearly molecules of **NOBOW** in the isotropic liquid or gas phase are "achiral" by the conventional chemical definition. That is, the compound **NOBOW** is not resolvable at room temperature. But, this material does form chiral liquid crystal supermolecular structures, as we have reported in detail[7]. The bow-phases are very similar to the well-known conventional SmC* system of Meyer. In order to aid in the following discussion, the tilt plane, layer plane, and polar plane of a small cube of SmC*/SmCP sample are illustrated in Figure 2. First, for all known smectic ferroelectric LCs there exists one two-fold axis of symmetry (singular points at the layer interfaces and the "middle" of the layers)—the polar axis. We define the tilt plane as the plane normal to the polar axis in an FLC. For antiferroelectrics, the tilt plane is defined as the plane normal to the polar axis in the ferroelectric state. The polar plane is defined as the plane containing the layer normal and the polar axis. For the bow-phase mesogens, we define the director as being along the imaginary bow-string of the molecular bows. Using these conventions, the geometry in the SmC* and SmCP systems is identical.

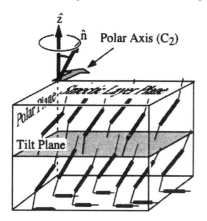

Figure 2. *The SmCP geometry.*

The smectic mesogenicity of **NOBOW** is actually quite complex and interesting, as follows. Upon melting from the chiral blue crystal phase, a liquid crystalline conglomerate is seen in 4μm transparent capacitor LC cells with low pre-tilt polyimide alignment layers, parallel-rubbed[16]. The alignment seen in the sample, seemingly directly upon melting, has the layers normal to the substrates (bookshelf alignment), with the tilt plane parallel to the substrates and the polar plane normal to the substrates. This is the same geometry observed in surface stabilized FLC (SSFLC) cells in the well-known bookshelf or quasi-bookshelf alignment. Note, however, that while the zenithal anchoring appears strong (the director is oriented parallel to the substrates), no azimuthal anchoring of the director is observed, resulting in a random focal conic texture by polarized light microscopy (PLM).

The macroscopically chiral domains of this conglomerate behave in a characteristically antiferroelectric and definitively chiral manner as evidenced by EO behavior upon application of a triangular driving waveform. Specifically, a smooth purple focal conic is observed in the absence of an applied field. This texture is "smectic A-like" (i.e. extinction brushes in cylindrical focal conic domains are oriented parallel and perpendicular to the crossed polarizer/analyzer, showing that the optic axis in the phase at zero field is along the layer normal). When a field is applied above a threshold of about 5 V/μm dramatic EO switching is observed, giving a green focal conic structure where the optic axis is rotated from the layer normal by about ±30°, depending upon the sign of the applied field. The observation of optic axis rotation either clockwise or counterclockwise depending upon the sign of an applied field normal to the "clock face," with heterochiral domains showing enantiomeric behavior, proves a chiral structure.

The structure of one of the enantiomers of this phase is illustrated in Figure 3, along with those of the two degenerate ferroelectric states. Symbolic representations of the bow-shaped structure indicate the molecular bows oriented with the polar plane as the plane of the page in the drawings at the top of the Figure, and with the tilt plane in the plane of the page in the drawings below. In both orientations, the layer normal is vertical; the layers are normal to the plane of the page and horizontal. Projected onto the polar plane, the director is tilted out of the plane, as is suggested by the heavier and lighter lines making up the molecular bows. In the projection in the tilt plane, the molecular bow plane is normal to the page. Experimentally, the substrates

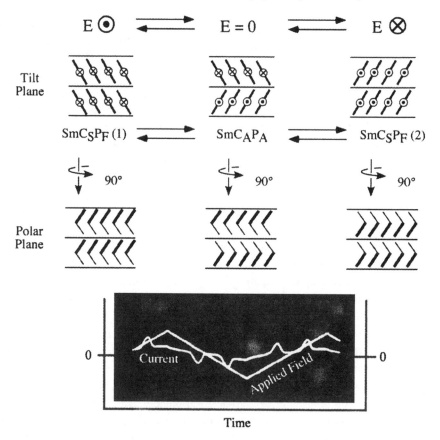

Figure 3.   *Structure of the SmC$_A$P$_A$ antiferroelectric phase and two SmC$_S$P$_F$ ferroelectric states accessed by application of a field. The typical antiferroelectric current response is indicated at the bottom of the Figure.*

in LC cells are parallel to the tilt plane. Thus, in the PLM experiments the system is being viewed as projected onto the tilt plane.

The drawings in Figure 3 represent a domain of negative chirality according to our convention (positive if $z \times n$ is pointing parallel to an imaginary arrow fitted to the bow). At zero field, the structure observed is antiferroelectric, denoted SmC_AP_A. SmC descriptor indicates a smectic structure with coherent, long-range tilt of the director relative to the layer normal. The SmCP notation conveys the unique identifying spontaneous breaking of nonpolar symmetry which occurs in the switching smectic bow-phases; polar orientation of the molecular bows within each layer. The subscripts refer to the relative structure of adjacent pairs of layers: C_A denotes *anticlinic* layer interfaces in the tilt plane, while P_A denotes *antiferroelectric* order for adjacent layer pairs in the polar plane. Note that in the SmC_AP_A ground state, the average optic axis is along the layer normal. In addition, the birefringence is lowered as viewed normal to the tilt plane by the anticlinicity—the conjugated aromatic system is substantially excited for light either parallel or perpendicular to the layer normal.

Application of a field above the switching threshold causes precession of the director in a set of alternate layers, providing a ferroelectric state. Two homomeric ferroelectric states, differing only in orientation, are accessed by application of fields of opposite sign. In Figure 3, these are denoted SmC_SP_F (1) and SmC_SP_F (2) (C_S refers to synclinic adjacent layer pairs in the tilt plane, and P_F refers to ferroelectric order in adjacent layer pairs in the polar plane). The chirality of layers does change during this kind of switching. All the layers shown in the Figure are of negative chirality.

Clearly, the switching causes the optic axis of the system to rotate off the layer normal by the SmC tilt angle of one sign or the other depending upon the sign of the applied field. Assuming the dipole moment of **NOBOW** molecules points antiparallel to the "arrow," (we use the physics convention that dipoles are pointing from negative to positive; the direction of the dipole moment was obtained with MOPAC using AM1), the domains with negative chirality have positive ferroelectric polarization (P along $z \times n$).

The classic antiferroelectric current response is given at the bottom of Figure 3. Note two current peaks during each "half-cycle" of driving. Starting at the bottom of the triangular driving waveform, the sample is switched into SmC_SP_F (1). As the field moves towards zero from its maximum value (which is above the switching threshold), the system switches to the antiferroelectric "ground state", releasing half the ferroelectric polarization of the ferroelectric state, which is seen as a positive current peak. This is due to the re-orientation of dipoles in half the layers as the sample switches to the SmC_AP_A phase. As the field crosses zero and continues to towards a minimum (maximum negative field), the second half of the polarization reversal current is seen when the system switches to the SmC_SP_F (2) state.

Supermolecular Diastereomers Exhibited by NOBOW

The SmC$_A$P$_A$ phase of **NOBOW** is metastable. Upon standing in the SmCP temperature range, the sample will spontaneously convert to a new phase exhibiting a very distinctive green stripe texture. The sample never seems to fully convert to this "green stripe" phase, however. Even upon cooling from the isotropic melt, about 10% of the sample exhibits the SmC$_A$P$_A$ texture. As described previously in detail[7], this phase is also antiferroelectric, with a structure denoted SmC$_S$P$_A$. The SmC$_S$P$_A$ antiferroelectric phase switches to the SmC$_A$P$_F$ ferroelectric state upon application of a field. The SmC$_S$P$_A$ phase and its corresponding ferroelectric state are macroscopic *racemates*, wherein adjacent layers possess heterochiral structures, these being macroscopically achiral.

The preceding discussion details the observation of six different supermolecular structures for the "B2 phase" of **NOBOW**, illustrated graphically in Figure 4. Bor-

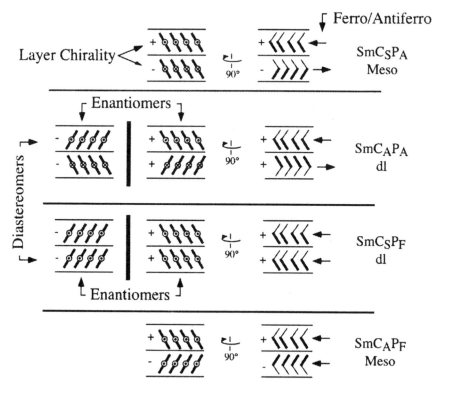

Figure 4.   *Illustration of the diastereomeric supermolecular structures exhibited by NOBOW. The graphics on the right are a symbolic representation of the structures projected onto the polar plane, while those on the left represent the structure projected onto the tilt plane. The only diastereomeric structures observed in the absence of an applied field are antiferroelectric, one racemate (meso) and one conglomerate (dl). The corresponding ferroelectric states are accessed by application of a field above the switching threshold.*

rowing from the language of molecular stereochemistry, we find it useful to consider these supermolecular stereoisomers. Two of the structures are supermolecular racemates (analogous to meso compounds), one antiferroelectric and one ferroelectric. The other four structures, two antiferroelectric and two ferroelectric, are supermolecular conglomerates (analogous to dl pairs), exhibiting macroscopically chiral domains of opposite handedness (supermolecular enantiomers).

While three distinct ferroelectric structures have been documented for **NOBOW** (one racemate and two enantiomers), these are not ferroelectric *phases*, being rather ferroelectric *states* resulting from application of a field to *antiferroelectric phases* (this distinction is clear and important). In fact, all of the electrooptically active bow-phases reported to date are apparently antiferroelectric. Quite obviously this state of affairs suggests a ferroelectric bow phase (or ferroelectric banana) as an attractive target for supermolecular stereocontrolled synthesis. Such a material, with an indefinitely stable net polar structure, could be useful as well. For example, the highest second order nonlinear optical susceptibility yet reported for a liquid crystal was seen in a ferroelectric state of **NOBOW**[17]. Capturing this structure in a thermodynamic phase in the absence of applied fields would be interesting.

Design and Synthesis of a Ferroelectric Banana

Intuition suggests that the antiferroelectric nature of the bow-phases is driven by a Coulombic free energy term. In fact, macroscopically, ignoring "edge effects," ferroelectric and antiferroelectric order in adjacent layer pairs have the same free energy. We suggest that the known bow-phases are antiferroelectric simply due to a free energy preference for *synclinic layer interfaces in the polar plane*. As shown in Figure 4, the antiferroelectric SmC_SP_A has synclinic layer interfaces projected in both the tilt plane and polar plane, while the antiferroelectric SmC_AP_A phase is synclinic in the polar plane. Consider that of all the known tilted smectics, the vast majority are synclinic SmC and SmC*.

Following this reasoning, if a bow-phase mesogen could be produced which causes a thermodynamic preference for *anticlinic* layer interfaces in the polar plane, then a ferroelectric phase should result. In fact no new paradigm is required to design such a system. Thus, one of the most important results in the FLC field since the end of the 1980s was the discovery of antiferroelectric smectic LCs[18]. In such materials, exemplified by the prototype chiral antiferroelectric methylheptyloxycarbonylphenyl octyloxybiphenylcarboxylate (**MHPOBC**), all layer interfaces are anticlinic, as shown in Figure 5 for chiral **MHPOBC** with positive **P**. Application of a field switches the sample to a ferroelectric state with all synclinic layer interfaces. Note that by our definition these illustrations show the director structure projected in the tilt plane.

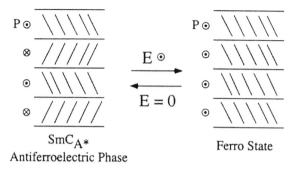

Figure 5. *Antiferroelectric phase and ferroelectric state of chiral **MHPOBC** with positive **P**, projected in the tilt plane.*

Importantly, it is well known that anticlinic layer interfaces are also obtained in a phase of *racemic* **MHPOBC**, proving that enantiomeric excess is not required to achieve the anticlinic structure, though racemic **MHPOBC** is neither ferroelectric nor antiferroelectric.

In the bow-phase system, the relationship between layer interface clinicity and ferro/antiferroelectricity is "reversed," as illustrated in Figure 6. The key defining characteristic here is the spontaneous polar order along the "arrows" of the molecular bows. Given this polar layer structure, the tilt plane of **MHPOBC** becomes the "bow plane", tilted from the polar plane by the SmC tilt angle (about 30° in the case of NOBOW). The tilt plane is now vertical, perpendicular to the plane of the page. In Figure 6, the structures of the ferroelectric and antiferroelectric $SmC_SP_{F/A}$ are illustrated projected onto this macroscopic bow plane. In these drawings the layers are *not* perpendicular to the page, a non-standard view. Nevertheless, these illustrations make a comparison with **MHPOBC** very simple. Synclinic layer interfaces provide an antiferroelectric structure while anticlinic layer interfaces provide a ferroelectric structure.

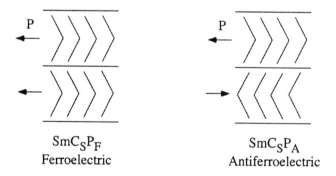

Figure 6. *Illustrations of ferroelectric and antiferroelectric SmC_SP bow-phase structures with the bow-plane parallel to the plane of the page.*

This idea was tested with the synthesis of a bow-phase analog of **MHPOBC**, similar to the classic NOBOW but possessing one racemic methylheptyloxycarbonyl tail. The structure of the new mesogen, termed **MHOBOW** (MethylHeptylOxycarbonyl-BOW), is given in Figure 7[19]. By PLM the material indeed can be seen to possess an enantiotropic bow-phase. The texture, however, is not that of one of the phases in the B2 motif, but rather the unmistakable, highly unusual "B7 phase."

The B7 texture was first described by Pelzl et al. in bow-phase mesogens possessing a nitro substituent at C1 of the resorcinol unit at the center of the core[20]. By PLM the B7 texture is characterized by the formation of twisted ribbons at the iso-B7 transition, and by the formation of beautiful focal conic domains containing unusual subtle stripes and other texture within the domains. Often the ribbons seem to anneal into focal conics with residual modulations from the ribbon structure. Double-helical intertwined twisted ribbon structures have been observed in the a different B7 bow-phase mesogen[21]. Of course the ribbons have a chiral supermolecular structure, and the samples are macroscopic conglomerates, with about half the ribbons being right handed, and half left handed, strongly suggesting a chiral structure for the phase, though the molecules are achiral. Recently, a bow-phase mesogen exhibiting the B7 texture was shown to be antiferroelectric[22].

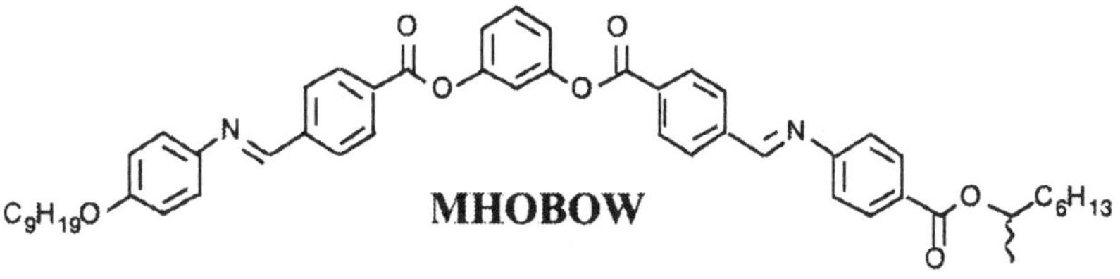

$$X \rightleftharpoons B4 \xrightarrow{130} B7 \xrightarrow{135} Iso$$

Figure 7.   *The structure and phase sequence of a bow-phase analog of MHPOBC: MHOBOW.*

The electrooptic behavior of the B7 texture of **MHOBOW** was studied in 4µm Displaytech cells[16]. The focal conic domains obtained by annealing the texture just below the Iso-B7 transition had a gold birefringence color suggesting relatively low birefringence. In addition, extinction brushes in cylindrical focal conic domains were parallel and perpendicular to the polarizer, showing that the optic axis was along the layer normal (SmA-like).

Application of a triangular electric field waveform of amplitude 10 V/µm produced a striking analog rotation of the optic axis in the focal conic domains of about ±10°. The optic axis rotation is approximately linear with applied field (similar to an electroclinic effect), and the sample is monostable, always returning to the SmA-like brush orientation at zero field. This analog response is chiral, with the sample behaving as a conglomerate. About half the domains rotate clockwise and half rotate counterclockwise for increasing positive field, and vice-versa for increasing field of opposite sign. Some domains seem to possess a decidedly smaller susceptibility, with optic axis rotation of about ±5° in response to a 10 V/μm triangular waveform.

Application of a field above a threshold value of about 12 V/μm causes a dramatic change in the texture. The gold focal conic domains, showing analog EO behavior, are seen to change to smooth SmC*-like bistable domains with a blue birefringence color indicating increased birefringence. Well-defined domain walls mediate the change when the applied field is just above the threshold. This "bistable blue" texture shows typical ferroelectric hysteresis in the EO switching, and behaves just as expected for a bistable SSFLC cell composed of chiral SmC* material, except it is, of course, a conglomerate. There is only one peak observed in each half-cycle of driving (the classic SSFLC ferroelectric response) in the bistable blue texture. The two bistable states of the conglomerate and the polarization current response of the cell are shown in Figure 8. To our knowledge, no other bow-phase material in the B7 texture has been seen to exhibit this type of ferroelectric response.

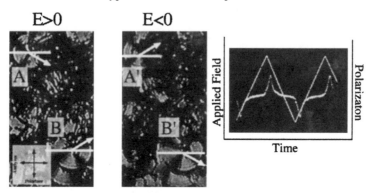

Figure 8.   *Left: Photomicrographs of a sample of **MHOBOW** in 4 μm transparent capacitor cells under drive. With a positive field applied, a pair of heterochiral focal conic domains can be seen (A and B). The director is tilted off the layer normal in these domains by about (±)30° as can be seen by the orientation of the extinction brushes. Reversal of the applied field switches the director in these domains in opposite directions about the layer normal (A' and B'). Right: The ferroelectric polarization response to an applied triangular driving field (1 kHz).*

Analog electrooptics in SmC* FLCs has been the subject of considerable study and interest in recent years. This desirable behavior has been achieved in SSFLC cells where the layers are uniformly tilted by the smectic C* tilt angle (Sony-mode V-shaped switching)[23]. Most recently, the analog response observed in "Thresholdless Antiferroelectric" (ThAFLC) cells[24] has been interpreted to be resulting from an unusual *alignment mode* in the SmC* phase.[25] Indeed, such SmC* cells are seen to show analog and bistable response in the same cell.[25]

While the details of the observed EO behavior of **MHOBOW** are still under investigation, this material has never been seen to exhibit EO properties indicative of an antiferroelectric structure. As suggested by the preceding discussion, the observation of an analog response in ferroelectric LCs is well known. The formation of a very robust bistable texture showing the characteristic ferroelectric current response appears to be unique to the B7 phase of **MHOBOW**. This EO evidence strongly suggests that a ferroelectric bow-phase, the target of the design effort, has indeed been obtained by incorporation of the **MHPOBC** tail into the bow-phase mesogen structure.

Furthermore, we speculate that in fact the **MHOBOW** "B7 phase" has the SmC_SP_F FLC conglomerate structure, precisely the target of the design effort, while other materials showing the antiferroelectric "B7 phase" are thermodynamic SmC_AP_A conglomerates. We suggest that the low birefringence, SmA-like gold focal conic texture occurs when the bow-plane of the sample is more or less parallel to the substrates (i.e. as shown in Figure 6, with the glass plates parallel to the plane of the page). In this geometry substantial excitation of the aromatic rings occurs for light polarized both parallel and perpendicular to the director, leading to two large indices of refraction and a small birefringence. Also, the director is oriented along the projection of the layer normal on the substrates, affording a SmA-like orientation of the extinction brushes. Of course the ferroelectric polarization is parallel to the substrates for this type of alignment—precisely the orientation leading to analog EO. Application of a field above the threshold of about 12 V/μm then causes an alignment change such that the bow plane, the ferroelectric polarization and smectic layers are now more or less normal to the substrates (bookshelf alignment), with the polar axis along the applied field. In this orientation the birefringence is much larger, since the aromatic rings are only excited for input light polarized parallel to the director.

Conclusions

A design effort aimed at formation of a ferroelectric bow-phase has led to the synthesis of the **MHPOBC** bow-phase analog **MHOBOW**. This structure was designed based upon the proposal that induction of anticlinic layer interfaces in the bow-phase polar smectic motif would stabilize a ferroelectric polar smectic structure. In the event, strong experimental evidence, based primarily upon electrooptic measurements, and aided by the chiral response of the system with bistable rotation of extinction brushes in cylindrical focal conic domains (the classic signature of FLC EO) suggests that indeed a ferroelectric bow-phase has been obtained.

Acknowledgements

The authors gratefully acknowledge support of this work by the Ferroelectric Liquid Crystal Materials Research Center (National Science Foundation MRSEC Award No. DMR-9809555).

References

[1] Meyer, R. B.; Liebert, L.; Strzelecki, L.; Keller, P. *J. Phys., Lett. (Orsay, Fr.)* **1975**, *36*, L69-L71.

[2] Putatively polar "polyphilic" SmA phases reported by Tournilhac et al. [Tournilhac, F. G.; Bosio, L.; Simon, J.; Blinov, L. M.; Yablonsky, S. V. *Liq. Cryst.* **1993**, *14*, 405-414.] have since been shown to possess a nonpolar bilayer smectic A structure: Shi, Y.; Tournilhac, F. G.; Kumar, S. *Phys. Rev. E* **1997**, *55*, 4382-4385.

[3] Soto Bustamante, E. A.; Yablonskii, S. V.; Ostrovskii, B. I.; Beresnev, L. A.; Blinov, L. M.; Haase, W. *Liq. Cryst.* **1996**, *21*, 829-839.

[4] Niori, T.; Sekine, T.; Watanabe, J.; Furukawa, T.; Takezoe, H. *J. Mater. Chem.* **1996**, *6*, 1231-1233.

[5] Pelzl, G.; Diele, S.; Weissflog, W. *Adv. Mater.* **1999**, *11*, 707-724.

[6] Weissflog, W.; Lischka, C.; Benne, L.; Scharf, T.; Pelzl, G.; Diele, S.; Kruth, H. *Proc. SPIE-Int. Soc. Opt. Eng.* **1998**, *3319*, 14-19.

[7] Link, D. R.; Natale, G.; Shao, R.; Maclennan, J. E.; Clark, N. A.; Körblova, E.; Walba, D. M. *Science* **1997**, *278*, 1924-1927.

[8] Walba, D. M.; Körblova, E.; Shao, R.; Maclennan, J. E.; Link, D. R.; Clark, N. A. *Antiferroelectric liquid crystals from achiral molecules and a liquid conglomerate*; in *Liquid crystal materials and devices*, Bunning, T., Chen, S., Chien, L.-C., Kajiyama, T., Koide, N. and Lien, S.-C., Eds.; Materials Research Society: San Francisco, 1999; Vol. 559, pp 3-14.

[9] The banana-phase nomenclature currently utilized in the field was established in connection with the symposium "Chirality by Achiral Molecules" held in Berlin in December 1997.

[10] Akutagawa, T.; Matsunaga, Y.; Yashuhara, K. *Liq. Cryst.* **1994**, *17*, 659-666.

[11] Heppke, G.; Moro, D. *Science* **1998**, *279*, 1872-1873.

[12] Pasteur, L. *C.R. Acad. Sci. Paris* **1848**, *26*, 535-539.

[13] (a) Stevens, F.; Dyer, D. J.; Walba, D. M. *Angew. Chem., Int. Ed. Engl.* **1996**, *35*, 900-901. (b) Walba, D. M.; Stevens, F.; Clark, N. A.; Parks, D. C. *Acc. Chem. Res.* **1996**, *29*, 591-597.

[14] (a) Weissbuch, I.; Lahav, M.; Leiserowitz, L. *J. Am. Chem. Soc.* **1997**, *119*, 933. (b) Nassoy, P.; Goldmann, M.; Rondelez, F. *Phys. Rev. Lett.* **1995**, *75*, 457-460.

[15] Huck, N. P. M.; Jager, W. F.; de Lange, B.; Feringa, B. L. *Science* **1996**, *273*, 1686-1688. Schuster had earlier implied a similar correspondence between the chiral nature of LC phases and enantiomeric excess of the molecules in the sample when he showed that a cholesteric LC doped with an enantiomerically enriched photoracemizable dopant molecules becomes "a compensated nematic" upon photoracemization: Zhang, M. B.; Schuster, G. B. *J. Phys. Chem.* **1992**, *96*, 3063-3067.

[16] The LC cells used in this work are commercially available from Displaytech, Inc. (http://www.displaytech.com).

[17] Macdonald, R.; Kentischer, F.; Warnick, P.; Heppke, G. *Phys. Rev. Lett.* **1998**, *81*, 4408-4411.

[18] (a) Takezoe, H.; Chandani, A. D. L.; Lee, J.; Gorecka, E.; Ouchi, Y.; Fukuda, A.; Terashima, K.; Furukawa, K.; Kishi, A., "What is the Tristable State?," Abstracts of the *Second International Conference on Ferroelectric Liquid Crystals*, Göteborg, Sweden, O 26 (1989). (b) Galerne, Y.; Liebert, L., "The Antiferroelectric Smectic O Liquid Crystal Phase," Abstracts of the *Second International Conference on Ferroelectric Liquid Crystals*, Göteborg, Sweden, O 27 (1989). (c) Chandani, A. D. L.; Gorecka, E.; Ouchi, Y.; Takezoe, H.; Fukuda, A. *Jpn. J. Appl. Phys.* **1989**, *28*, L 1265-L 1268. (d) Chandani, A. D. L.; Ouchi, Y.; Takezoe, H.; Fukuda, A.; Terashima, K.; Furukawa, K.; Kishi, A. *Jpn. J. Appl. Phys.* **1989**, *28*, L 1261-L 1264.

[19] Walba, D. M.; Körblova, E.; Shao, R.; Maclennan, J. E.; Link, D. R.; Glaser, M. A.; Clark, N. A. *Science* **in press.**

[20] Pelzl, G.; Diele, S.; Jákli, A.; Lischka, C.; Wirth, I.; Weissflog, W. *Liq. Cryst.* **1999**, *26*, 135-139.

[21] Lee, C. K.; Chien, L. C. *Liq. Cryst.* **1999**, *26*, 609-612.

[22] Heppke, G.; Parghi, D. D.; Sawade, H., "A laterally fluoro-substituted "Banana-Shaped" Liquid Crystal Showing Antiferroelectricity," in the abstracts of the *7th International Conference on Ferroelectric Liquid Crystals*, pp 202-203 (Poster PC05) (1999).

[23] Nito, K.; Takanashi, H.; Yasuda, A. *Liq. Cryst.* **1995**, *19*, 653-658.

[24] Inui, S.; Iimuro, N.; Suzuki, T.; Iwane, H.; Miyachi, K.; Takanishi, Y.; Fukuda, A. *J. Mater. Chem.* **1996**, *6*, 671-973.

[25] Rudquist, P.; Lagerwall, J. P. F.; Buivydas, M.; Gouda, F.; Lagerwall, S. T.; Clark, N. A.; Maclennan, J. E.; Shao, R.; Coleman, D. A.; Bardon, S.; Bellini, T.; Link, D. R.; Natale, G.; Glaser, M. A.; Walba, D. M.; Wand, M. D.; Chen, X.-H. *J. Mater. Chem.* **1999**, *9*, 1257-1267.

Chapter 21

Control of Structure Formation of 1,3,5-Triazines through Intermolecular Hydrogen Bonding and CT-Interactions

D. Janietz[1], D. Goldmann[1], C. Schmidt[2], and J. H. Wendorff[2]

[1]Department of Chemistry and Institute of Thin Layer Technology, Potsdam University, Kantstr. 55, D-14513 Teltow, Germany
[2]Institute of Physical Chemistry and Centre of Material Science, Philipps University, Hans-Meerwein-Str., D-35032 Marburg, Germany

Sheet-like 2,4,6-triarylamino-1,3,5-triazines substituted with long peripheral alkoxy groups are presented. Triazines containing six lipophilic side chains exhibit columnar mesophases due to a segregation of the polar central molecular part from the non-polar aliphatic molecular segments. The corresponding trialkoxy derivatives are non-mesomorphic in their pure states. The triazines form hydrogen bonded aggregates with alkoxybenzoic acids. Depending on the lateral substitution pattern of both complementary components the molecular recognition allows the variation of hexagonal lattice constants, the variation of the two-dimensional lattice symmetry and even the induction of columnar phases. In contrast, CT-interactions with flat nitrofluorenone based electron acceptors give in certain cases rise to lamellar mesophases. The control of structure formation is caused by certain specific anisometric rigid conformations of the triazines which become favored through the attractive interactions with the complementary molecules.

Introduction

The rigid anisometric geometry of molecules is known to be a major factor in controlling the formation of thermotropic liquid crystalline phases (1). Rod-like mesogens predominantly form smectic phases (2) whereas molecules possessing, on the average, a flat rigid core surrounded by a certain number of long flexible side chains preferably organize to columnar mesophases (3). On the other hand, it has become apparent that self-organization with formation of liquid crystalline structures can be achieved by directed intermolecular forces between identical or different individual molecules leading to non-covalently bonded anisometric associates (4-6).

282

It is, for example, well established that electron-rich systems such as triphenylene ethers or radial multialkynylbenzene compounds form charge-transfer (CT) complexes with rather flat but non-liquid crystalline electron acceptors such as 2,4,7-trinitrofluoren-9-one (TNF) (7). CT-interactions may cause a stabilization as well as an induction of columnar mesophases (7-9). Except for a few cases (8,10), columnar phases formed by binary mixtures of complementary donor and acceptor molecules are hexagonal (7,8,11,12) or nematic columnar (12-15).

Columnar mesomorphic structures may also arise from intermolecular hydrogen bonding (4). Examples are primary benzamides attached with three long alkoxy chains (16), inositol ethers (17,18), derivatives of cyclohexane-1,3,5-triol (19-21), 1,3-diacylamino-benzene compounds (22,23), mixtures of 2,6-diacylaminopyridines with complementary uracil derivatives (24-26). A common feature of these systems is that the flat mesogenic unit is only formed by the association of two or more molecules via hydrogen bonding, i.e., single molecules do not fulfil the criteria of a disc-like anisometric geometry.

A flat anisometric shape, on the one hand, should give rise to molecules having the capacity to form columnar phases as single components. On the other hand, the combination with a molecular recognition site located in the inner core region supplementary should enable control of structure formation through the docking of a complementary component via hydrogen bonding (side-by-side interactions). If the center of the molecules additionally comprises an electron-rich polar group, than manipulations of the phase type should be possible for the same molecule through associations with electron-poor counterparts via charge transfer interactions perpendicular to molecular periphery (face-to-face interactions). A suitable approach consists in "open-sided" core systems (27) as presented schematically in figure 1.

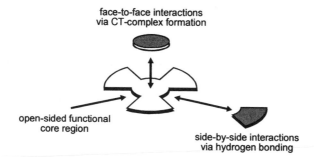

Figure 1. Control of structure formation of functional sheet-like mesogens via directed interactions with complementary molecules.

To realize this concept we have synthesized the 2,4,6-triarylamino-1,3,5-triazines (melamines) **1** and **2** containing long peripheral alkoxy chains (scheme 1). We will summarize here the thermal properties of compounds **1** and **2** in their pure state, the structure formation in binary mixtures with complementary alkoxy substituted aromatic acids as well as mesomorphic assemblies of mixtures with flat electron deficient molecules.

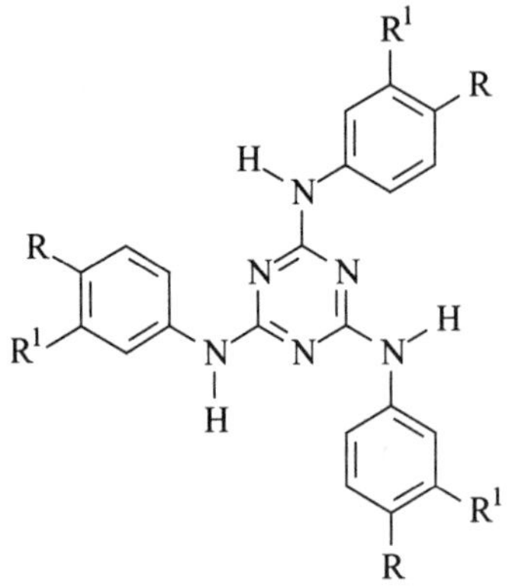

compound	R	R^1
1a	OC$_{10}$H$_{21}$	OC$_{10}$H$_{21}$
1b	OC$_{12}$H$_{25}$	OC$_{12}$H$_{25}$
2a	OC$_{10}$H$_{21}$	H
2b	OC$_{12}$H$_{25}$	H

Scheme 1. Chemical structure of the alkoxy substituted 2,4,6-triarylamino-1,3,5-triazines 1 and 2.

Thermal and structure forming properties of the pure melamines 1 and 2

The six-fold alkoxy substituted triarylmelamines **1** exhibit a hexagonal columnar mesophase with an aperiodic intracolumnar stacking of the molecules (*28*). Contrary, the triazines **2** substituted with only three peripheral flexible chains do not show mesomorphic behaviour (*29,30*). The phase transition data are given in table 1.

Table 1. Phase transition temperatures (°C) and lattice constants (Å) of the 2,4,6-triarylamino-1,3,5-triazines 1 and 2. Phase transition enthalpies (kJ mol^{-1}) are given in brackets.

Compound	Cr		Col$_{hd}$		Iso	a_{hex} (Å)
1a (*28*)	●	70.4 (14.03)	●	156.6 (2.44)	●	30.2
1b (*28*)	●	54.4 (30.08)	●	90.8 (3.24)	●	32.8
2a (*29*)	●	111.5 (42.12)	-		●	
2b (*29*)	●	105.2 (58.20)	-		●	

Cr: crystalline, Col$_{hd}$: hexagonal columnar disordered, Iso: isotropic.

The three phenyl substituents of the triazines **1** and **2** are attached to the central heterocyclic nucleus via secondary amino groups and exhibit significant conformational mobility. Thus, not only the flexible alkyl chains, but also the core region of the compounds can adopt different conformations. The result is a lack of inherent molecular planarity. No specific anisometric geometry is provided by the central part of the molecules. For this reason, the columnar mesophases found for the melamines

1 in their pure state cannot be attributed to a conventional rigid disc-like molecular shape.

The triarylmelamines **1** are characterized by a distinct polar region in the center of the molecules surrounded by lipophilic alkyl chains. Therefore, we conclude that the main driving force for mesomorphic behaviour of the triazines **1** is a micro-segregation (*31*) of the polar central molecular part from the non-polar aliphatic chains. The intercolumnar distances determined for the hexagonal columnar phases by X-ray are somewhat smaller than the molecular diameter of radial two-dimensional conformers of compounds **1** assuming all-trans arranged alkyl chains (figure 2a). Therefore, it is most reasonable to suggest that the columns are piled up from the polar triphenylaminotriazine cores, preferably exhibiting flat radial conformations, with a two-dimensional random arrangement of the flexible alkyl groups (*32*). Thus, cylindrical aggregates are formed with the polar groups located in the center surrounded by a shell of the lipophilic chains (figure 2b).

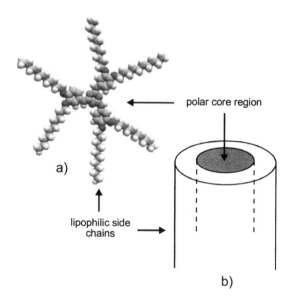

Figure 2. *Columnar mesophases formed by the six-fold alkoxy modified melamines 1. a) Two-dimensional radial conformers that become favored (exemplified for the triazine 1a; Cerius 2, force field Dreiding 2). b) Accumulation of the polar groups in the center of cylindrical aggregates.*

The parallel alignment of these aggregates gives rise to the observed two-dimensional hexagonal lattice symmetry. However, considering the non-liquid crystalline melamines **2** grafted with only three side chains it follows that the appearance of mesomorphic properties strongly depends on the number of lipophilic chains.

Variation and induction of columnar mesophases through intermolecular hydrogen bonding

The triarylmelamines 1 and 2 are characterized by a nitrogen heterocycle with a three-fold substitution with exocyclic secondary amino groups. These structural features are similar to those of alkylamino substituted 1,3,5-triazine compounds and 2,6-diacylaminopyridine derivatives which have been described to form mesomorphic complexes with aliphatic or aromatic carboxylic acids (33-35). Therefore we expected that the aminotriazine core of the melamines should promote attractive interactions with complementary molecules via hydrogen bonding.

Table 2. Phase transition temperatures (°C) and lattice constants (Å) of the equimolar mixtures of the triazines 1 with the alkoxy substituted benzoic acids 3-5. Transition enthalpies (kJ mol^{-1}) are given in parenthesis.

Sample		Phase transition data				Lattice constants
1a/3a (36)	Cr	32.4 (33.44)	Col$_{hd}$	68.4 (3.38)	Iso	a$_{hex}$ = 32.4
1a/3b (36)	Cr	34.5 (13.42)	Col$_{hd}$	75.1 (3.90)	Iso	a$_{hex}$ = 33.2
1a/3c (36)	Cr	18.7 (15.41)	Col$_{hd}$	73.6 (3.49)	Iso	a$_{hex}$ = 34.4
1a/4a	Cr	70.5 (4.44)	Col$_{rd}$	94.1 (4.07)	Iso	a = 31.0; b = 29.3
1a/4b (36)	Cr	62.2 (9.27)	Col$_{rd}$	88.7 (4.91)	Iso	a = 35.8; b = 30.1
1a/4c	Cr	80.4 (12.61)	Col$_{rd}$	98.2 (9.45)	Iso	a = 44.7; b = 27.2
1a/5a	Cr	34.9 (11.96)	Col$_{hd}$	79.8 (3.80)	Iso	a$_{hex}$ = 32.6
1a/5b	Cr	33.1 (7.03)	Col$_{hd}$	78.98 (3.79)	Iso	a$_{hex}$ = 33.0
1a/5c	Cr	33.1 (5.54)	Col$_{hd}$	79.1 (5.34)	Iso	a$_{hex}$ = 34.4
1b/5a	Cr	34.7 (19.57)	Col$_{hd}$	80.9 (3.86)	Iso	a$_{hex}$ = 32.6
1b/5b	Cr	33.7 (14.71)	Col$_{hd}$	80.8 (3.63)	Iso	a$_{hex}$ = 32.6
1b/5c	Cr	36.3 (18.84)	Col$_{hd}$	83.8 (3.34)	Iso	a$_{hex}$ = 32.8

Cr: crystalline, Col$_{hd}$: hexagonal columnar disordered, Col$_{rd}$: rectangular columnar disordered, Iso: isotropic.

We investigated the thermal properties of the triarylamino-1,3,5-triazines 1 in mixtures with the alkoxy substituted benzoic acids 3-5 (table 2). The acid components were systematically modified with respect to the number, the positions and the chain lengths of the alkoxy substituents. All investigated binary mixtures show complete miscibility with formation of an enantiotropic mesophase only at a molar ratio of the components of 1:1. The phase transition temperatures of the equimolar mixed systems **1/3-5** are summarized in table 2.

The equimolar compositions of the melamine **1a** with the non-mesogenic 3,5-dialkoxy substituted benzoic acids **3** exhibit a hexagonal columnar disordered (Col_{hd}) phase (*36*). The hexagonal lattice constants are increased compared with those of the pure melamine **1a**. Furthermore, an increase of the intercolumnar distances is observed as a function of increasing alkoxy chain lengths of the benzoic acids **3**. Thus, the number of methylene groups of the 3,5-dialkoxy substituted benzoic acids allows control over the spacings between the hexagonally arranged columns.

The non-liquid crystalline 3,4-dialkoxybenzoic acids **4** were applied to study the dependence of mesomorphic properties on the positions of lateral flexible chains of the acid component (*27,36*). The X-ray diffractograms of the binary mixtures **1a/4** (figure 3) are characterized by a (010) reflection that appears beside the (100), (200) and (300) reflections in the small angle region and by a diffuse halo at larger scattering angles which is caused by the liquid-like arrangement of the flexible alkoxy side chains. These characteristics give evidence of a rectangular columnar disordered (Col_{rd}) structure.Thus, the 3,4-dialkoxy substitution pattern of the benzoic acid component gives rise to a change of the columnar mesophase structure of the melamine **1a** from a hexagonal to a rectangular lattice in mixtures with aromatic acids.

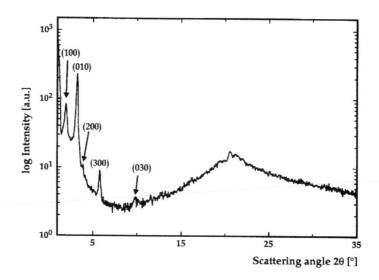

*Figure 3. Wide angle X-ray diffractogram of the equimolar mixture of the triazine **1a** with the 3,4-bis(hexadecyloxy)benzoic acid **4c**.*

The 4-alkoxybenzoic acids **5a-c** are known to exhibit calamitic (SmC and N) mesophases (*37,38*) in their pure state due to a dimerization via hydrogen bonds. The X-ray diffractograms of all equimolar mixtures of the melamines **1** with the benzoic acids **5** show sharp (100) and (110) reflections in the small angle region that correspond to lattice spacings $d_{100} : d_{110} = 1 : 1/3^{1/2}$ (figure 4). Furthermore, all samples **1/5** show a broad halo in the wide angle region. Thus, all binary mixtures **1/5** display a hexagonal columnar disordered (Col_{hd}) structure (*27*). Similar to the mixed systems **1a/3** the hexagonal lattice constants (table 2) of the mixtures **1a/5** are enhanced compared to the pure triazine **1a** and increase with the increasing number of methylene groups of the acid component. Contrary, the intercolumnar distances of the 1:1 mixtures **1b/5** and of the pure triarylmelamine **1b** are of the same order of magnitude.

It has to be emphasized that neither the thermal nor the X-ray investigations indicate the appearance of a calamitic phase characteristic for the benzoic acids **5** in their pure state. Hence, association of the acids **5** with the aminotriazines **1** completely frustrates the tendency of the aromatic acids to form dimers.

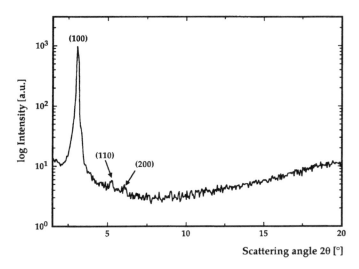

*Figure 4. WAXS diagram of the binary mixture of the melamine **1a** with 4-undecyloxybenzoic acid **5b**.*

The non-liquid crystalline three-fold dodecyloxy substituted triarylmelamine **2b** was used to investigate the influence of the number of peripheral substituents of the triazine component on the structure formation in mixtures with aromatic carboxylic acids. The equimolar mixture of **2b** with the calamitic 4-octyloxybenzoic acid **5a** was found to exhibit an enantiotropic hexagonal columnar disordered (Col_{hd}) phase (*30*). The hexagonal lattice constant of the columnar phase induced here by interaction of the triazine **2b** with the acid **5a** is of the same order of magnitude as those found for the binary mixtures of the triazine **1b** bearing six dodecyloxy groups with the aromatic acids **5**.

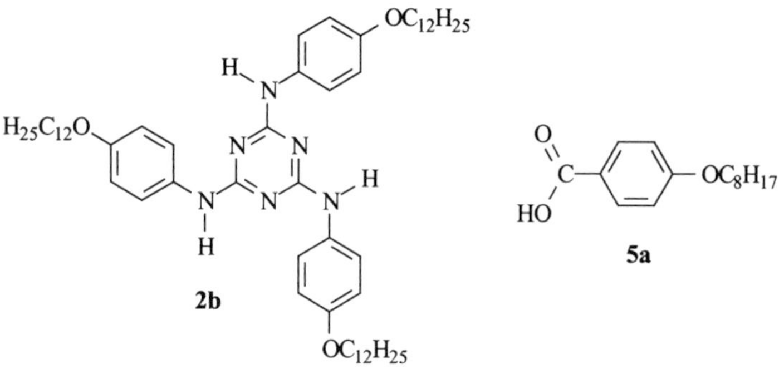

2b/5a: Cr 35.9 Col$_{hd}$ 80.4 Iso; a$_{hex}$ = 32.7 Å

*Scheme 2. Phase transitions temperatures (°C) and lattice constant for the induced hexagonal columnar phase of the 1:1 mixture **2b/5a**.*

FT-IR spectroscopic investigations performed after annealing of the binary mixtures in the isotropic state confirm association of the complementary molecules via hydrogen bonding between the amino substituted triazine core and the acid component (scheme 3). For example, the v(C=O) absorption that appears at 1695 cm^{-1} for the pure benzoic acid **3a** is shifted to 1665 cm^{-1} in case of the appropriate equimolar mixture with the melamine **1a** (*36*). The two absorption bands at 2640 and 2520 cm^{-1} arising from overtones and combinations of the OH-bending and CO-stretching vibrations of the carboxylic acid **3a** are not visible in the spectrum of the mixed system **1a/3a**. The γ(OH) absorption of the benzoic acid **3a** peaking at 930 cm^{-1} disappears in the spectrum of the binary mixture **1a/3a**.

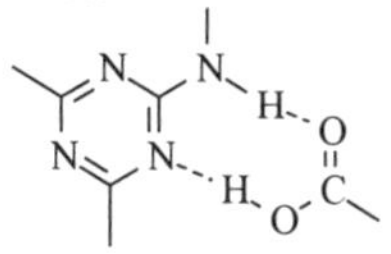

*Scheme 3. Schematic representation of the hydrogen bonding between the aminotriazine core of the melamines **1** and **2** and the carboxylic group of the benzoic acids **3-5**.*

Depending on the lateral substitution pattern of both components, the molecular recognition allows the variation of the hexagonal lattice constants, the variation of the two-dimensional lattice type and even the induction of a hexagonal columnar phase. The control of columnar mesophases of 1,3,5-triazines in this way, however, is largely limited to the intercolumnar parameters.

Considering the three hydrogen bonding sites of the melamines **1** and **2** it still remains unclear why the control of structure formation is restricted to equimolar compositions of the complementary aminotriazine and acid molecules.

Molecular mechanics simulations (Cerius 2; force field Dreiding 2) reveal that a peripheral attack (side-by-side interaction) of the acid component to the inner functional core region of the aminotriazines 1 and 2 gives rise to a flat anisometric hydrogen bonded aggregate only, if the triphenylaminotriazine cores preferably adopt a non-symmetric conformation (30). It follows that a disc-like central core already originates from the docking of one equivalent of the aromatic acid whereas two molecular recognition sites of the triazines remain free due to steric hindrance (figure 5). The formation of the acid-amine associates most likely results in stiffening of the molecular core and the introduction of highly polar hydrogen bonding facilitates microphase separation. Thus, micro-segregation along with the flat shape that arises for associates of the melamines 1 and 2 and the alkoxybenzoic acid 3-5 may explain the modulations and inductions of columnar mesophases of the appropriate equimolar mixtures quite well.

Figure 5. Molecular model (Cerius 2, force field Dreiding 2) of the hexagonal columnar phase forming hydrogen bonded complex between the melamine 1a (left) and the 3,5-bis(decyloxy)benzoic acid 3a (right).

Induction of lamellar mesophases through charge-transfer interactions

Considering the electron-rich polar core region of the triarylmelamines 1 and 2 we expected that structural modifications or even the induction of mesomorphic assemblies should be possible through charge-transfer (CT) interactions in mixtures with flat electron acceptors such as 2,4,7-trinitrofluoren-9-one (TNF) 6 and 2,4,7-trinitrofluoren-9-ylidene malodinitrile 7 (scheme 4).

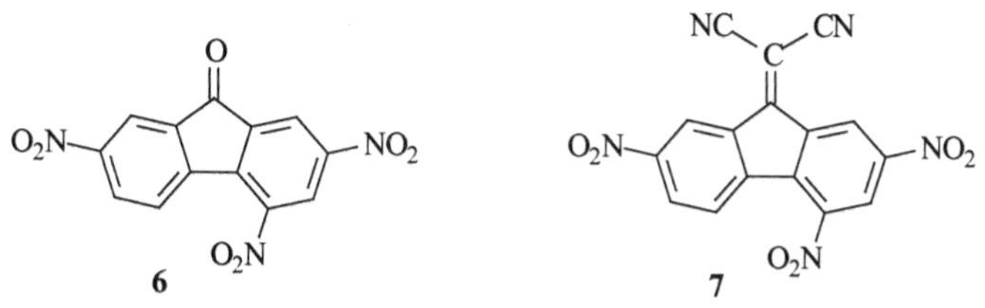

Scheme 4. Chemical structures of the nitrofluorenone based electron acceptors 6 and 7.

Binary mixtures of the six-fold alkoxy substituted triazines 1 with the acceptors 6 and 7 exhibit an enantiotropic mesophase at least at an equimolar ratio of the components (32). The scattering diagrams of the donor-acceptor complexes derived from the melamines 1 and TNF 6 reveal that a smectic layer structure exists. The layer spacings amount to 36.6 Å in case of the CT-complex 1a/6 and 41.5 Å for 1b/6 (table 3). Contrary, the X-ray diffraction pattern of the dodecyloxy substituted triazine 1b and the acceptor 7 give evidence that the molecules are organized on a two-dimensional rectangular lattice.

Table 3. Phase transition temperatures (°C) and lattice constants (Å) for equimolar mixtures of the melamines 1 and 2 with the nitrofluorenone based acceptors 6 and 7. Transition enthalpies (kJ mol⁻¹) are given in parenthesis.

Sample		Phase transition temperatures				a	b
1a/6 (32)	Cr	68.1 (21.19)	SmA	110.4 (1.91)	Iso	36.2	-
1b/6 (32)	Cr	68.6 (27.05)	SmA	115.0 (2.33)	Iso	41.5	-
1b/7 (32)	Cr	93.7 (17.72)	Col$_r$	147.2 (2.48)	Iso	36.3	27.6
2b/6 (30)	Cr	135.1 (20.11)	Col$_r$	154.9 (3.28)	Iso	38.5	32.7
2a/7	Cr	101.8 (3.59)	SmA	170.1 (2.97)	Iso	32.9	-
2b/7 (30)	Cr	98.2 (7.49)	SmA	168.4 (3.25)	Iso	32.2	-

Cr: crystalline, SmA: smectic A, Col$_r$: rectangular columnar, Iso: isotropic.

Doping of the triarylamino-1,3,5-triazines 2 with the electron acceptor derivatives 6 and 7 leads to the induction of enantiotropic mesomorphism (30). The equimolar mixture of compound 2b and TNF 6 exhibits a CT-induced rectangular columnar phase whereas the mixed systems of 2a,b and the dinitrile 7 form a smectic layer structure with only a short range order within the layers. The layer spacings are somewhat smaller compared with those of the CT-complexes 1a,b/6 (table 3).

Similar as for CT-complexes of, i.e. triphenylene ethers (39,40), the dark brown color, which appears instantaneously when solutions of the triarylaminotriazines 1 or 2 are combined with either TNF 6 or with the acceptor 7, is indicative for complex formation due to charge-transfer interactions. The brown color remains for the residues after evaporating the solvent. Certainly, the three alkoxyphenylamino groups are responsible for the electron donating properties of compounds **1** and **2** which compensate for the electron deficiency of the six-membered nitrogen heterocycle.

The origin for charge-transfer complex formation involving nitrofluorenone based acceptors like TNF **6** is the π-π interaction perpendicular to the planes of the aromatic cores of the donor and the acceptor (*41-43*). Therefore, an intercalated structure is most reasonable with the phenyl rings of the donor components **1** and **2** and of the acceptors **6** and **7** arranged closely face-to-face.

The best contact is possible for conformers of the triarylmelamines with an almost planar arrangement of the aromatic rings (*44*). However, it seems rather unlikely that the lamellar mesophases induced by the CT-interactions are accomplished with a flat radial conformation of the triphenylaminotriazine cores since the formation of smectic phases is predominantly caused by a rod-like anisometric geometry of the molecules. We, therefore, are forced to the conclusion that the lamellar mesophases originate from a two-dimensional linear conformation of the triarylmelamines **1** and **2** which becomes favored during the CT-complex formation. One such non-symmetric conformation with an almost flat geometry of the triphenylaminotriazine core is shown in figure 6 along with a molecular model of TNF **6**.

*Figure 6. Molecular models (Cerius 2, force field Dreiding 2) of a two-dimensional linear conformer of the triarylmelamine **1a** (left) and of the acceptor 2,4,7-trinitrofluoren-9-one **6** (right).*

Comparing the molecular models we find that the core region of the melamines including the phenyl groups and the acceptor TNF fit quite well in shape and size. This further supports our assumption of a close face-to-face arrangement.

It seems that one equivalent of the electron acceptor is required to change the melamine conformation to more rod-like giving rise to the smectic phases. Therefore, the fact that mesomorphic behavior is restricted to an equimolar composition of the mixtures, probably, may be explained in terms of the formation of molecular 1:1 complexes (*45,46*).

The layer spacings determined by X-ray for the smectic phases of the six-fold alkoxy substituted compounds **1** in mixtures with TNF **6** are slightly smaller than the maximum molecular dimensions for linear conformers of the melamines **1a,b** (**1a**: 41 Å; **1b**: 46 Å; Cerius 2, force field Dreiding 2) (*32*). This may indicate a slight interdigitation of the flexible alkyl chains separating the polar groups of neighboring layers or it may be explained by the fact that the larger space requirement of the mixed donor-acceptor polar regions necessitates a better space filling of the aliphatic regions by a disordered packing of the flexible alkyl groups. The smaller layer periodicities found for the CT-complexes of **2/7** may arise from a stronger degree of interdigitation or from a more distorted arrangement of the lipophilic chains. A structural model for the CT-induced lamellar mesophases is presented in figure 7a.

It is most likely that in case of the equimolar compositions **1b/7** and **2b/6**, displaying at least a two-dimensional regular structure, the CT-interactions impose a flat linear conformation of the triazine molecules as well. The lattice parameters *a* determined by X-ray are in reasonable agreement with the molecular length of the melamines evaluated by molecular mechanics simulations. Assuming the simplest molecular arrangement, namely a columnar structure on a face-centered rectangular array, the lattice parameters *b* would correspond to the lateral side-by-side distance per repeat unit. The problem is, however, that such a face-centered structure would lead to systematic extinction of X-ray reflections which is not observed (*32*). Therefore, we have to assume that the molecules are systematically shifted from the face-centered positions (*47*) as shown schematically in figure 7b.

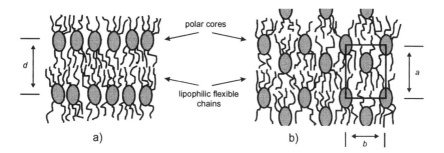

Figure 7. Schematic models for the mesophases of the melamines 1 and 2 in binary mixtures with the electron acceptors 6 and 7 (shown for the case of the six-fold alkoxy modified compounds 1). a) CT-induced lamellar phases. b) Rectangular columnar mesophase of the binary mixtures 1b/7 and 2b/6.

On average, the sterical conformations of the core region of the triarylaminotriazines **1** and **2** are the same for the CT-complexes with both acceptors **6** and **7**. Thus, the rectangular columnar phases of **1b/7** and **2b/6** can be considered as resulting from a destruction of the CT-induced lamellar phases.

Conclusions

The 1,3,5-triazines **1** and **2** are the first examples of functionalized mesogens (**1**) or premesogenic molecules (**2**) that promote intermolecular hydrogen bonding with complementary molecular species as well as charge-transfer complex formation with electron acceptors. For the first time the induction of smectic liquid crystalline structures via donor-acceptor interactions was found for electron donors which exhibit a columnar phase in their pure state. The columnar phases of the pure triarylmelamines **1**, the molecular recognition directed control and induction of columnar mesomorphic assemblies in mixtures with aromatic acids as well as the CT-induced lamellar phases in mixtures with electron- deficient molecules are attributed to segregation effects due to distinct polar and non-polar molecular regions. Certain specific anisometric conformations become favored during the process of self-organization.

References

1. Demus, D. *Liq. Cryst.* **1989**, *5*, 75.
2. Gray, G.W.; Goodby, J.W. *Smectic Liquid Crystals – Textures and Structures*; Leonard-Hill: Glasgow – London, UK, 1984.
3. Destrade, C.; Foucher, P.; Gasparoux, H.; Nguyen Huu Thin; Levelut, A.-M.; Malthete, J. *Mol. Cryst. Liq. Cryst.* **1984**, *106*, 121.
4. Paleos, C.M.; Tsiourvas, D. *Angew. Chem.* **1995**, *107*, 1839.
5. Praefcke, K.; Singer, D. In *Handbook of Liquid Crystals*; Demus, D.; Goodby, J.; Gray, G.W.; Spiess, H.-W.; Vill, V., Eds.; Wiley – VCH: Weihheim, Chichester, Brisbane, Singapore, Toronto, 1998; Vol. 2b, 945-967.
6. Diele, S.; Pelzl, G.; Weissflog, W.; Demus, D. *Liq. Cryst.* **1988**, *3*, 1047.
7. Bengs, H.; Ebert, M.; Karthaus, O.; Kohne, B.; Praefcke, K.; Ringsdorf, H.; Wendorff, J.H.; Wüstefeld, R. *Adv. Mater.* **1990**, *2*, 141.
8. Ebert, M.; Frick, G.; Baehr, C.; Wendorff, J.H.; Wüstefeld, R.; Ringsdorf, H. *Liq. Cryst.* **1992**, *11*, 293.
9. Praefcke,K.; Holbrey, J.D. *J. Incl. Phenom. Mol. Recog. Chem.* **1996**, *24*, 19.
10. Janietz, D.; Festag, R.; Schmidt, C.; Wendorff, J.H. *Liq. Cryst.* **1996**, *20*, 459.
11. Ringsdorf, H.; Wüstefeld, R.; Zerta, E.; Ebert. M.; Wendorff, J.H. *Angew. Chem.* **1989**, *101*, 934.
12. Janietz, D.; Praefcke, K.; Singer, D. *Liq. Cryst.* **1993**, *13*, 247.
13. Praefcke, K.; Singer, D.; Kohne, B.; Ebert, M; Liebmann, A.; Wendorff, J.H. *Liq. Cryst.* **1991**, *10*, 147.
14. Praefcke, K.; Singer, D.; Langner, M.; Kohne, B.; Ebert, M.; Liebmann, A; Wendorff, J.H. *Mol. Cryst. Liq. Cryst.* **1991**, *215*, 121.
15. Bengs, H.; Karthaus, O.; Ringsdorf, H.; Baehr, C.; Ebert, M.; Wendorff, J.H. *Liq. Cryst.* **1991**, *10*, 161.
16. Beginn, U.; Lattermann, G. *Mol. Cryst. Liq. Cryst.* **1994**, *241*, 215.
17. Marquard, P.; Praefcke, K.; Kohne, B.; Stephan, W. *Chem. Ber.* **1991**, *124*, 2265.

18. Praefcke, K.; Marquardt, P.; Kohne, B.; Stephan, W.; Levelut, A.-M.; Wachtel, E. *Mol. Cryst. Liq. Cryst.* **1991**, *203*, 149.
19. Lattermann, G.; Staufer, G. *Liq. Cryst.*, **1989**, *4*, 347.
20. Lattermann, G.; Staufer, G. *Mol. Cryst. Liq. Cryst.*, **1990**, *191*, 199.
21. Ebert, M.; Kleppinger, R.; Soliman, M.; Wolf, M.; Wendorff, J.H.; Lattermann, G; Staufer, G. *Liq. Cryst.*, **1990**, *7*, 553.
22. Matsunaga, Y.; Terada, M. *Mol. Cryst. Liq. Cryst.*, **1986**, *141*, 321.
23. Pucci, D.; Veber, M.; Malthete, J. *Liq. Cryst.*, **1996**, *21*, 153.
24. Brienne, M.J.; Gabard, J.; Lehn, J.-M.; Stibor, I. *J. Chem. Soc., Chem. Commun.* **1989**, 1868.
25. Fouquey, C.; J.-M. Lehn, J.-M.; Levelut, A.-M. *Adv. Mater.* **1990**, *2*, 254.
26. Lehn, J.-M. *Makromol. Chem., Macromol. Symp.* **1993**, *69*, 1.
27. Janietz, D. *J. Mater. Chem.* **1998**, *8*, 265.
28. Goldmann, D.; Janietz, D.; Festag, R.; Schmidt, C.; Wendorff, J.H. *Liq. Cryst.* **1996**, *21*, 619.
29. Goldmann, D. Ph.D thesis, Potsdam University, GERMANY, 1998.
30. Janietz, D.; Goldmann, D.; Schmidt, C.; Wendorff, J.H. *Mol. Cryst. Liq. Cryst.* **1999**, *332*, [2651]/141.
31. Tschierske, C. *J. Mater. Chem.* **1998**, *8*, 1485.
32. Goldmann, D.; Janietz, D.; Schmidt, C.; Wendorff, J.H. *Angew. Chem.* **1999**, submitted.
33. Paleos, C.M.; Tsiourvas; D., Fillipakis, S.; Fillipaki, L. *Mol. Cryst. Liq. Cryst.* **1994**, *242*, 9.
34. Kato, T.; Nakano, M.; Moteki, T.; Uryu, T., Ujiie, S. *Macromolecules* **1995**, *28*, 8875.
35. Kato, T.; Kubota, Y.; Nakano, M.; Uryu, T. *Chem. Lett.*, **1995**, 1127.
36. Goldmann, D.; Dietel, R.; Janietz, D.; Schmidt, C.; Wendorff, J.H. *Liq. Cryst.* **1998**, *24*, 407.
37. *Flüssige Kristalle in Tabellen;* Demus, D.; Demus, H.; Zaschke, H. Eds.; Deutscher Verlag für Grundstoffindustrie: Leipzig, GERMANY; 1973.
38. Gray, G.W.; Jones. B. *J. Chem. Soc.*, **1953**, 4179.
39. Kranig, W.; Boeffel, C.; Spiess, H.W.; Karthaus, O.; Ringsdorf, H.; Wüstefeld, R. *Liq. Cryst.* **1990**, *8*, 375.
40. Green, M.M.; Ringsdorf, H.; Wagner, J.; Wüstefeld, R. *Angew. Chem.* **1990**, *102*, 1525.
41. Diederich, F.; Philp, D.; Seiler, P. *J. Chem. Soc., Chem. Commun.* **1994**, 205.
42. Markovitsi, D.; Bengs, H.; Ringsdorf, H. *J. Chem. Soc. Faraday Trans.* **1992**, *88*, 1275.
43. Markovitsi, D.; Pfeffer, N.; Charra, F.; Nunzi, J.-M.; Bengs, H.; Ringsdorf, H. *J. Chem. Soc. Faraday Trans.* **1993**, *89*, 37.
44. Neumann, B.; Joachimi, D.; Tschierske, C. *Liq. Cryst.* **1997**, *22*, 509.
45. Park, J.W.; Bak, C.S.; Labes, M.M. *J. Am. Chem. Soc.* **1975**, *97*, 4398.
46. Homura, N.; Matsunaga, Y.; Suzuki, M. *Mol. Cryst. Liq. Cryst.* **1985**, *131*, 273.
47. Lattermann, G.; Schmidt, S.; Kleppinger, R.; Wendorff, J.H. *Adv. Mater.*, **1992**, *4*, 30.

Author Index

Subject Index

RETURN TO: CHEMISTRY LIBRARY

100 Hildebrand Hall • 642-3753

LOAN PERIOD 1	2	3
	1-MONTH USE	
4	5	6

ALL BOOKS MAY BE RECALLED AFTER 7 DAYS.
Renewable by telephone.

DUE AS STAMPED BELOW.

FORM NO. DD 10
3M 3-00

UNIVERSITY OF CALIFORNIA, BERKELEY
Berkeley, California 94720–6000